INFORMATION BEYOND BORDERS

Information Beyond Borders

International Cultural and Intellectual
Exchange in the Belle Époque

Edited by

W. BOYD RAYWARD

*University of New South Wales, Sydney, Australia and
University of Illinois at Urbana-Champaign, USA*

ASHGATE

Published by
Ashgate Publishing Limited
Wey Court East
Union Road
Farnham
Surrey, GU9 7PT
England

Ashgate Publishing Company
110 Cherry Street
Suite 3-1
Burlington, VT 05401-3818
USA

www.ashgate.com

British Library Cataloguing in Publication Data
A catalogue record for this book is available from the British Library

The Library of Congress has cataloged the printed edition as follows:
Information beyond borders : international cultural and intellectual exchange in the Belle Époque / [edited] by W. Boyd Rayward.
        pages cm
    Includes bibliographical references and index.
    ISBN 978-1-4094-4225-7 (hardback) – ISBN 978-1-4094-4226-4 (ebook) –
    ISBN 978-1-4724-0212-7 (epub) 1. Information services–History–19th century.
2. Information services–History–20th century. 3. Documentation–History.
4. Information society–History. 5. Intellectual cooperation–History. 6. International cooperation–History. 7. Communication–Technological innovations–History.
8. Globalization–History. 9. Culture and globalization–History. 10. Language, Universal–History. I. Rayward, W. Boyd, 1939-

    ZA3157.I53 2013
    001.09'034–dc23

                                                                        2013020327

ISBN 9781409442257 (hbk)
ISBN 9781409442264 (ebk – PDF)
ISBN 9781472402127 (ebk – ePUB)

MIX
Paper from
responsible sources
FSC
www.fsc.org   FSC® C013985

Printed in the United Kingdom by Henry Ling Limited,
at the Dorset Press, Dorchester, DT1 1HD

Contents

List of Figures

Notes on Contributors

Alistair Black is a Professor in the Graduate School of Library and Information Science, University of Illinois at Urbana-Champaign, USA. He received a BA in Modern History and MA in Social and Economic History from the University of London and a Postgraduate Diploma in Information Studies and his PhD from the Polytechnic of North London (now London Metropolitan University). After working in public and academic libraries, in 1990 he became a lecturer at Leeds Metropolitan University where he was appointed professor in 2001. He was Chair of the Library History Group of the Library Association, 1992–9, and of the International Federation of Library Associations (IFLA) Section on Library History, 2003–7. He was editor of *Library History*, 2004–8 and North American editor of *Library and Information History*, 2009–13. He is co-editor of the journal *Library Trends*. He is author of *A New History of the English Public Library* (Leicester University Press 1996) and *The Public Library in Britain 1914–2000* (the British Library 2000). He is co-author with Dave Muddiman of *Understanding Community Librarianship: The Public Library in Post-Modern Britain* (Ashgate 1997); and with Dave Muddiman and Helen Plant of *The Early Information Society in Britain, 1900–1960* (Ashgate 2007); and with Simon Pepper of *Books, Buildings and Social Engineering: Early Public Libraries in Britain from Past to Present* (Ashgate 2009). He is co-editor of Volume 3 (covering 1850–2000) of the *Cambridge History of Libraries in Britain and Ireland* (2006). His recent research is on the history of corporate libraries and staff magazines, and the design of public libraries in the 1960s.

Volker Barth is Assistant Professor of Modern History at the University of Cologne. He was educated in Germany and France. In 2004 he received a co-tutelle PhD from the Ludwig-Maximilians-University (LMU), Munich, and the École des Hautes Études en Sciences Sociales (EHESS), Paris. His teaching and research positions include the Université Paris VII–Denis Diderot, the German Academic Exchange Service (DAAD) and the Bureau International des Expositions (BIE), Paris. His current research project is on the cooperation of globally operating news agencies between the mid-nineteenth century and the interwar period, in particular on the emergence and establishment of a world information order that determined the global flow of news items in that period. His publications include: *Mensch versus Welt. Die Pariser Weltausstellung von 1867* (Wissenschaftliche Buchgesellschaft 2007); "The Micro-history of a world event: Intention, perception and imagination at the Exposition universelle de 1867," *Museum & Society* 6 (2008); "Medien, Transnationalität und Globalisierung,

1830–1960: Neuerscheinungen und Desiderata," *Archiv für Sozialgeschichte* 51 (2011); "Making the Wire Speak: Transnational Techniques of Journalism, 1860–1930," in *Global Communication Electric. Actors of a Globalizing World* edited by Michaela Hampf and Simone Müller-Pohl (forthcoming).

Julie Carlier is the research coordinator of the interdisciplinary Ghent Centre for Global Studies (Ghent University, Belgium, 2013). She obtained a Masters degree in History at Ghent University (2003) and in European Studies at the Institut d'Études Politiques de Paris (Sciences-Po), 2005. Her PhD dissertation from Ghent University in 2010 was entitled, "Moving beyond boundaries. An entangled history of feminism in Belgium, 1890–1914." Her main research interests include gender, transnational history and the history of masculinities. Recent publications include, "An entangled history of ideas and ideals. Feminism, social and educational reform in children's libraries and children's literature in Belgium before the First World War" with co-author Christophe Verbruggen in *Paedagogica Historica*, 45 (2009); "Forgotten transnational connections and national contexts. An entangled history of the political transfers that shaped Belgian feminism, 1890–1914," *Women's History Review*, 19 (2010); "Réseaux transnationaux et la construction d'une identité féministe collective: transferts oubliés, connexions et conflits dans le cas de la Belgique (vers 1890–1918)", in *Au-delà et en deçà de l'État. Le genre entre dynamiques transnationales et multi-niveaux* edited by Bérengère Marquez-Perreira, Petra Meier and David Paternotte (Louvain-la-Neuve: Academia-Bruylant, 2010).

Mary Carroll is Associate Course Director for the School of Information Studies Charles Sturt University, Australia. She was educated in Australia and her PhD is from the Charles Sturt University. She was an Early Career Research Fellow at RMIT University, Melbourne, and a teacher and lecturer at Victoria University, Melbourne. Much of her research has focused on the history of libraries and of information organisation and on the history of education for the information professions internationally. In 2012/2013 she undertook the *Public Spaces – Private Places: Private RTOs and Information Infrastructure* research project funded by the National Centre for Vocational Education Research. This research explores the relationship between public libraries and private training providers in Australia. As well as publications on aspects of professional education, her recent publications include *Defining Difference: The 'noble and less noble' traditions of education and training in Australia (Advances in Librarianship* 2011*)* and with Sue Reynolds as co-author "'Disaffection in the library': Shaping a living centre of learning," *History of Education Review* (2012).

Heather Gaunt is Curator of Academic Programs (Research) at the Ian Potter Museum of Art, the University of Melbourne. She received her BMus, BA, and Graduate Diploma of Art Curatorial Studies from the University of Melbourne, Australia, and her PhD from the University of Tasmania, Australia. She has worked in a variety of areas in the museum sector in Australia during the past two decades. Her research in book and library history examines the historical development of Australiana collections in Australia's major "State" public libraries from the mid-nineteenth century to the mid-twentieth century in the contexts of nationalism and the development of historical consciousness in specific communities. Recent research and publications have explored the use of art museum environments and collections to teach visual observation and empathy skills in medical and dental students. Among her recent publications are: "'A native instinct of patriotism': nationalism in the Australian public library from Federation to the 1930s", *Library History* (2008); "Social memory in the public historical sphere: Henry Savery's 'The Hermit in Van Diemen's Land' and the Tasmanian Public Library", *Library and Information History* (2009); and "'How to encourage our literature': Australian fiction in the Australian Public Library before the Second World War", *Australian Literary Studies* (2012).

Frank Hartmann is Professor in the Faculty of Art & Design, Bauhaus-University Weimar, Germany. He was educated in Austria and his doctorate in philosophy is from the University of Vienna where he received his postdoctoral qualification (Habilitation) in the new interdisciplinary field of media philosophy. He has taught at several European Universities. Among international appointments is a guest professorship at the Universidade de São Paulo, Brasil. Among a number of books, textbooks, book chapters and articles on media history, theory, and philosophy are: *Bildersprache: Otto Neurath Visualisierungen* (2006); *Globale Medienkultur: Technik, Geschichte, Theorien* (2006); Multimedia (2008); and he edited *Vom Buch zur Datenbank: Paul Otlets Utopie der Wissensvisualisierung* (2012).

Charles van den Heuvel is Head of Research for the History of Science and Scholarship at the Huygens Institute for the History of the Netherlands of the Royal Netherlands Academy of Arts and Sciences. He holds the chair, Digital Methods and Historical Disciplines, in the University of Amsterdam. He studied Art History and Archaeology at the University of Groningen (PhD 1991). Together with Wijnand Mijnhardt (Utrecht University) he currently leads the project: Circulation of Knowledge and Learned Practices in the 17th Century Dutch Republic: a Web-based Humanities Collaboratory on Correspondence. His research has dealt with the history of architecture, fortification and town planning, history of cartography, and digital humanities. In this connection see especially his *'De Huysbou', A reconstruction of an unfinished treatise on architecture, town planning and civil engineering by Simon Stevin* (Royal Dutch Academy of Sciences 2005). More recently aspects of his research deals with the history of information science. Recent publications include studies on Paul Otlet, Frits Donker Duyvis, the universe of knowledge metaphor (with Richard P. Smiraglia) and on visualizations of knowledge. His article "Facing Interfaces: Paul Otlet's Visualizations of Data Integration" with co-author W. Boyd Rayward appeared in the *Journal of the American Society for Information Science and Technology* (*JASIS&T* 2011).

Fabian de Kloe is a teaching assistant and member of the Graduate School of the Faculty of Arts and Social Sciences of Maastricht University, the Netherlands. His research examines internationalist claims and initiatives by scientists in the twentieth century. He explores their ideological functions and their relationship with the broader cultural and political contexts in which they were articulated. His PhD dissertation investigates early twentieth century attempts by scientists from France, Germany and the US to create and promote a scientific international language that was meant to facilitate the transfer of scientific knowledge across national boundaries. He is author of "Arenas of truth making: the HIV/AIDS correlation as a case of contested facticity in science, politics and journalism," *Krisis: Journal for Journal for Contemporary Philosophy* (2012).

Markus Krajewski is Associate Professor of Media History in the Faculty of Media at Bauhaus University Weimar. During the academic year 2008/09 he was a fellow at the Humanities Center at Harvard University and Visiting Professor in the History of Science Department. He is author of *Paper Machines: About Cards & Catalogs, 1548–1929* (The MIT Press 2011); *Der Diener: Mediengeschichte einer Figur zwischen König und Klient* (2010) [forthcoming as *The Servant. Media History of a Figure between King and Client*, Yale University Press]; and of *Restlosigkeit: Weltprojekte um 1900* (2006) [forthcoming as *For the Rest: World Projects and Notions of Globality around 1900*, University of Minnesota Press, 2014]. He is editor of *Projectors. Knowledge Production in the Pre-form of Failure* (in German, Kulturverlag Kadmos, 2005). His current research projects include the media and cultural history of servants, epistemology of the peripheral, and the history of exactitude in scholarly and scientific contexts. He has also developed and maintains the bibliographical software *synapsen–a hypertextual card index* (www.verzetteln.de/synapsen.).

Daniel Laqua is Senior Lecturer in European History at Northumbria University in Newcastle upon Tyne. He previously worked at University College London, where he also presented his doctoral thesis. His research deals with transnational activism in nineteenth- and twentieth-century Europe. He is particularly concerned with different forms of internationalism ranging from cultural exchange to political varieties (for instance socialist and pacifist movements). He is the author of *The Age of Internationalism and Belgium, 1880–1930: Peace, Progress and Prestige* (Manchester University Press, 2013) and the editor of *Internationalism Reconfigured: Transnational Ideas and Movements Between the World Wars* (I.B. Tauris, 2011). He is co-editor with Christophe Verbruggen and Gita Deneckere of an issue of *Revue belge de philologie et d'histoire/Belgisch tijdschrift voor filologie en geschiedenis* (2012) entitled "Beyond Belgium: Encounters, Exchanges and Entanglements, 1900–1925." His research has appeared in periodicals such as *Labour History Review*, *Critique Internationale*, the *Journal of Global History* and *The International History Review*.

Dave Muddiman retired in 2008 as Principal Lecturer in the School of Information Management, Leeds Metropolitan University. He was educated at Leicester University, UK and gained postgraduate degrees from the Polytechnic of North London and the UK Open University, subsequently working in public libraries until 1988. His early academic career and related research and publications focused on the sociology and recent history of the public library service. In 2000 he led a major UK government funded research project entitled *Open to All? The Public Library and Social Exclusion*. In the late 1990s his research turned to the history of the information society, in particular in the period 1870–1945. Recent publications have dealt with H.G. Wells, J.D. Bernal, the early history of ASLIB and the Imperial Institute among other topics. He is co-author with Alistair Black and Helen Plant of *The Early Information Society: Information Management in Britain before the Computer* (2007 Ashgate). In addition to part-time teaching and research, he is currently Book Reviews Editor of the *Journal of Librarianship and Information Science* and a contributor to the latest edition of the *Encyclopaedia of Library and Information Sciences* (Taylor and Francis).

W. Boyd Rayward is Emeritus Professor in the Graduate School of Library and Information Science of the University of Illinois and in the School of Information Systems, Technology and Management of the University of New South Wales. He was educated in Australia and the US. His PhD is from the University of Chicago. He has held professorial and decanal positions in the University of Chicago and the University of New South Wales. He is the recipient of the Research in Information Science Award of the American Society for Information Science and Technology. His recent research examines the history of the international organization of knowledge with studies of Paul Otlet's ideas in relation to hypertext, the internet, the World Wide Web and the beginnings of modern information science, of a number of utopian schemes of knowledge organization including H.G. Wells's idea of a world brain, and of the implications of digitisation and networking for libraries and museums. Among his publications are: *The Universe of Information: The Work of Paul Otlet for Documentation and International Organization* (VINITI 1975) translated into Russian and Spanish; *International Organisation and Dissemination of Knowledge: Selected Essays of Paul Otlet* (Elsevier 1990). He edited *European Modernism and the Information Society* (Ashgate 2008) and contributed chapters to *Paul Otlet (1868–1944) Fondateur du Mundaneum: Architect du savoir, Artisan de paix*, edited by Jacques Gillen (Les Impressions Nouvelles 2008) and, *Mundaneum: Archives of Knowledge* (University of Illinois 1910). *"Mondothèque: A multimedia desk in a global Internet"* (with Charles van den Heuvel) was issued in 2011 in the series "Science Maps as Visual Interfaces to Digital Libraries."

Sue Reynolds is a Senior Lecturer and Program Director for the Master of Information Management program in the School of Business IT and Logistics at RMIT University in Melbourne Australia. She was educated in Australia and the US. Her PhD is from Charles Sturt University in New South Wales. Her doctoral thesis won the American Library Association Phyllis Dain Library History Dissertation Award in 2011. Among her previous positions were appointments in Victoria University, Melbourne and San Jose State University, California. Her current research in library and information history is a project with Mary Carroll on the similarity of relationships between libraries and communities as they were in the nineteenth century and are developing in the twenty-first century, particularly with regard to the organization of resources. Among her publications are: "A Nineteenth Century Library and its Librarian: Factotum, Bookman or Professional" *The Australian Library Journal*, 56 (2007); "Cultural Record Keepers: The Library of the Supreme Court of the Colony of Victoria, Australia" in *Libraries & the Cultural Record* 45 (2010); "The Library of the Supreme court of the Colony of Victoria: 'Continuities … Discontinuities …, and … Original Contributions'" in *Ireland and the Irish Antipodes: One World or Worlds Apart? Papers delivered at the 16th Australasian Irish Studies Conference, Massey University, Wellington, New Zealand*, edited by Brad and Kathryn Patterson (Anchor Books, 2010); with Mary Carroll, "Collaborators or Competitors? The Roles of School Libraries, Classroom Libraries, Teachers and Teacher librarians in Literacy Development" in *Synergy* 10 (2012); and *Books for the Profession: The Library of the Supreme Court of Victoria* (Australian Scholastic 2012).

Paul A. Servais is Professor in the Department of History of the Catholic University of Louvain where he took his PhD in 1979. He has held professorial and decanal positions in the Catholic University of Louvain, the École des Hautes Études en Sciences Socials (EHESS, Paris), the École Nationale des Chartes (Paris), and the University of Laval (Québec). His research examines the history of the network of international Orientalists and how knowledge and representations about the East in the Western countries were constructed between the 18th and 20th centuries. Among his publications relevant to this book are, as editor, *De l'Orient à l'Occident et retour. Perceptions et représentations de l'Autre dans la littérature et les guides de voyage (*Série Rencontres Orient-Occident 9, Louvain-la-Neuve, 2006) and *Christianisme et Orient, 17e–21e siècles* (Série Rencontres Orient-Occident 12; Louvain-la-Neuve, 2010). Other publications include "Sylvain Lévi, Louis de la Vallée Poussin et Etienne Lamotte, une généalogie orientaliste?," in *Sylvain Lévi (1863–1935): Etudes indiennes, histoire sociale* (Turnhout 2007); "Images de Chine: Le Père *Léon* Dieu *à* l'ombre de la Grande Muraille," in *Images et paysages mentaux des 19e et 20e siècles de la Wallonie à l'Outre-mer …"* edited by L. Courtois, J.P. Delville et al. (Editions Academia 2007); and *"Une Chine de papier: Charles de Harlez de Deulin: des études éraniennes aux études chinoises,"* in *Chine – Europe – Amérique. Rencontres et échanges de Marco Polo à nos jours*, edited by Shenwen Li (Presses de l'Université Laval 2009).

Geert J. Somsen teaches history of science in the History Department and the Science and Technology Studies Program of Maastricht University, the Netherlands. He was educated in the Netherlands and the US. His PhD is from the Utrecht University. He has been a visiting scholar at the universities of Uppsala, Harvard and Cambridge, UK. His research deals with scientific planning in the Netherlands and the UK, the propagandist role of science journalists, the uses of neutrality by scientists from small countries, and the campaigns for a scientific World State by H.G. Wells. Current work focuses on A.D. White's ideas of science and arbitration and George Sarton's internationalist history of science. Among his publications are: *Neutrality in Twentieth-Century Europe: Intersections of Science, Culture, and Politics after the First World War*, edited with Rebecka Lettevall and Sven Widmalm (Routledge 2012); "A History of Universalism: Conceptions of the Internationality of Science, 1750–1950," *Minerva* (2008); and "De Politiek van de Wetenschapsjournalistiek. De Socialistische Agenda van de Britse 'Scientific Journalists'", (*The Politics of Science Journalism. British Scientific Journalists' Socialist Agenda*) in *Leonardo voor het Publiek. Een Geschiedenis van Wetenschaps- en Techniekcommunicatie* edited by Frans J. Meijman, Stephen Snelders and Onno de Wit (Amsterdam: VU University Press).

Jan Surman received his PhD in History from the University of Vienna in 2012. His dissertation was on "Habsburg University 1848–1918: Biography of a Space." He was associated with the PhD program, The Sciences in Historical Context, at the University of Vienna and was visiting scholar at the Max Planck Institute for History of Science (Berlin), the Center for Austrian Studies at the University of Minnesota, the Institute for the History of Science of the Polish Academy of Sciences (Warsaw), and the Centre for the History of Sciences and Humanities (Prague). His publications include *Dwa życia Ludwika Gumplowicza. Wybór tekstów* co-edited with Gerald Mozetič (Oficyna Naukowa 2010) and *The Nationalization of Scientific Knowledge in the Habsburg Empire, 1848–1918* coedited with Mitchell Ash (Macmillan 2012). He is currently a Leibniz–DAAD Research Fellow at the Herder Institute in Marburg, Germany.

Wouter Van Acker since 2013 is lecturer in Contemporary and Australian Architecture at the Griffith School of Environment (Architecture), Griffith University, Queensland, Australia. He obtained his Masters and PhD degrees in Architecture at Ghent University and was a post-doctoral Researh Fellow in the Department of Architecture and Urban Planning in Ghent University. His research explores the interdisciplinary relations between architecture, knowledge, the city, and art in the late 19th and early 20th century. Among his publications are: "Internationalist utopias of visual education: the graphic and scenographic transformation of the Universal Encyclopaedia in the work of Paul Otlet, Patrick Geddes and Otto Neurath" *Perspectives on Science* (2011); *Universalism as Utopia. A Historical Study of the Schemes and Schemas of Paul Otlet (1868–1944)* Doctoral Dissertation (2011); "Information and Space. Analogies and Metaphors," a special issue of *Library*

Trends co-edited with Pieter Uyttenhove (2012); "A Tale of Two World Capitals – the Internationalisms of Pieter Eijkman and Paul Otlet," co-authored with Geert Somsen *Revue Belge de Philologie et d'Histoire/Belgisch Tijdschrift voor Filologie en Geschiedenis* (2012)); and *Gent 1913: Op het breukvlak van de Moderniteit* co-edited with Christophe Verbruggen (Snoeck 2013).

Christophe Verbruggen is Professor in the research unit for Social History since 1750 in Ghent University, where he obtained his PhD. He has held fellowships at the Centre de Sociologie Européenne in Paris (2008) and the Institute for Historical Studies, Texas University at Austin (2009). His field of research is the social history of intellectuals and cultural institutions and he is currently studying transnational dynamics and relationships between intellectuals (1880–1930) and the development of Virtual Research Environments for the study of transnational history. Among his publications are a book on literary sociability (*Schrijverschap tijdens de Belgische belle époque. Een sociaal-culturele geschiedenis*, 2009) and articles in *Paedagogica Historica*, *History of Science* and with Lewis Pyenson as co-author in *Isis*. He is co-editor with Daniel Laqua and Gita Deneckere of *Beyond Belgium. Encounters, Exchanges and Entanglements, 1900–1925* (a 2012 special issue of *Revue Belge de Philologie et d'Histoire – Belgisch Tijdschrift voor Filologie en Geschiedenis*).

Nader Vossoughian is an Associate Professor of Architecture at the New York Institute of Technology. He studied philosophy, cultural studies, and German literature at Berkeley, Swarthmore, the Albert-Ludwigs-Universitaet (Freiburg), and the Humboldt University (Berlin) before receiving his MPhil and PhD in the History and Theory of Architecture from Columbia University. From 2008–2010, he was a Researcher at the Jan van Eyck Academie in Maastricht. In 2012, he was a Visiting Scholar at the Canadian Centre for Architecture. He has curated exhibitions at Stroom den Haag, the MAK Center for Art and Architecture at the Schindler House in Los Angeles, the AIA Center for Architecture in New York, and the Museum of Applied Arts in Vienna. His research centers on the relationship between architecture, politics, and knowledge in 20th century modernism. He is author of *Otto Neurath: The Language of the Global Polis* (2008, paperback edition 2011). He has contributed to *European Modernism and the Information Society*; edited by W. Boyd Rayward (Ashgate 2010); *Otto Neurath's Economics in Context* edited by Elisabeth Nemeth, et al. (Vienna Circle: Springer 2007); and *Josef Frank 1885–1967 – Eine Moderne der Unordnung*, edited by Iris Meder (Pustet Anton 2008). His most recent publications include "On Ernst Neufert's *Architects' Data*: Architecture, Standardization, and the Language of Total War," which is forthcoming in the journal *Grey Room*.

Acknowledgements

These chapters represent a selection from the papers that were initially presented at a Colloquium, "Transcending Boundaries in the Period of the Belle Époque: organising knowledge, mobilising networks and effecting social change." This took place in the Mundaneum in Mons, Belgium, in May 2010. Special recognition should be accorded to the staff of the Mundaneum in setting up and managing the colloquium so successfully, especially the Director, Mme Charlotte Dubray for her support of the idea and the chief of the Archives Service, Stéphanie Manfroid and the archivist, Jacques Gillen for their never-failing help.

The papers chosen for publication here have been further developed, revised and reviewed.

The Colloquium, Transcending Boundaries is the fourth in a series that began in 2002:

Architecture of Knowledge: The Mundaneum and European Antecedents of the World Wide Web, Mundaneum, Mons, Belgium, 24–25 May 2002. An account of the discussions and papers is given in, Charles Van den Heuvel, W. Boyd Rayward, and Pieter Uyttenhove, "L'Architecture du savoir: une recherche sur le Mundaneum et les précurseurs européen de l'Internet," *Transnational Associations / Associations transnationales,* Issue 1–2, June 2003, pp. 16–28.

European Modernism and the Information Society: Informing the Present, Understanding the Past, a colloquium organised by Professor W. Boyd Rayward at the University of Illinois, Urbana, Illinois, USA, 6–8 May 2005. A selection of the papers was revised and published as *European Modernism and the Information Society.* Edited by W. Boyd Rayward. Aldershot, Hants: Ashgate, 2008.

Analogous Spaces: Architecture and the Space of Information, Intellect, Action, a colloquium organised by Pieter Uyttenhove and Wouter van Acker at the University of Ghent, Belgium, 1 May, 2008. A selection of papers from this colloquium were published as Wouter Van Acker and Pieter Uyttenhove (eds), "Information and Space: Analogies and Metaphors," *Library Trends*, 61(2): Fall 2012.

Introduction
International Exhibitions, Paul Otlet, Henri La Fontaine and the Paradox of the Belle Époque

W. Boyd Rayward

From the close of the year 1811 intensified arming and concentrating of the forces of Western Europe began, and in 1812 these forces—millions of men, reckoning those transporting and feeding the army—moved from the west eastwards to the Russian frontier, toward which since 1811 Russian forces had been similarly drawn. On the twelfth of June, 1812, the forces of Western Europe crossed the Russian frontier and war began, that is, an event took place opposed to human reason and to human nature. Millions of men perpetrated against one another such innumerable crimes, frauds, treacheries, thefts, forgeries, issues of false money, burglaries, incendiarisms, and murders as in whole centuries are not recorded in the annals of all the law courts of the world, but which those who committed them did not at the time regard as being crimes ... The people of the west moved eastwards to slay their fellow men, and by the law of coincidence thousands of minute causes fitted in and co-ordinated to produce that movement and war (Tolstoy, *War and Peace*).[1]

The Proud Tower built up through the great age of European Civilisation was an edifice of grandeur and passion, of riches and beauty and dark cellars. Its inhabitants lived, as compared to a later time, with more self-reliance, more confidence, more hope; greater magnificence, extravagance and elegance; more careless ease, more gaiety, more pleasure in each other's company and conversation, more injustice and hypocrisy, more misery and want, more sentiment including false sentiment, less sufferance of mediocrity, more dignity in work; more delight in nature, more zest. The Old World had much that has since been lost, whatever may have been gained (Tuchman 1967, 544).

From the middle of the nineteenth century to 1914 the daily life of Europeans was radically transformed by an astonishing number of innovations. This period of innovation, historically unique in its speed and global reach, led Charles Péguy to say that "the world changed more between 1880 and 1914 than since the Romans" (Matin and Giget, 2001, 30).

---

1   Tolstoy, Leo. *War and Peace* (Book Nine, Chapter 1). Translated by Louise and Aylmer Maude [1922–23 pp. 458, 460] http://books.google.com.au/books?id=jhZzwKsi0 OsC&pg=PT1&dq=Aylmer+maude&hl=en&sa=X&ei=TYsYUciOCYSkkwXc0IHgAQ& output=reader.

On June 28, 1914 the heir apparent to the Austrian Empire, the Archduke Franz Ferdinand and his wife, were paying a state visit to Sarajevo in Bosnia. As they drove by in an open carriage, they were assassinated by a member of a group of Serbian conspirators. Slowly, inexorably, despite all efforts to avert it, a situation arose like that described in Tolstoy's account of Napoleon's invasion of Russia. Day by day, for five agonising weeks, thousands of minute causes fitted in and co-ordinated like the slowly engaging gears and wheels of a huge engine to produce a movement that ended in the outbreak of a general European war. The war rapidly escalated until, except for a handful of neutral states, it involved the entire world. The four years of violence and invasion that followed the echo around Europe of those shots in that remote corner of the Austro-Hungarian empire produced not only devastation, destruction, mutilation and death on a scale so vast that even Tolstoy could hardly have imagined them, but also political and social upheaval on a scale no less vast. The old world of European emperors and empires disappeared, though the sun continued not quite yet to set on the British Empire, and a recognisably modern world emerged from the ruins of the War. This was a world, however, so insecurely, so precariously ordered by the peace treaties of 1919 and 1920 and by subsequent national and international developments that the armistice of November 11, 1918 that was supposed to end the "War to End all Wars" in hindsight did nothing of the sort. It had taken a hundred years from the Congress of Vienna for the general peace of Europe to be shattered by the first World War. It took a mere twenty years for an unstable, uneasily maintained peace sustained at the end by misleading documents, one merely a flimsy note waved in the air before a cheering crowd by the British Prime Minister, Neville Chamberlain, to be broken by Nazi Germany's invasion of Poland on September 1, 1939. And once more thousands of minute causes fitted in and co-ordinated to produce the movement that led to a new world war even more terrible in its conduct than the war that had preceded it.

Yet in those last decades of the nineteenth century and the first years of the new century, the period known as the *belle époque*, Europe seemed relatively secure, prosperous, stable, full of possibilities. General world peace had been disturbed, for example, by the Boer War (1889–1902), the Boxer rebellion in China (1900–1901), the brief Russo-Japanese War of 1904–05, some failed German gun-boat diplomacy in Morocco in 1905–6 and again in 1911, and by never-ending tensions in the Balkans all of which excited intense European interest. But most of these disturbances were relatively short, localised and distant. There was nothing, at least until the outbreak of the Balkan Wars of 1912 and 1913, to suggest that rising nationalisms, the emerging pattern of international alliances, the rivalries of imperial expansion, an accelerating international arms and naval race fuelled by dramatic innovations in weapons design and the construction and movement of shipping, the spread of socialism, the advent of communism and of anarchism with its spectacular assassinations along with the social and industrial unrest that fuelled the development of these movements, could not be controlled.

Looking back, we may think that it was inconceivable that so exciting, so eventful a period culturally and intellectually with its ever-strengthening international orientation could end in the disaster of world war. It was a period of efflorescence in the arts and sciences. This was the period of Rutherford, Bohr, Einstein, Mme Curie, the Solvay Conferences on Physics, Freud, Proust, Gertrude Stein and her circle, of Oscar Wilde, of James Joyce, of the *International Catalogue of Scientific Literature* and other great bibliographic enterprises, of the *Carte du Ciel*, of Stravinsky and the Diaghilev ballets, of the triumph at last by Wagner in the opera houses of Europe, the advent of Debussy and of Webern, the creation of the Nobel Prizes, the spread of the influence of the Art and Crafts Movement and of Art Nouveau into almost every nation of Europe and beyond.

It was the period of a widespread development of interest in international arbitration as the basis for settling disputes between nations to ensure the maintenance of peace. This led in 1889 to what became grand annual Universal Peace Congresses in the different cities of Europe, the United Kingdom and the USA. In 1891 in part as a result of the early Peace Congresses, the International Peace Bureau was set up, the headquarters for the International Union of Peace Societies. The Bureau won the Nobel Peace Prize in 1910.[2] The year 1889 also saw the formation of the Interparliamentary Union which also met almost annually, for a number of years in association with the Peace Congresses. Despite the hypocrisy, scorn and reluctant participation of many of the official governmental representatives, The Hague Peace Congresses of 1899 and 1907 seemed to have had important successes in terms of the international conventions eventually concluded for the pacific settlements of disputes, the creation of regulations governing the conduct of war, and the specification of the rights and obligations of neutral powers (Scott 1913; Tuchman 1967, Ch. 5; Cooper 1991).

A relatively young French Jew of thirty-one, Julien Broda, who was to become a well known philosopher, a man of letters and survivor of the Second World War, reflecting on what he remembered of this period observed: "We were sincerely persuaded in 1898 that the era of wars was over. For fifteen years from 1890 to 1905 men of my generation really believed in world peace" (quoted in Tuchman 1967, p. 272).

It was moreover a period in which governments increasingly came together to create official multinational treaties such as those for managing standardised international systems of weights and measures, posts, the telegraph and railways, and establishing, for example, the prime meridian for providing a common global reference point for determining time and cartographic location for navigation. It was, that is to say, a period of emerging and strengthening infrastructures of communications, of technologies and agreements that allowed people, ideas and capital to flow relatively unhindered and with a previously unheard of rapidity across all the land and sea borders of Europe and the Western World. Progress, industrial and scientific development, global trade, social amelioration, hard-won political liberalism, a flowering in the fine and liberal arts, dynamic peace movements and

---

2   http://www.nobelprize.org/nobel_prizes/peace/laureates/1910/.

above all, internationalism in all of its ramifications seemed to typify the age. But the end did come violently, seemingly abruptly, completely.

In his memoirs of growing up in Vienna in this period, Stefan Zweig captures something of the pre-War sense of peaceful, comfortable progress.

> When I attempt to find a simple formula for the period in which I grew up, prior to the First World War, I hope that I convey its fullness by calling it the Golden Age of Security … In its liberal ideals, the nineteenth century was honestly convinced that it was on the straight and unfailing path toward being the best of all worlds. Earlier eras, with their wars, famines, and revolts, were deprecated as times when mankind was still immature and unenlightened. But now it was merely a matter of decades until the last vestige of evil and violence would finally be conquered, and this faith in an uninterrupted and irresistible "progress" truly had the force of a religion for that generation. One began to believe more in this "progress" than in the Bible, and its gospel appeared ultimate because of the daily new wonders of science and technology. In fact, at the end of this peaceful century, a general advance became more marked, more rapid, more varied (Zweig 1943, 1, 3)

Zweig goes on to mention not only new inventions such as electricity illuminating the city's streets and the presence of "horseless carriage" but advances in plumbing, hygiene, social welfare, and the extension of the franchise. "Sociologists and professors," he said, "competed with one another to create healthier and happier living conditions for the proletariat" (Zweig 1943, 3–4). He also stressed that "There is hardly a city in Europe where the drive towards cultural ideals was as passionate as it was in Vienna." It was an epicurean city too in which life was to be enjoyed, not only in terms of food and wine but for its music, dancing, and theatre. Indeed, the ordinary Viennese, he said, opened their morning newspapers not for world news but for news of the theatre. "It was," he concluded, "wonderful to live here, in this city which hospitably took up everything foreign and gave itself so gladly …" (Zweig 1943, 13–14).

Technology, Commerce, Culture and Knowledge: The Belle Époque on Display

Much of what Zweig says of Vienna could equally be said of most of the great cities of the world at this time for, to echo Zweig, they too took up everything foreign and gave of themselves so gladly. We see this reflected dramatically in the great Expositions or World's Fairs that were so much a feature of this period. They constituted not so much periodic snapshots or even changing kaleidoscopes but detailed, encyclopedic, museum-like representations of the world as it was thought to be at each of times in which they were held. Ephemeral like fireworks, they sprang up, sometimes as many as three or four a year, in the major and not so major centres across Europe, the US and even as far away as the British dominions. In

1888, for example, international expositions appeared in Barcelona, Copenhagen, Melbourne and Brussels (Poirier, 1958 98). In these magnificently engineered environments, epitomised perhaps by the Eiffel Tower of 1889, rose that Proud Tower of European civilisation to which Barbara Tuchman refers so elegiacly and which was brought down by the First World War.

The Expositions drew enormous admiring crowds from all over the world. Millions marvelled at innovations and discoveries that heralded the appearance of a new kind of modern life that was rapidly developing and expanding before their astonished eyes. The presence of these multitudes of visitors reflected the ease and speed with which people and goods could be transported by the network of railways that, as the century progressed, ever-increased in power, capacity and comfort as they ever-more tightly interlinked every corner of the European continent. The oceans were equally hospitable. Developments in maritime technologies, especially steam turbines, iron and then steel-framed ships and submarines, while reflecting and driving national naval rivalries, also led to the building of commercial liners and cargo ships that became ever faster, safer, larger and, some of them, refrigerated. As various national commercial shipping lines became established and in competition, the long journeys of goods and people to and from the new and old worlds, to and from the metropolitan centres and the colonial peripheries, to and from India, the Orient and Australasia became more regular, cheaper and in a sense shorter and more profitable. And these vessels could be used to lay the undersea cables that allowed information in the form of news and of various personal and official kinds to be transmitted almost instantaneously from one continent to another.

At the international expositions in often architecturally spectacular if temporary pavilions, the nations celebrated historical anniversaries, proclaimed the brilliance of their achievements, and revealed the magnitude of their nationalist aspirations in science, technology, industry, commerce, agriculture, empire-building, architecture and the arts—in effect, every sphere of activity and daily life. The number of innovations that appeared in this period, observed Martin and Giget, was "staggering." "Essentially what we have around us today had been created or had its basis in this period" (Matin and Giget 2001, 30–31). They mention the automobile, the bicycle, the lift or elevator, the telegraph and telephone, the typewriter and the sewing machine among many other machines and devices, all making their appearance both as products for exhibition and later as services—available to the public at the Expositions themselves in the case of the telephone and telegraph. And successive expositions revealed how these sorts of invention while constantly being improved seemed to move ever more quickly from idea, to prototype, to industrial production, to international availability and distribution in a diffusion process that signaled the advent of our modernist society and economy. Uniting propaganda, entertainment, and information, these universal encyclopedic exhibitions also provided "elaborate projects for the improvement of social conditions in the areas of health, sanitation, education and welfare." They were "promissory notes," observed Tony Bennett, "that the engines of progress would be harnessed for the general good" (Bennett 1995, 82).

To take an example. The Paris Exposition of 1889, over which the Eiffel Tower loomed, saw 32 million people pass through its gates. This Exposition was to be

> the greatest in every sense of the word of any that have preceded it in France and abroad. Never will buildings so vast have been devoted to an exposition. Never will the manufacturers of any country in the world have had at their disposal a hall comparable to the Palais des Machines … Never will an engineer have dared to raise a tower three hundred metres high. Never moreover will so great a number of exhibitors be brought together from so varied a number of countries. Never, finally, will more attractions and entertainments have been promised to the innumerable visitors from all points of the globe already indicating that they will be attending.[3]

Its organisers pointed out that there had been international congresses of various kinds held on the occasion of past expositions, but these congresses had been few and uncoordinated. In 1889, the centennial year of the French Revolution, the Minister had decided that the physical splendours of the Exposition would be matched intellectually by a series of congresses and conferences that would cover all the branches of human knowledge. These gatherings would help create international agreement in such matters as "weights, measures and currencies, the application of sanitation regulations, the preparation of comparable statistics, and the conduct of great scientific work either in collecting comparable data on questions on which differences in climate, race and temperament do not permit the adoption of uniform solutions or in determining the relative state of science in the different parts of the world" (Picard 1891, 327–8). Over 70 international conferences are listed with reports of their discussions and resolutions taking up one or more volumes.[4]

The Paris Exposition revealed in a way that could not be clearer, both explicitly by the organisers and implicitly by what was finally achieved "on the ground," the fundamental paradox implicit in these great expositions. They embodied a universalist aspiration that was harnessed to goals of nationalist prestige. Beneath the excitement and glitter was a powerfully competitive ranking and display of the relative power and progress of the nations in all of the transformative developments of modern life as it was then emerging.

The Universal Exposition of Brussels, 1897

This is no less clear in Brussels almost ten years later. On the initiative of King Leopold II, King of the Belgians, the fifteenth section of the 1897 Universal Exposition of Brussels, was designated a separate International Colonial Congress by the Exposition's organising committee. The congress had a large attendance

---

3   *Bulletin Officiel de l'Exposition universelle de 1889*, vol 1; Paris: Bureaux. p. 3.

4   For a list of proceedings volumes see Le Conservatoire Numérique des Arts & Métiers. Publications des Expositions nationales et universelles (par auteurs) http://cnum. cnam.fr/RUB/fcata_expo.html).

of scholars, politicians and colonial administrators. It was held not where the rest of the exposition was taking place in the Parc du Cinquantenaire, more or less in the centre of Brussels, but in the impressive, newly reconstructed Palais des Colonies in the Parc de Tervuren. The two sites were linked by a special tramway, a "monorail" (although it seems that the monorail "had five rails in fact"; Legrand 1898, 298) and a grand boulevard constructed for the occasion at the instigation of the King. Only one colony, however, was involved in the Palais des Colonies, the Belgian Free State which was neither Belgian nor free but the private fiefdom of Leopold II, knowledge of the horrors of whose "reign" there were only beginning to emerge (Hochschild 1998).

Opening the congress, the Belgian Foreign Minister articulated in a rhetoric of civic virtue the rationalisations that led what we now accept as the economic exploitation and social oppression of native peoples by the colonial powers in their competitive quest for new territories. The Congress's aim, he said, was "to study in a common aspiration for progress, the serious problems that were being experienced by the government and administration of the new continents to which the old nations must offer the benefits of civilisation." Science along with "A legion of intrepid sailors" and "heroic travellers" had led to "civilisation's conquest." He went on to observe aphoristically: "In effect to colonise is essentially to civilise. Universal history proves this axiom. ... Colonisation is therefore a manifestation of progress; it is also an expression of human solidarity. It brings together inferior races with those who are more advanced. It teaches them how to improve, to ennoble their conditions of existence" (De Favereau, 1887, 38).

These words help reveal an aspect of the ironies inherent in the doctrine of modernist social and economic progress in this *fin-de-siècle* period of the great international expositions for, outside the Palais des Colonies in the Parc de Tervuren, were the huts of nearly 300 Congolese men and women of whom only 6 it is said died of influenza from the effects of the temperamental weather of that Belgian summer (Esgain 2001, 127). "The blacks [les noirs] were installed in villages imitating those of the Congo and were surrounded by native African plants." Rather than maps and posters of the principal tribes, the visitor was provided with "a living Congo ethnography" (Braun 1987, 19, 138). The native peoples in their huts and with their hints of "fierce and cannibal customs" fascinated a public of over a million persons who visited them in the course of the six months of the Exposition. They came to witness of the primitiveness of their modes of life, to experience a sense of the almost non-human otherness that they represented and to glory in the benevolent, improving power of their Belgian master.

Inside the Palais des Colonies in a beautiful art nouveau hall built of Congolese wood and intended to represent the jungle, were stuffed animals and exhibits of art and other cultural, botanical and ethnographic objects from the Congo. But there were also separate galleries for exhibits of imports and exports designed to emphasise its potential benefits to be had from the commercial exploitation. especially in terms of collecting and processing its products such as wood, ivory, cocoa, coffee and above all rubber. A separate gallery called a *musée commercial*, was devoted to the sorts of merchandise believed to be desired by the indigenous populations. Here was provided information about packaging methods that could be used to help avoid damage and theft in the transport of these sorts of goods to the colony by ship (Legrand 329; Braun 1897, 155). The exhibits in the Palais des Colonies and the Palais itself became the basis of the Musée du Congo created the next year and now known as the Musée Royale d'Afrique Centrale.[5]

Away from this outpost of Empire, in the Parc du Cinquantenaire the rest of the International Exposition, the opening of which was delayed by torrential rains, unfolded in the many pavilions in the gardens and in the exhibition galleries of the great wings of the Palais du Cinquantenaire. The 1897 Exposition, undertaken and funded privately but with major support from the State, was to be in fifteen sections such as Fine Arts; Industrial and Decorative Arts; Hygiene; Lighting and Heating and Their Applications; Electricity; and the Material, Processes and Products of Industrial Manufacture. The fourteenth section dealt with the Congo and the 13th with Congresses and conferences, though this enumeration was later varied and congresses and conferences were no longer regarded as forming a special section but as a component of the many ancillary exhibitions, music competitions, concerts, and festivals that would take place on the occasion of the Exposition (Legrand 1898, 27). Thirty nations participated and provided 10,000 exhibits for the six million visitors to the Exposition (Slate, n.d.).

An innovative feature of the Exposition was the setting up of an International Competition of Science and Industry. The idea was that the various sections should raise important questions and problems that might be addressed in the displays and help guide future progress. Cash prizes were offered for the best solutions or answers demonstrated by the various processes, products and machines being exhibited. Exhibitors were expected to indicate how their exhibits responded to particular questions and the exhibits were assembled in the international galleries according to the particular phenomena involved. Only exhibits not in competition were to be displayed in national galleries. The importance of this aspect of the Exposition is indicated by the fact that 885 special prizes were awarded at the Exposition, by far the highest number going to French exhibitors (Legrand 1898, 10, 27, 34).

Bibliographic Internationalism: An Epiphenomenon of Science and Scholarship

A section of the Sciences was devoted to the disciplines concerned with observation and experiment without necessarily having industrial or commercial applications. This section consisted of six divisions and was entered somewhat ludicrously amid cannons, artillery shells and torpedoes through an area shared with the Military Arts. As well as exhibits related to the traditional disciplines of astronomy and

5   "History of the RMCA," http://www.africamuseum.be/about-us/museum/history.

meteorology, physics, chemistry, biology, geology and anthropology, there was a division related to Bibliography. This last division had the goal of "collecting, classifying and cataloguing all the production of the human intellect." It presented an outline of the Dewey Decimal classification. This seemed to Legrand a very simple method that he hoped might be "adopted by all. In this way not only the scholar or the professional person but the ordinary reader would be able to find immediately the list of works that dealt with the questions that interested him (Legrand 1898, 312).

The bibliographic exhibit in the Section of the Sciences had been organised by Paul Otlet and Henri La Fontaine representing the International Office of Bibliography, a semi-official agency of the Belgian government that they had founded in 1895 following an International Conference on Bibliography in September that year. Legrand's observations reflected the aspirations that they had invested in the special development of the Decimal Classification as the tool by means of which a universal bibliographic catalogue on cards (Répertoire Biliographique Universel) would be organised within the Office of Bibliography. The Office was also to be the headquarters for a related International Institute of Bibliography, a loosely affiliated group of individuals from all over the world interested in the kinds of bibliographical problems being explored by the Office of Bibliography and its founders. The idea was that members of the institute would meet from time to time (the IIB's pre-War international conferences were held in 1895, in 1897 at the International Exposition of Brussels, in 1900 at the Universal Exposition of Paris, 1908, and in 1910 at the Universal Exposition of Brussels and continued after the War).

One of the problems presented by the construction of the Universal Catalogue was how to reproduce cards quickly and in many copies so that the whole or parts of the universal repertory could be distributed throughout the world. Following the Exposition procedure, Otlet and his colleagues offered a prize of 500 francs for a machine or process that would enable the rapid and economical printing of from 50 to 100 copies of cards with the plate created for each card storable for later reuse.[6] Nothing came of this competition and the problem was referred to the International Bibliographical Conference that was held in August that year as one of the conferences associated with the Exposition.

This conference was well attended with a number of representatives from the US as well as from the countries of Europe. Its discussions included a wide-ranging canvassing of important bibliographic developments from around the world; the problems involved in developing the Decimal classification as the central tool for organising a universal cooperatively compiled universal catalogue; and issues related to international cooperation more generally. And of course a committee was established to examine the problem of finding the most economical and practical methods for printing bibliographic cards. The delegates were taken on visits to

---

6 "Faits et Documents: Exposition Bibliographique de Bruxelles," *Institut International de bibliographie Bulletin* 1 (1895–6): 120–21.

the installation of the International Office of Bibliography to examine the work being undertaken there for the Universal Catalogue, to the bibliographic exhibit that had been prepared for the Science section of the Exposition and, among other excursions, at night to the Parc du Tervuren ablaze with electrical lights.

One of the conference's resolutions was to offer congratulations to the Swiss authorities for their support of the Concilium Bibliographicum in Zurich. Behind these few words of tribute lie an important exercise in scientific internationalism. The Concilium Bibliographicum had been set up in 1895 by a brilliant, multi-lingual American zoologist, Herbert Havilland Field, to prepare and distribute card bibliographies initially on Zoology but expanding subsequently to cover related subject areas such as anatomy, physiology, and palaeontology. In developing his bibliographic project, Field had decided that bibliographies on cards were an important innovation that would avoid the delays and inconveniences of published bibliographies. The use of cards enabled the maintaining currency and access to a systematically organised, always up-to-date, cumulated record of a science. He had consulted Melvil Dewey about the usefulness of the Decimal Classification for the subject arrangement of such bibliographies. He had also held discussions about the need for them with various scientists in the US and Europe and raised the matter at a number of European scientific conferences. The third International Congress of Zoology held in Leiden in 1895 agreed to support his plan for a special bibliographical agency for Zoology and with subventions from the Canton and City of Zurich, Field set up the Concilium Bibliographicum in Zurich in the autumn of 1895 (Ward 1921). As he prepared to begin work, Field consulted closely with Otlet and La Fontaine in Brussels. To secure Field's collaboration, they agreed to use the standard American 3 × 5 inch (or 125 × 75 mm) card for their bibliographic work. Independently of Field, Otlet had already established an agreement with Melvil Dewey in the US to develop a European version of the Decimal Classification that would be suitable for the minute classification required for bibliographies. Field for his part undertook not only to supply references on cards prepared at the Concilium Bibliographicum to the Universal Catalogue being compiled in Brussels but also to develop the tables of the Decimal Classification related to the sciences of interest to the Concilium Bibliographicum. In addition to the card bibliographies, Field also used his cards to prepare regular printed bibliographical supplements to two of the major scientific journals, the *Zoologische Anzeiger* and the *Anatomische Anzeiger* (Rayward, 1975, ch. V) Thus Field and the Concilium Bibliographicum represent a kind of pivot in the creation of an international information infrastructure involving elements from the US, Switzerland, Germany and Belgium for managing flows of scientific bibliographic information originating in books, journals and research reports from all around the world.

One reason that the Bibliographical Conference in Brussels was held in 1897, apart from the international context that the Exposition provided for it, was not to have a meeting in potential conflict with a meeting in London that the Royal Society had called in 1896 to begin to discuss what became the monumental *International Catalogue of Scientific Literature*. Otlet, La Fontaine and a colleague represented Belgium at this and subsequent meetings and tried to influence the Royal Society's representatives to adopt the Decimal Classification for this work. A delegation from the Royal Society visited Brussels but remained unconvinced and drew up a special classification that was continually refined and revised during the course of the publication of the *International Catalogue*. This began to appear annually from 1900 in 17 broad subject areas covering the scientific literature from 1900 to 1914. Ultimately the catalogue comprised 238 volumes, the last of which was not published until 1916. Governed by an international convention and financed by a loan from the Royal Society, the catalogue involved a complex machinery of international cooperation. Like spokes to a wheel, 32 countries set up Regional Bureaux to collect references to the most important scientific literature published in their country or region, to transcribe the references in a highly stylised format onto standardised slips of paper and forwarded the slips to London where they were collated, checked and published. The currency of the material in each annual volume of the catalogue depended on the efficiency of the bureaux and the speed of rail and sea transport of the packets of slips from the outlying areas to the centre in London. There were regional bureaus set up, for example, in the distant former British colonies that after 1901 had become the states of NSW, Victoria, Queensland, South Australia and Western Australia of the newly independent, federated Commonwealth of Australia. Nearer the centre, the International Office of Bibliography in Brussels became the Regional Bureau for Belgium. Otlet and La Fontaine represented Belgium at the various meetings of the convention governing the catalogue until it went into liquidation in the early 1920s.

There was yet another bibliographical venture of this kind, *the Répertoire bibliographique des sciences mathématiques*. The idea for this had been broached as early as 1885 by the Société mathématique de France. Eventually with the cooperation of some 50 mathematicians in 16 countries the repertory appeared between 1894 and 1912. It was published on long narrow cards with a classification number "of a rare complexity"[7] at the top and up to ten entries for the books and journal articles in that category listed on each card. The repertory was designed to cover the period from 1800 through 1900, not to be a current index. It eventually comprised the analysis of an estimated 20,000 works. Its classification and the use of cards represented a carefully considered experiment in the diffusion of mathematical information and, until Field had persuaded Otlet and La Fontaine otherwise, they considered using the format of these cards for their own bibliographical work.

What is of interest is that these vast, sometimes overlapping, sometimes competitive bibliographical enterprises, whose compilation depended on various

---

7 The phrase "rare complexité" was used to describe the classification system in the French original of this article http://poincare.univ-nancy2.fr/digitalAssets/12594_ bibliographie_mathematique_ideale.pdf.

forms of international cooperation, should emerge at much the same time in this period of rapid scientific development. They may be seen as a useful epiphenomenon of science and scholarship that testifies to the rapidity of the growth in size, complexity, specialisation and fragmentation of the world of contemporary scientific knowledge. They suggest how pressing the imperative had become that this world be brought under some form of integrated, coordinated, international control that would ensure orderly cumulation, ease of access and the avoidance of duplicative research. The discussions that led to the creation of the various Classification systems in their turn suggest the extent to which groups of scientists and scholars struggled to achieve agreement on how their disciplines were constituted conceptually, to formulate the criteria for inclusion and exclusion when identifying the components and relationships of various subject areas and to how to create effective notations to represent the affiliations of established and emerging fields and subfields. As Rollet and Nabonnand point out:

> During the second half of the 19th century, a large number of disruptions affected science: an unprecedented increase of research and a growing specialisation in every domain; the organisation of institutions in networks; an institutionalisation of research in most European countries via the creation of academies, learned societies and universities … finally the considerable acceleration of the internationalisation of science (Rollet and Nabonnand 2003, 9)

They also refer to the exponential growth of the scientific literature in this period which stimulated so much of the bibliographical activity discussed above.

Some sense of the growing international traffic of scientists, scholars, officials, and others is given by the increasing number of international meetings that were held in this pre-War period. In the contemporary *Annuaire de la Vie International* for 1910–11, Table VII shows that in the period from 1860 to 1879 there were 246 of these meetings. In the period 1880 to 1899 this number had risen to 819 and in the ten year period 1909 to 1909 the number was 1,070 (*Annuaire de la Vie Internationale 1910–11*). How were the transnational flows of information that resulted from participation in these meetings to be organised? The answers to such a question could only be tentative but it nevertheless required and received immediate practical attention.

The Chapters in this Volume

The authors of the chapters in this book in terms of their affiliations come from Melbourne and Sydney in Australia, from the USA via Leeds, from Vienna, from Switzerland via Paris, one has come from Vienna via Moscow, others have come more directly from England, the Netherlands, Germany, and Austria. Several have transcended the invisible but powerfully restrictive internal borders that can divide a single nation like Belgium. This is a form of intellectual voyaging that is intended to explore the ways in which for a relatively limited historical period societies create and manage knowledge, how networks of personal contacts and of publication emerge to instantiate knowledge for distribution and use, and how boundaries between disciplines and epistemic cultures are transcended as we strive to arrive at new kinds of understanding of our past. In bringing together scholars from a variety of disciplines, research practices, linguistic backgrounds and different levels of experience the book itself is intended to represent a kind of physical exemplum of information transcending a range of boundaries.

The chapters in this volume do not pretend to be a systematic or comprehensive study but to assess through the various subjects they explore the implications of the increasing, diversifying flow of information (and the documents in which it was recorded and the people who created, expressed and used it) across and beyond the borders of the states of Europe. The chapters deal with the period of the so-called *belle époque*, approximately from 1880 to the outbreak of the First World War. As discussed above, this was a period of enormous, exhilarating, disturbing change that was a response in part to the emergence of new communications technologies and information infrastructures. These allowed the relatively easy international movement of people, goods, influence and ideas to create what is recognisably a new kind of globalising information society, but one cut short by the movement of deeper historical forces that led to world war.

Information History

Our hope is that the chapters offer an approach that is unusual and suggestive of what more might be done along the lines it describes. All societies depend in both obvious and less obvious ways on information and the technical and institutional infrastructures by means of which it is produced and disseminated and access to it facilitated or withheld. Our hope is that these chapters suggest that an information history lens is valuable in examining an important historical period and the issues it encompasses. As global trade begins to develop how do the companies involved create and retain some kind of constructive identity across the various international sites in which they establish themselves? In an innovative chapter Alistair Black offers one answer to the question in his chapter: "An Information Tool for Dismantling barriers in Early Multinational Corporations: The Staff Magazine in Britain before World War I." As global trade in the period grew in volume, importance financially and in the competitiveness of national business interests, the idea gained ground not only that acquiring the right kind of information or intelligence might well secure a competitive edge for an enterprise but that the provision of such information or intelligence might itself be a commercial proposition. These ideas led to the formation of commercial museums in most of the major centres of Europe. In his chapter, "From Display to Data: The Commercial Museum and the Beginnings of Business Information, 1870–1914," Dave Muddiman explores "the origins, functions and significance of commercial museums," concentrating on the examples of the Musée Commercial in Brussels (founded in 1882), the Imperial Institute in London (1892) and the Commercial Museum of Philadelphia (1894).

International Organisation and Pacifism

An important document both in itself but also as a source of contemporary data about international organisation is the *Annuaire de la Vie Internationale* in its two editions of 1908–09 and 1910–11. In his chapter in this book, "Alfred H. Fried and the Challenges of 'Scientific Pacifism' in the Belle Époque," Daniel Laqua discusses the background to the development of this example of transnational information exchange and the eventual outcome of the collaborations it involved with Fried, Paul Otlet, Henri La Fontaine and others involved in setting up in Brussels the Central Office of International Institutions in 1907 and the Union of International Associations in 1910. The *Annuaires* represented an interest in documentation shared by all of the protagonists involved in that the volumes originated in an extensive, extremely detailed, formal survey of international organisations undertaken by the Central Office of International Institutions. The survey was intended to cover all aspects of the foundation, range of activities and in some cases dissolution of international official and non-official "unions, associations, institutes, commissions, bureaux, offices, conferences, congresses, expositions, publications" that then existed—the subtitle of the volume. Laqua discusses the difficulties experienced in preparing the two editions of the *Annuaire*, the light that they throw the pacifist movement in the period, the contributions of Fried and La Fontaine to this movement, and the stresses and strains that appeared in the relationships between them until the final rupture that occurred when war broke out. Laqua also discusses the Dutchman, Pieter Eijkman's alternative approach to documenting the international movement that resulted in his "l'internationalisme medical" (1910) and L'internationalisme scientifique" (1910).

The curious story of Eijkman is told by Geert Somsen in his chapter "Global Government through Science: Pieter Eijkmans Plans for a World Capital." Somsen outlines Eijkman's scientific and medical background and discusses his attempts to harness the pacifist movement, especially the 1907 meeting of the Peace Congress in The Hague in support of his idea of creating a World Capital there. Eijkman battled to convince the Dutch authorities that nationalism and internationalism were not mutually antithetical, that arbitration was the highest form of civilisation, and that a World City would enhance the country's status internationally. Though he failed, his arguments nevertheless suggest something of the paradox in the ideological movements that swept across Europe in the period of the belle époque. Influenced by Fried, Eijkman believed that internationalisation was a natural, inevitable process. He argued that an international government based on the certitudes of science was possible and should be part of the institutions of his World City.

A case study of the evolution of one of the scholarly associations of the kind that were grist to the mill of the *Annuaire de la Vie Internationale* was the Congrès international des orientalistes. In his chapter "Scholarly Networks and International Congresses: The Orientalists before the First World War," Paul Servais follows the international congresses of orientalists from city to city across Europe from the first congress in 1873 to that of 1912, the last before the War. He discusses the physical features of the various congresses, their reception, the changing numbers of participants and what this may have meant, the locations of the meetings in the different European cities and above all the subjects that were included for discussion (or excluded) and how the categories of conference sections and so on were constructed. Referring to the hypotheses of Edward Said, he points out the essential Eurocentrism of the congresses and how various scientific and political, "specifically imperialist" priorities were entangled in them, but also how, because of the information that they produced and disseminated, they transcended these limitations in important ways.

Both Otlet and La Fontaine's interest in documentation, pacifism and internationalism were strongly influenced by their membership in Brussels of a number of organisations concerned with the emerging social sciences. In his chapter, "Sociology in Brussels, Organicism and the Idea of a World Society in the Period before the First World War," Wouter van Acker discusses the contributions to the social sciences of the Société d'études sociales et politiques, the Institut des Sciences Sociales (later the Institut de Sociolgie Solvay) and the Institut des Hautes Études of the Université Nouvelle. Active in these settings were leading Belgian social theorists and later politicians, strongly socialist in their political orientations. The meetings of the various groups provided a forum for publications, reports and discussions on issues of political reform and economic and social welfare. Van Acker suggests that the idea of organicism, the comparison of society to a biological organism, was an important aspect of Brussels sociology at the time. It was also the basis for the active participation of the Belgian sociologists in the Institut International de Sociologie that had been founded in Paris in 1893 by René Worms. Van Acker shows how the theory of organicism influenced Otlet and la Fontaine' ideas about international organisation and underpinned their development in association with their colleague Cyrille van Overbergh of, and their rationale for, the Union of International Associations.

In their chapter "Laboratories of Social Thought: The Transitional Advocacy Network of the Institut International pour la Diffusion des Expériences Sociales and its Documents du Progrès," Christophe Verbruggen and Julie Carlier deal with an organisation that was part of the "nebula of reform movements" in Europe in the period under review. Created by Rudolf Broda, the authors describe the Institute as "an enlarged and locally rooted transnational advocacy network, interconnecting scientific expertise and social activism." Its fundament aims were to "encourage intellectual cooperation and the dissemination of social expertise as the engine of social progress and peace." They analyse the national locales and their influence on the sharing of content between the several journals in France, England, Germany and elsewhere that were affiliated both formally and informally with the Institute. Broda believed that female emancipation and enfranchisement was an aspect of social progress that would contribute to the obsolescence of war. In this context, Verbruggen and Carlier focus special attention on the involvement of the Institute with the rise of the Belgian feminist movement. Leading feminists, Belgian and non-Belgian, gave lectures at the Institut des Haut Études of the Université Nouvelle which, a centre for innovation in the study of the social sciences, became a major site for the creative transnational entanglements of those involved in Broda's Institute (including Broda himself) with the Belgian feminist movement.

The Problem of Language and International Communication

One of the outcomes of the increasing internationalisation of scientific and other forms of communication during this period in terms of the formation of international organisations, the growing number of international meetings, and the proliferating quantity and range of publications was a recognition in the scientific and scholarly community that national languages presented serious obstacles to mutual understanding in areas which should be universal in outlook. If information were to be shared transnationally, if the natural, spontaneous social processes of global integration, ideas about which animated so many of the social theories of the time, were to be allowed to develop freely, new forms of language were necessary.

Two chapters in this volume deal with the phenomenon of the development of international auxiliary languages that were a feature of this period. They are complementary in subject, emphasis and points of view. Markus Krajewski points out that in 1900 there were about 250 "planned or artificial languages." In his chapter, "Organizing a Global Idiom: Esperanto, Ido and the World Auxiliary Movement," Krajewski provides an overview of the development of perhaps the major artificial languages of the period, Volapük, Esperanto and its derivative Ido. He discusses the roles of the "devisers" of these languages, how the languages were received, the characteristics that were thought necessary for an auxiliary language to be effective, their particular importance in the pacifist movement, and how they supplanted each other. He analyses the involvement of the famous German chemist, Willhelm Ostwald, in the popularisation of Esperanto and his conversion to the use of Ido. In any explanation of the paradoxes of the belle époque, the role Ostwald played in the international auxiliary language movement is emblematic. With the outbreak of war he abandoned his interests in Ido as an international language with a pacifist orientation. Instead, Ostwald worked on the formation of a simplified German that could be used in the countries newly occupied by German forces in the early period of the War. This was Weltdeutsch, a language "which could be learned and used by everyone with little effort."

Fabian de Kloe's "Beyond Babel: Esperanto, Ido and Louis Couturat's Pursuit of an International Scientific Language," concentrates on the work of the Frenchman, Louis Couturat, in the development of Esperanto and Ido. He discusses the increasing numbers of international meetings that brought together people from different national language practices in mutual incomprehension, the growth in the literature of science and the number of languages in which it was published, the problems that this posed for translation, and the kinds of solution that might be had first from Esperanto and then from Ido as, in Couturat's view, a more linguistically precise and neutral development from Esperanto. Couturat was a committed pacifist and wrote in defence of the idea of "the progress of civilization towards a peaceful world state." Ido, he believed, in its logical structure and its linguistic characteristics could have maximum international impact "in advancing the principles of justice and neutrality."

The intense struggles for national identity and the languages in which this should be expressed and, in particular, the bearing that linguistic decisions would have on the conduct and publication of science is the subject of Jan Surman's chapter, "Divided Space—Divided Science? Closing and Transcending Scientific Boundaries in Central Europe between 1860 and 1900." Universities and academies in the Austro-Hungarian empire faced language and identity issues that tighter linguistic constraints in the Russian and German empires obviated. In the Hapsburg empire the lingua franca was German, an international language in the sense that it was used in so many scientific publications and in many German-speaking universities and research institutions who trained those who were to become academic staff throughout the Universities and Academies of central Europe. But with rising nationalist movements its use was challenged in the Polish, Ruthenian, and Czech communities. As the stitching that held the polygot Hapsburg empire together began to unravel and local Slavic languages gained territorial supremacy, there were paradoxically increasing efforts in these territories to transcend the limitations that the use of these languages imposed on the conduct and reporting of science. These efforts occurred not only because of the universalist orientation of science but also because of the national prestige that could accrue from highly visible participation both personally and in publications in the international scientific community. Such participation required the use of one of the "international" languages, French, German, English and to a lesser extent Italian.

Communications Technologies and News

One of the features of the period of the *Belle époque* was the speed, development and global spread of communications infrastructures. One of the most important of these was the telegraph and the submarine telegraph cables that connected the continents so that, in the words of one of the contemporaries quoted by Frank Hartmann, the ocean became "a highway of thought." Hartman discusses the implications of the new world of global telecommunications inaugurated by the submarine cables in his chapter, "Of Artifacts and Organs: World Telegraph Cables and Ernst Kapp's Philosophy of Technology." The new infrastructure changed both "the way people communicated and their view of the world." He discusses Ernst Kapp's idea that there was "a specific relationship between biology and technology": tools as an extension of the body (hammer and fist); mechanical systems paralleling organ systems (the railway and the circulatory system; the telegraph and the nervous system). He explains Kapp's theory of unconscious organic projection reproducing Kapp's illustration of the visually remarkably similar section of the submarine telegraph cable and the cross section of a nerve fibre. He stresses Kapp's interest in analysing objectively, not romanticising, the changes that were occurring as the new technology spread across the globe and in finding an appropriate conceptual framework for them, thus becoming an early media theorist.

The advent of the telegraph and the cable had an immediate effect on the creation of agencies for the dissemination of news. In his chapter, "The Formation of Global News Agencies, 1859–1914," Volker Barth discusses the history of the Agence Havas founded in Paris as early as 1835, the creation of the Reuter Agency in London in 1851 just after a telegraph line had been laid under the English channel (though Reuter had earlier resorted to a pigeon service to bridge a gap in the telegraph line from Berlin to Paris), Wolffs Telegraphisches Burö in 1848, and the American Associated Press also in 1848. Barth analyses the "cartel treaties" that created "diversely structured national zones of influence within a global communications network" for each of the agencies. He describes their modes of operation, critiques their ethos of objectivity, neutrality, accuracy and factuality in terms of the political and cultural contexts within which each operated and the various techniques that were evolved to establish these characteristics. Finally, he discusses the demise of the cartel idea as competitor agencies emerged in the 1920s and 1930s.

Willhelm Ostwald's international interests were nowhere better reflected than in his participation in the period before the outbreak of war in Die Brücke, the Bridge. In his chapter, "Collecting Paper: Die Brücke, the Bourgeois Interior, and the Architecture of Knowledge," Nader Vossoughian, quoting Ostwald, indicated that Die Brücke's name was derived "from its goal of using a specially constructed organ to unify harmoniously and effectively separate intellectual undertakings that emerge on isolated islands." The Bridge was in effect a kind of virtual organisational technology. Vossuoghian suggests that those who were interested in Die Brücke hoped "to coordinate and control the production of information from the bottom up. That is, they wanted to manage how it circulates in and between offices, schools, government agencies, and private citizens, and not just in scientific laboratories." One of the most important outcomes of the work of Die Brücke was the idea of international standard paper formats, especially Ostwald's "world formats" that influenced today's A-Series formats. This leads Vossoughian into a discussion based on the book, *Raumnot und Weltformat,* published by Die Brücke and written by one of its co-founders, Karl Bührer, of the ways in which the commitment to these formats led to a reformulation of the idea of the collector and the nature of the areas, the rooms, where the collections were housed. Vossoughian analyses images of Freud's consulting room and study and makes comparisons between Die Brücke and the German Werkbund, concluding ultimately that "standard paper formats anticipate the advent of 'collecting machines' such as the Internet."

Information, Classification and National Identity

The impact of these changes in communications form the background for Heather Gaunt's paper "'In the Pursuit of Colonial Intelligence': The Archive and Identity in the Australian Colonies in the Nineteenth Century." She stresses the importance of information to the six colonies on that remote continent some 12,000 nautical miles and more from Europe and the mother country. She analyses how the systems and patterns of the acquisition of information developed and changed as the speed and capacity of sea travel increased and when the submarine cable eventually reached the northern tip of Australia from Britain in 1871 and proceeded overland to connect the southern colonies. She discusses the emergence of formal programs of publication exchange as a way both of participation by the colonies in the international flow of knowledge but also as a way to build up the physical intellectual capital represented by library collections. The range of European, US and Imperial institutions with which these distant Australian colonial Boards of Exchange established and maintained contact was extraordinary. In the second part of her chapter, Gaunt reverses the direction of her discussion and analyses the need in the colonies of information about their own recent past in order to recreate their histories and to begin to formulated and shape their identities. The new historical imperative of the time was for formal, objective documentation. Gaunt discusses among other issues the problem of the local loss or destruction of records. To recreate them involved transcription projects especially dealing with colonial records held in London. Gaunt concluded that at work here was the idea that through collecting the documentation internationally and writing the local histories, it was possible for "the new nation to write itself into the wider histories of the British Empire, and indeed the world." These histories "offered a vehicle to project a formalised Australian identity to a global audience."

International developments in the transfer of information played out in a curious way in the Melbourne Public Library in 1910. Melbourne was the capital city of the State of Victoria and indeed of the Commonwealth of Australia until Canberra was built. Metropolitan and modern, it was a "large, complex, multifaceted urban centre" with "a world class public library." Mary Carroll and Sue Reynolds in their chapter "The Great Classification Battle of 1910: A Tale of 'Blunders and Bizarreries' at the Melbourne Public Library" discuss the conflicts that arose in the library over the question of how its collections should be classified—by the home-grown system currently in operation or the Dewey Decimal system from the US or the Universal Decimal Classification developed in Brussels by Otlet and La Fontaine. With the increasing availability, speed and affordability of sea travel in this period, it had become not uncommon for officials and others to spend time in exploring what European countries had to offer them. Libraries and librarians were no exception. Morris Miller, one of the protagonists in the classification debate, on leave from the Library, visited a number of European libraries and also the installation of the International Office and Institute of Bibliography in Brussels. Its catalogue left him "spellbound." He also met the Chief Librarian of the Melbourne Public Library, a proponent of the Dewey system, in Scotland where they had amicable discussions. Back in Melbourne however, the classification controversy boiled over into formal hearings before a "Library Staff Disaffection Committee" established by the Board of Trustees. The national–international classification battle was won by the American Decimal Classification.

The completed full edition of the Universal Decimal Classification was published in 1905 as the *Manuel du Repertoire Bibliographique Universel*. It was a huge volume of over 2,000 pages incorporating tables that had been collaboratively developed by scientists and scholars throughout Europe including the classification tables for a range biologically related subjects prepared by Field and his staff in the *Concilium Bibliographicum*.[8] The appearance of this huge volume attracted the attention of Johan Zaalberg in the Netherlands who consulted Otlet about the use of the Universal Decimal Classification for the management of the administrative records of municipal governments, something about which Otlet had already written at some length. In his chapter, "Dynamics of Networks and of Decimal Classification Systems, 1905–35," Charles van den Heuvel discusses the increasing Dutch interest in the classification that began with Zaalberg in the early 1900s. His focus is ultimately on Frits Donker Duyvis who was to become a central figure in managing the development of the UDC after World War I— the full French revision that was achieved 1929–31, the translations into other languages, eventually English, and transformative revisions mooted by Duyvis as late as 1951. How to secure a workable concordance between the Dewey system and the UDC, a kind of basic international harmonisation of the two systems, had been a tendentious problem between the Belgians and the Americans even as the first edition of the UDC was being developed in the 1890s. The problem became even more acute after its publication in 1905. Duyvis played a major diplomatic role in post-War negotiations between the Americans and Otlet and La Fontaine in Brussels about concordance between the two editions and the publication of an English edition of the UDC. These discussions see-sawed across the Atlantic between Dewey's Lake Placid Club in upper New York State and Brussels and lasted well into the period after the Second World War. Van den Heuvel also discusses a curious development of the UDC by a colleague of Otlet's in the 1930s. Walter Théodore Glineur suggested that the UDC be modified in a variety of ways to create what he called the Decimal Classification of the Human Senses [Classification Décimale d'après les Sens Humains]. He changed the name later to the Decimal Classification of Consciousness (Classification Décimale de la Conscience—CDC) to make the parallel with the UDC more apparent. Nothing came of Glineur's initiative.[9]

But the outcome of all the discussions and speculations of the time was never in doubt. In one of his great set pieces, reflecting on the British fleet's visit to Kiel, Churchill observed: "the old world in its sunset was fair." He captured the sense of what had been achieved and what might be lost as world peace hung in the balance in the fateful days of late June 1914.

---

8   For an account of the development of the material that was incorporated into the volume and of the volume itself, see Rayward, 1975, Chapter V.

9   It might well be noted that the UDC continues to be in use in 130 countries around the world and has been published as a whole or in part in 40 languages. "UDC Consortium: About the Universal Decimal Classification" http://www.udcc.org/about.htm.

The world on the verge of its catastrophe was very brilliant. Nations and Empires crowned with princes and potentates rose majestically on every side, lapped in the accumulated treasures of the long peace. All were fited and fastened—it seemed securely—into an immense cantilever. ... Would Europe have thus marshalled, thus grouped, thus related, unite into one universal and glorious organism capable of receiving and enjoying in undreamed of abundance the bounty which nature and science stood hand in hand to give.

The German Emperor, out sailing when the news of the assassination of the Archduke Franz Ferdinand arrived, "came on shore in noticeable agitation, and that same evening, cancelling his other arrangements, quitted Kiel" (Churchill 1923, pp. 198–9). By the first days of August, the world was at War.

Conclusion

The chapters of this book reveal aspects of the complexity of aspirations and the realities that far-sighted, well-meaning, imaginative individuals faced in the great melting pot of the *belle époque* period. As people, publications, objects and ideas— as information—in all its many formats and carriers moved ever more freely and quickly to and fro across the boundaries of the European states and beyond them to animate international conferences and international expositions; to stimulate local adaptations of international artistic and craft styles flowing out of England, France and Belgium; to influence the creation and disciplinary orientations of new international scholarly societies and institutes; to lay the foundation of new kinds of international information infrastructures for the management of the escalating volume and increasing fragmentation of the literature of science and scholarship, there were always tensions at work. These arose from a profound, powerful, inescapable, nationalism that was deeply rooted in the individual psyches of peoples, even in those individuals who consciously aspired to transcend the languages and cultures, the social and political institutions in which it was expressed. We see it colouring the attitudes and behaviour of some of the most deeply committed pacifist figures discussed in this book as the First World War began to draw ever closer. It everywhere informed the political and diplomatic crises of the times and the alarm that they created over the possibility of belligerent outcomes. These crises escalated in number and severity as the period drew inexorably towards the last years of peace. Perhaps in part because of the energy that such tensions produce, the *Belle époque* remains, especially in Europe but not only there, an enormously exciting and productive period scientifically, technologically, intellectually, socially and artistically. It provided a forge in which was shaped an emergent modern world and the information systems, formats, institutions, and modalities of communication on which this new world was to depend.

References

Annuaire de la Vie Internationale. 1910–11, *Annuaire de la Vie Internationale: Unions, Associations, Instituts, Commissions, Bureaux, Offices, Conférences, Congrès, Expositions, Publications*, Volume II. Bruxelles: Office Central des Associations Internationales.

Bennett, Tony. 1995. *The Birth of the Museum: History, Theory, Politics*. London: Routledge.

Brown, James Scott. (ed.) 1913. *The Hague Convention and Declarations of 1899 and 1907*. New York: Oxford University Press.

Churchill, W.S. (1923). *The World Crisis*. Toronto: Macmillan.

Conférence Bibliographique Internationale. 1897. "Compte rendu sommaire des déliberations," *Institut Internationale de Bibliographie Bulletin* 2: 253–63.

Cooper, Sandi E. 1991. *Patriotic Pacifism: Waging War on War in Europe, 1815–1914*. New York and Oxford: Oxford University Press.

Esgain, Caroline. 2001. "Pavillon Colonial," in *La Belle Europe: Le Temps des Expositions Universelles, 1851–1913*. Bruxelles: Musees Royaux d'Art et d'Histoire.

Favereau, Paul de. 1897. "Discours inaugurale." *Compte Rendu, Congrès International Colonial: Exposition Internationale de Bruxelles, 1897*. Bruxelles: Imprimerie des travaux Publics, pp. 37–41.

Hochschild, Adam. 1998. *King Leopold's Ghost: A Story of Greed, Terror, and Heroism in Colonial Africa*. New York: Houghton Mifflin.

Martin, Christophe and Giget, Marc. 2001. "Pavillion des Innovations," in *La Belle Europe: Le Temps des Expositions Universelles, 1951–1913*. Bruxelles: Musées Royaux d'Art et d'Histoire, pp. 30–47.

Otlet, Paul. 1990 (1891–2). "Something about Bibliography," in *International Organisation and Dissemination of Knowledge. Selected Essays of Paul Otlet*. Edited and translated by W. Boyd Rayward. Amsterdam: Elsevier, pp. 11–24.

Picard, Alfred. 1891. *Exposition universelle international de 1889 à Paris. Rapport Général*; t.1. Paris: Imprimerie Nationale (Le Conservatoire numérique des Arts & Métiers. Publications des Expositions nationales et universelles (par auteurs) http://cnum.cnam.fr/RUB/fcata_expo.html).

Rayward, W. Boyd. 1975. *The Universe of Information: The Work of Paul Otlet for Documentation and International Organisation*. FID Publication 520. Moscow: All-Union Institute for Scientific and Technical Information of the USSR Academy of Sciences.

Rollet, Laurent and Nabonnand, Philippe. 2003. "An Answer to the Growth of Mathematical Knowledge? Le Répertorie bibliographique des science mathématiques." *European Mathematical Society Newsletter* 47 (March): 9–14.

Slate, Paul F., "World's Fair of 1897," *Historical Dictionary of Brussels*, http://brussels.enacademic.com/822/World%27s_Fair_of_1897.

Tuchmann, Barbara. 1967. *The Proud Tower: A Portrait of the World before the War, 1890–1914*. New York: Bantam Books.

Ward, Henry B. 1921. "Herbert Haviland Field," *Science*, N.S.54: 424–8.

Zweig, Stefan. 1943 (reproduced 1964). *The World of Yesterday*. Lincoln, Nebraska: University of Nebraska Press.

Chapter 1

Of Artifacts and Organs: World Telegraph Cables and Ernst Kapp's Philosophy of Technology

Frank Hartmann

The Nascence of Global Media

By 1850 the electric telegraph had spread all over England, Europe and the populated areas of North America. Now began the ambitious endeavor of "wiring the abyss," to put it in the words of science fiction writer Arthur C. Clarke (1992). This was to use submarine telegraphy first to connect England with the continent, and second the two continents of Europe and North America—or Newfoundland with Ireland, to be exact. In fact there was no abyss at all. The idea of laying a transatlantic cable was based on recent oceanographic findings showing: "There is at the bottom of the sea, between Cape Race in Newfoundland and Cape Clear in Ireland, a remarkable steppe, which is already known as the telegraphic plateau" (Maury 1855, 317). The oceanic submarine cables, whose successful functioning required much trial and error from the engineers side, not only meant the generation of enormous revenue, but also laid foundation for the first global telecommunications hegemony of the British Empire (Hugill 1999).

The structure of the new global telecommunications order was based on old trade routes with connections and nodes at geopolitically important locations, with London eventually as the logical center of the colonial world. As this new kind of technology brought the world closer together than as never before, it created the new task of finding international standards and technical codes of exchange. This led to the first *International Telegraphy Congress* in Paris in 1865. But even in 1851 with the emergence of international World fairs such as the "Great Exhibition of the Works of Industry of all Nations" in the Crystal Palace in London, new technologies like photography and telegraphy were being popularized through demonstrations in this new environment, though not Charles Babbage's new calculating machines, much to the discomfort of their inventor who reported on the *Great Crystal Palace Exhibition* (Babbage 1851).

In these years of building an international network of communication cables, the notion of "communication" itself changed significantly from personal, direct communications to its modern, mediated form. Tele-technologies now were conceived of as extensions of mankind and as forming an environment for a culture of modern fluid aesthetics. We see this stated by Charles H. Cooley, the first sociologist of communication:

> By communication is here meant the mechanism through which human relations exist and develop—all the symbols of the mind, together with the means of conveying them through space and preserving them in time. It includes the expression of the face, attitude and gesture, the tones of the voice, words, writing, printing, railways, telegraphs, telephones, and whatever else may be the latest achievement in the conquest of space and time (Cooley 1909, 61).

What would this new infrastructure mean for human communication and thought? Could it be related to the new and much discussed theory of evolution by Charles Darwin? Was there some kind of co-evolutionary relationship between technology and biology? Could technology fit into a process of forming an intelligent network for which later the term *World Brain* would be employed?

The aggregation of technologically mediated experience in the nineteenth century came as a challenge to the philosophy of mind that had originated with Immanuel Kant and Georg W.F. Hegel just a few decades before. In 1855, a report in *Scientific American* could state triumphantly: "English telegraph engineers deserve great credit for the boldness and enterprise they have exhibited in laying down so many ocean lines. They have made the ocean a highway of thought."[1] However, in the mainstream academic traditions of the times there was little interest in reflecting theoretically on the epistemological impact of the new technology of the telegraph with its capacity to build a "highway of thought" all around the world.

Unlike most academics, however, German geographer and philosopher, Ernst Kapp, had indeed been reflecting on the shift towards a global form of communication through "universal telegraphic" technology for which he used the term *Weltcommunication*. Moving away from his orientation as a left-wing Hegelian, he created a form of materialistic philosophy for which he first used the term *Philosophie der Technik* (Philosophy of Technology). He proposed that there was a specific relationship between biology and technology which he called *Organprojektion* (organ projection), thus anticipating by about a century the notion of media technology as "extensions of man" (cf. McLuhan 1964).

Biographical Sketch

Ernst Kapp (1808–1896) had an unusual career. After academic studies in philology, he took up a post teaching at the gymnasium in Minden where he became fascinated by the new science of geography of Alexander von Humboldt and Carl Ritter. It was especially Ritter who influenced Kapp to think of geography in a physiological way in which elements of the earth were considered to be like

---

1   "European Sub-Marine Telegraph." *Scientific American*, 10 (May 19) 1855: 285.

inter-related organs. He developed these ideas in his two volume *Vergleichende allgemeine Erdkunde* which he published in 1845. This *General Comparative Geography* anticipated "what might today be called an environmental philosophy" (Mitcham 1994, 21).

Kapp was less an academic thinker than an unruly intellectual and libertine. In 1849 he published a treatise against political despotism which, following the revolutions of 1848 in the German states, made him an inconvenience to the authorities. He was prosecuted for sedition. As a political dissenter, he sought refuge in the United States where he joined a German pioneer settlement in central Texas. There he built a house and worked as a farmer. Nearby he also began to operate a spa called Badenthal where he offered "Dr. Ernst Kapp's Water-Cure." This involved hydropathy treatment and gymnastic exercises.[2] To understand Kapp's philosophy of technology, Carl Mitcham stresses the fact that during his years in Texas, he "led a life of close engagement with tools and machinery." In 1865 he paid a visit to Germany again. But suffering from bad health, on medical advice he decided against traveling back to Texas and died in 1896 in Düsseldorf.

After his arrival in Germany Kapp had taken up his academic interests again. "He revised his philosophical geography and then undertook, through reflection on his frontier experience, to formulate a philosophy of technology in which tools and weapons are understood as different kinds of *organ projections*" (Mitcham 1994, 23). The outcome of his studies was published 1877 in Germany as *Grundlinien einer Philosophie der Technik. Zur Entstehungsgeschichte der Cultur aus neuen Gesichtspunkten* [Fundamentals of a philosophy of technology; The genesis of culture from a new perspective].[3] As the subtitle indicates, this philosophical approach to technology tackled the origins of culture in a new way. What Kapp proposed was the concept of a relationship between biology and technology in which the latter was discussed as a bold yet unconscious externalization of human nature which gave rise to culture and civilization in general.

Technology as New Culture

As mentioned above, this philosophy of technology was published only a decade after the transatlantic cable came into full operation. The submarine lines had not only changed international relations, but they had also created a world of telecommunications that changed the way in which human beings experienced their existence. We see in this that the information society was already beginning to emerge and mature.

---

2   Cf. *Handbook of Texas Online* http://www.tshaonline.org/handbook/online/articles/fka01 (accessed December 16, 2012).

3   Referred to generally here as *Grundlinien*.

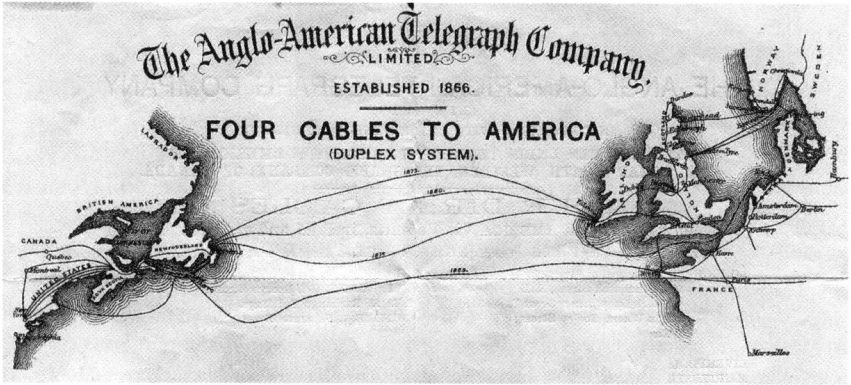

Figure 1.1   1880: Anglo-American Telegraph Company North Atlantic map[4]

Some decades before the advent of this technological phenomenon, Johann Gottfried Herder had published his *Ideen zur Philosophie der Geschichte der Menschheit* [*Outline of a Philosophical History of Mankind*, 1784–1791]. In this work he suggested that mankind was still a project in the making, subject to underlying organizational and existential changes. But while he mentioned the role of arts and crafts in the histories of different peoples, it was not until the mid-19th century with the rise of new media technologies, such as photography, sound recording and the electric telegraph that human cognition and the idea of communication itself began to change: "The far could now speak to the near, and the dead could now speak to the living" (Peters 1999, 138).

Thus technologically-induced change led to new instruments which were thought of either as potentially dangerous prostheses or as auxiliary organs which help to conquer space and time. This was observed by Alexander von Humboldt in his popular Berlin *Kosmos*-lectures (1845–62) which Kapp quoted:

> Creating new organs (tools for observing) increases the mental, often also the physical strength of humans. In its closed circuit, electricity carries thoughts and volition faster than light. Forces, the gentle powers in elemental nature, such as the tender cells of organic tissues, yet still escaping our senses, are recognized, used, and determined for higher purposes. They are aligned in the unforeseeable array of means, which draws us closer to the mastery of particular domains of nature and the lively cognition of the world as a whole (Humboldt in: Kapp 1877, 105).

Beginning with Hegel's reflection on the technical as a kind of instrumental reason in which mind externalizes itself to form an applied system of means, the concept of technology in the continental philosophical tradition stands for the extension

---

4   http://atlantic-cable.com/Maps/index.htm (accessed December 16, 2012).

of subjective skills to industrial machines and apparatuses. It can be said that to free himself from the physical world, man developed master–servant relationships with one part supplying the needs of the other (cf. Hegel 1807, chapter IV). To do so—following Carl Mitcham's interpretation—the servants

> must undertake technological work, and through work realize their own inherent dignity [...]. Slaves can transform the world, which is thus less noble than they are. From such realization comes the drive for technological progress that can free the slave too from the physical environment and create the idea of a new society of free and equal citizens. In the spirit of this analysis, Kapp's history is not the necessary unfolding of [Hegel's] *Absolute Idea*, but the differential record of human attempts to meet the challenges of various environments—to overcome dependence on raw nature. This requires the colonization of space (through agriculture, civil engineering, etc.) and of time (through systems of communication, from language to telegraph). The latter, in its perfected form, would constitute a 'universal telegraphics' linking world languages, semiotics, and inventions into a global transformation of the earth and a truly human habitat (Mitcham 1994, 22–3).

In *Grundlinien*, Kapp argued that the constructions of machines is not an arbitrary process but follows an unconscious process which he described as an "externalization." Mechanical engineering displayed morphological parallels to the human organism: first, analogies of tools with body parts are obvious (e.g., a hammer as the extension of the fist); and second, mechanical systems resemble organic arrangements (e.g., the railroad system as an extension of the circulatory system and the telegraph of the nervous system). These parallels do not follow a conscious pattern and can be observed only after the fact. Engineers constructing tunnel supports probably have never seen the longitudinal cut of a femur, yet unconsciously they follow nature's construction principles. For Kapp it was

> the intrinsic relationship that arises between tools and organs, and one that is to be revealed and emphasized—although it is more one of unconscious discovery than of conscious invention—is that in the tool the human continually produces itself. Since the organ whose utility and power is to be increased is the controlling factor, the appropriate form of a tool can be derived only from that organ. A wealth of spiritual creations thus springs from hand, arm, and teeth. The bent finger becomes a hook, the hollow of the hand a bowl; in the sword, spear, oar, shovel, rake, plow, and spade one observes sundry positions of arm, hand, and fingers, the adaption of which to hunting, fishing, gardening, and fields tools is readily apparent (Kapp 1877, 44–5, transl. by Mitcham 1994, 23–4).

Kapp was thus inverting the Cartesian tradition of regarding the organic as a mere mechanism by considering mechanical artifacts as unconscious projections of organic potentiality. As mechanical tools are unconscious projections of the osteomuscular apparatus and instruments are eventually projections of human organs, the international cable networks can be explained as projections of the nervous system. The various uses of the new technologies, he believed, would lead to a morphology of cultures and rewriting cultural history would be a necessity in terms of a co-evolutionary process. Humans unconsciously apply the form and functionality of their bodies to the objects they create. They only become conscious of this *post factum*. "This creation of mechanisms based on organic models, and also the understanding of organisms by means of mechanical devices, as well as the accomplishment of the principles of organ projection for fulfilling the objectives of human productivity, is the thesis of this work" (Kapp 1877, Preface).

Technology and the Unconscious

This attempt to reconcile technology and biology makes Kapp a philosopher, not of industrial society, but of the nascent information age. Traces of his *Grundlinien* can be found in the concepts that were developed by 20th century philosophers, among others ranging from Sigmund Freud—for whom man, with the help of his "auxiliary organs," has become "a kind of prosthetic God" (Freud 1962, 39)—to Martin Heidegger's questions concerning technology and to Marshall McLuhan's concept of the media as extensions of man. Also Kapp's notion of the "universal telegraphic," mentioned by Mitcham, invokes the idea of telecommunication networks and anticipates the world being online. At the time of the publication of the *Grundlinien*, telegraph networks had already developed far beyond the experimental period. Extended internationally, they were redefining geopolitics and global communications (Hugill 1999). With the maturing of the telegraph, communications boomed in administrative and business contexts. An incredible number of technical inventions and applications, to which Humboldt had earlier referred to rather hypothetically, were created in the period of the *belle époque*. Society changed, and as Hegel had stated earlier in his *Lectures on the Philosophy of History*, "the technological appears, when the necessity exists."[5]

New tools and instruments (and their relative mechanics) differed greatly from those that were previously available and so the concept of *techné* (from Greek: to create or manufacture something) needed revision as well as the concept of *organ* (from Greek: implement, tool, bodily organ, musical instrument) which is similar to the "érgon" meaning "work" (Kapp 1877, 60). As the hand is concerned with an object, so does the object relate to its maker, the human hand. For Kapp, the mechanical extension of the hand in a tool is not simply a lengthening of the arm such as a pole would lengthen it, but more along the lines of Humboldt's thinking, that it was a means to increase or enhance the experience of manual operations.

---

5 "[...] das Technische findet sich ein, wenn das Bedürfnis vorhanden ist." Hegel 1970, 491).

Kapp's description of organ projection involves the projection of an unconscious or unrecognized development of human capacities. Hands, arms, teeth are at the beginning of all culture. Historical artifacts such as cave discoveries and archaeological excavations signaled a new history of human development. For him technical innovations were to be considered the result of performance, not simply of invention. As argued later by French paleoanthropologist, André Leroi-Gourhan (1964–5), Kapp had already stated that new inventions were not the result of reflection about improvement but were the result of a long-term process of optimization in tiny steps. That was the remarkable new basis for what he considered "the cultural historical rationale."

It is not easy to understand "unconscious organic projection." It would be simplistic to think of it as an extension of the human in the way a prosthesis is a replacement of a body part. And by no means should it be thought of as resemblance, as a romantic idea. Rather the idea should be understood in context and in relation to the separation of organic and inorganic technology mentioned above. Kapp was not so much concerned with defining technical media in terms of the human body as with putting the concept of the cultural on new grounds that emphasized the co-evolution of culture and technology. Organic life creates stable forms, such as bones, because these are functionally necessary. According to Kapp, science takes on new significance by means of scientific exploration such as discovering, with the help of a microscope, the proximity of natural and technical forms. The internal structure of the femur, to quote Kapp's example, was not an irregular tangle of fibers and cavities as was previously thought, but instead proved to be a well-organised structure whose push and pull lines corresponded to the stress points on a bone in a manner that can be found also in structural steel work (Kapp 1877, Chapter VI). This is a kind of congruence which is not necessarily conscious to engineers when they construct technical artifacts. On the ordinary level of human perception, this dimension of the technology, which would be a feedback of nature upon itself, is imperceptible. What is involved is the Hegelian idea of consciousness revised: tools and technology become a manifestation of the depths of the "unconscious."[6] The most prominent example of such a manifestation—on a level that was never reached before—was the electric telegraph system, which had been implemented globally during Kapp's lifetime.

To draw parallels between nature and technology was not uncommon at the time. Kapp compares the depiction of nerves—referring to the cellular pathology of the Berlin physician, Rudolf Virchow and to the holistic physiology of Carl Gustav Carus—to the new technology of telegraph cables. Among the many illustrations in the *Grundlinien*, we find the cross section of a transatlantic telegraph cable compared to that of a nerve fiber (Figures 1.2 and 1.3). The nerves in the animal body carried signals in a manner similar to telegraph cables, which he stated had become the growing "nerves of Mankind" (Kapp, 1877, 141).

---

6 Cf. Kapp, 1877, *Grundlinien*, p.123 where Kapp draws on Eduard von Hartmann (1869).

Figure 1.2   Submarine telegraph cable section (illustration in Kapp 1877, 141)

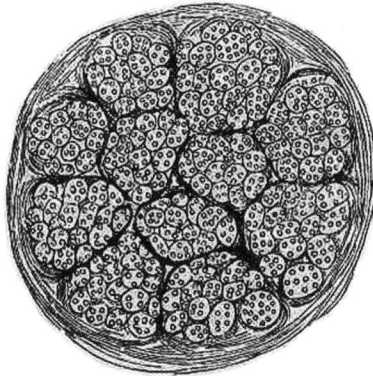

Figure 1.3   Cross section of a nerve fiber (illustration in Kapp 1877, 142)

In the 19th century, this common parallelism indicated a rising awareness of natural science and the developing media culture that sought new ways of connecting emerging technical realities to changing relationships with previous knowledge. The theory of organic development reflected the influence of Charles Darwin's theory of evolution and the search for philosophically based explanation for mechanical–industrial practices. In the discussion of these issues by Franz Reuleaux in his theory of mechanical engineering, for example, attention is focused not on ideas but on what is external to human nature: language, politics, technology (Reuleaux, 1875). Reuleaux was a German engineer and professor at the Berlin Royal Technical Academy, whose book on theoretical kinematics is quoted in Kapp's *Grundlinien* (Kapp 1877, 167–208). He stressed the existence of a "fundamental morphologic law" which could be found by examining ("nachspüren") how natural form follows

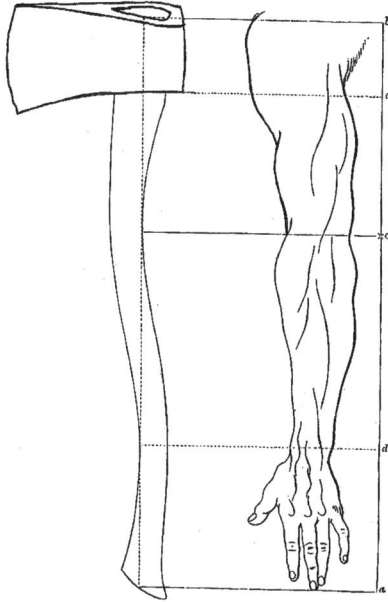

Figure 1.4   The American axe and the human arm (illustration in Kapp 1877, 242)

function and not ideas (Reuleaux 1875, 257). Hence, technological artifacts are not to be considered as results of theoretical reflection but of antecedent craftsmanship. Kapp's reference here is "an old backwoodsman" from Texas who, in a practical demonstration, showed him the superiority of an American axe which unlike the straight German axe fitted the organic form of the human arm (Fig. 1.4) (Kapp 1877, 241). The transatlantic cables may be considered to be a further, rather more grand example of this idea. They were constructed and laid without any crucial knowledge of the underlying physics or mathematics related to cable engineering. Relevant electromagnetic theory was not formulated until decades later by Oliver Heaviside (1950 [1887]). And it took about a hundred years later still in the epistemology of Michel Foucault (2002), for formal, philosophical justification to be set aside in favor of historical *a priori*, and for it to be accepted that the nature of the human subject as well as discursive practices also depend on technical preconditions.

World Communication

In the *Grundlinien*, the Hegelian "world spirit" appears in the new ontology of global media technology or *Weltcommunication* (world communication), a term Kapp coined for the macro-spherical technology which he described as a non-human consciousness which might affect human self-awareness as a conscious being.[7] This turns Hegel upside down by stressing the "externals" of human nature. When Kapp describes tools as an evolutionary development from the hand as an *agile brain*, he understands the use of the phrase as a mediological gesture: it allows the transmission of the knowledge of its application. That is to say, this incorporated knowledge of how to do things becomes detached from its original "handiness" and may be transferred to a machine's function or an apparatus's program. Thus, the hand actually might be seen as a "reflexive" part of human existence. According to Kapp, because the hand was the "material reason for human self-confidence" and remained so for art and science "in the transcendentalist sense" (Kapp 1877, 151), the growing system of world communication would not only form a new foundation of culture but a new form of *World Brain*, a term popularized by H.G. Wells (1938).

It remains important to mention that in all of this, Kapp was not expressing a type of technical determinism but rather presenting a basis for cultural morphology. Not only the usage of tools but also the technology of storage, transfer, and processing of data and information can be seen as advanced activity that liberates culture from nature. Technology, in the broadest sense of the word, consists of the evolution of a chain of events that is transferred from one human to the next, from generation to generation and from culture to culture. This also shows how biological organisms react to technology, how the kinetic of the human gesture changes to the indirect movement of the machine that is further developed and automated. The liberation of memory through the invention of writing and printing is followed in succession by the development of micro-electronics, computer systems, and the current global network of knowledge resources. Thus too, photography, telegraphy, and now computers, as inventions which can manipulate symbols beyond the linguistic code, have played a co-evolutionary role in the development of culture.

Technology is a comprehensive phenomenon unfolding throughout history, permanently reacting to the challenge of more complex forms of existence. New uses became possible through the application of electricity and electromagnetism. The telegraph was much more than just a technological improvement: it changed the way people communicated and it changed their view of the world. During the era of the *belle époque*, when telecommunication technology started to develop into a world project, Ernst Kapp neither romanticized the idea of a technologically unified mankind nor exactly predicted mankind's unification into some kind of a *Global Village*. But he was committed to analyzing the changes that were occurring and trying to find a conceptual framework for them, which—for his time—is surprisingly fitting. Less a visionary and more a forgotten analyzer of the intrinsic relationship between tools and organs, Kapp demystified the advance of modern technology and *avant la lettre* thus offered a defense for the need for a cultural theory of technology … or even a media philosophy.

---

7   "Von der Beschaffenheit nichtmenschlicher, Bewusstseine' macht der Mensch sich Vorstellungen nur auf Kosten der Integrität seines Selbstbewusstseins" (Kapp 1877, 161).

References

Babbage, Charles. 1851. *The Exposition of 1851; or Views of the Industrie, the Science, and the Government of England.* London: John Murray.

Clarke, Arthur C. 1992. *How the World was One: Beyond the Global Village.* London: Gollancz.

Cooley, Charles Horton. 1909. *Social Organization: A Study of the Larger Mind.* New York: Scribner's Sons.

Foucault, Michel. 2002. *The Archaeology of Knowledge* [1969]. London, New York: Routledge.

Freud, Sigmund. 1962. *Civilization and its Discontents* [1930]. New York: Norton & Co.

Hartmann, Eduard von. 1869. *Philosophie des Unbewußten: Versuch einer Weltanschauung* [Philosophy of the Unconscious: Towards a Philosophy of life]. Berlin: Carl Duncker.

Heaviside, Oliver. 1950. *Electromagnetic Theory: Complete and Unabridged Edition of Vols 1–3.* New York: Dover Publications.

Hegel, Georg W.F. 1807. *Phänomenologie des Geistes* [Phenomenology of Spirit]. Bamberg und Würzburg: Goebhardt.

Hegel, Georg W.F. 1970. *Vorlesungen über die Philosophie der Geschichte* [1837] (Lectures on the Philosophy of History). Werke Band 12, Frankfurt am Main: Suhrkamp.

Herder, Johann Gottfried. 2002. *Ideen zur Philosophie der Geschichte der Menschheit* [1784–1791] (Reflections on the Philosophy of the History of Humanity) edited by Wolfgang Proß. München: Hanser.

Hugill, Peter J. 1999. *Global Communications since 1844: Geopolitics and Technology.* Baltimore: The Johns Hopkins University Press.

Kapp, Ernst. 1877. *Grundlinien einer Philosophie der Technik. Zur Entstehungsgeschichte der Cultur aus neuen Gesichtspunkten.* Braunschweig: Westermann.

Leroi-Gourhan, André. 1964/65. *La geste et la parole*, 2 vols. Paris: Albin Michel.

Maury, Matthew Fontaine. 1855. *The Physical Geography of the Sea and its Meterology.* New York: Harper.

McLuhan, Marshall. 1964. *Understanding Media: The Extensions of Man.* New York: McGraw-Hill.

Mitcham, Carl. 1994. *Thinking Through Technology: The Path between Engineering and Philosophy.* Chicago: University of Chicago Press.

Peters, John Durham. 1999. *Speaking Into the Air: A History of the Idea of Communication.* Chicago: University of Chicago Press.

Releaux, Franz. 1875. *Theoretische Kinematik. Grundzüge einer Theorie des Maschinenwesens* Braunschweig: Vieweg [The Kinematics of Machinery: Outlines of a Theory of Machines, translated by A.B.W. Kennedy, London: Macmillan and Co. 1876].

Standage, Tom. 1999. *The Victorian Internet: The Remarkable Story of the Telegraph and the Nineteenth Century's Online Pioneers*. London: Phoenix.

Wells, H.G. 1938. *World Brain*. London: Methuen & Co.

Chapter 2

The Formation of Global News Agencies, 1859–1914

Volker Barth

In 1835 Charles Havas founded the first modern news agency in Paris. Born in 1783 in Rouen, Havas could already look back on a tumultuous career when he first decided in 1832 to open a translation office in the French capital. On a daily basis he transcribed—in his mind—the most important foreign press articles into French and sold them to interested newspapers. What Havas offered first French and soon international papers was not a genuine journalistic activity—journalism as a profession only emerged at the end of the nineteenth century (Requate 1995)—but rather a service specifically developed for the press.

The cases of Bernhard Wolff and Julius Reuter, who both worked for a short period within the *Agence Havas* before opening their own agencies, were quite similar. Reuter gained his first business experience in Aachen where he bridged the gap in the Berlin–Paris telegraph line by the means of a pigeon service before setting up a news agency in London in 1851, only a few months after the first undersea telegraph line across the English Channel came into service. Only Bernhard Wolff, who in 1848 first founded the *National Zeitung* before opening Wolff's Telegraphisches Bureau, had some, if limited, journalistic experience.

The founding fathers of the first three European news agencies were more concerned with media business than with public opinion. Their entrepreneurial idea was as simple as it was brilliant: They wanted to compile the "pure facts" of every day's most important events and sell them to interested newspapers; these could then process them according to their journalistic convictions. The news agencies emerging in the mid-nineteenth century, promised politically and personally unbiased news reports. The agencies based their activities on the idea of objective news, in which objectivity neither indicated a journalistic maxim, nor an insight into the history of ideas and even less a philosophical conviction. Primarily it described a business model which was to guarantee the trustworthiness of an agency's news and promised to maximize the number of potential customers. With this commitment to objectivity the idea was that the agencies' news reports could be used by newspapers of the most diverse views and opinions.

This business model proved to be extremely successful and in the second half of the nineteenth century a substantial number of regional, national and service-specific news agencies came into existence. Nevertheless, the first four news agencies remained the uncontested market leaders well into the twentieth century:

the *Agence Havas* (1835), *Wolff's Telegraphisches Bureau* (1848), *Reuters* (1851)
and the US *Associated Press* (1848). The last had its headquarters in New York City
and in the 1890s was absorbed by the Western Associated Press with headquarters
in Chicago before later moving back to New York (Schwarzlose 1990). The four
agencies cooperated over several decades within a sophisticated contract network,
the so-called cartel treaties. These treaties divided the global news market along
the lines of national borders. The constitution of a planetary news system was thus
characterized by a tense relationship between diversely structured national zones
of influence within a global communications network.

The period from the middle of the nineteenth century up to the 1930s marked
the founding phase of a "world news order" the impacts of which are still noticeable
today (Siebold 1984; Meier and Schanne 1979, 213; Weaver and Wilhoit 1984).
Havas, Reuters, Wolff and the *Associated Press* were the decisive protagonists in this
development. They invented, developed, and professionalized the global exchange
of news. In the 1980s about 80 per cent of all published news came from the four
biggest agencies (Reuters, AP, AFP, UPI – The *Agence France Presse* (AFP), founded
after World War II, is the successor of *Havas*). The ancestors of the US agency,
United Press International (UPI), reach back to the first decade of the twentieth
century (Siebold 1984, 45). In 1932, only two years before the end of the news cartel,
Wolff's Telegraphisches Bureau, the smallest of the four agencies, served no fewer
than 1,300 newspapers in Germany and another 900 news services around the world
with the aid of 42 German and 21 foreign branches (Basse 1991, 235).

Despite those impressive numbers very little research has been conducted
on the four agencies. A few business histories mainly retrace the biographies of
the respective company founders (For *Wolff* see Basse (1991) and Wilke 1991;
for *Reuters* see Read 1999 and Storey 1951; for *Havas* see Frédérix 1959 and
Lefebure 1992; for the *Associated Press* see Schwarzlose 1979; Schwarzlose 1989;
Schwarzlose 1990; Blondheim 1994; and Gramling 1940). In addition, these works
are concerned with the spectacular expansion of the agencies and the ways in which
this was closely linked to revolutionary developments in news technology. The
stages: pigeons, railways, electric telegraph, and undersea cables in conjunction
with the audacity of a handful of entrepreneurial geniuses in these accounts tell
the story of irresistible expansion (Boyd-Barrett and Palmer 1980; UNESCO
1953; Desmond 1978; Meier and Schanne 1980). Despite the cartel treaties, the
interaction and the interdependence of the agencies are rarely mentioned and are not
systematically integrated into the analysis. Only the British and French examples
have partly been analysed within a comparative framework (Palmer 2002; Palmer
1983; Palmer 1976; Wilke 2000; Wilke 1984; Wilke and Rosenberger 1991).

For useful new research three new emphases seem necessary. First, the four
news agencies—in order to comprehend their monopoly and manner of operation—
must be put into a comparative perspective, an approach that already has proved
useful with regards to telegraph companies (Winseck and Pike 2007; Headrick
1991). Second, the news reports themselves, specifically their production and their
transmission, must be taken into consideration. Third, objectivity as the credo of

the agencies seems to be a promising point of focus for studying international cooperation between the agencies and the production process of individual news items. The claim to objectivity suggests the value of investigating them according to their professed self image.

In this paper, I will deal with the history and the modes of operation of modern news agencies. I then analyse the claim of objectivity with regard to its economical, political, and cultural underpinnings, before going on to describe various techniques which were intended to generate objectivity within the day to day business of the agencies. Finally, I will discuss the reasons for the dissolution of the news cartel.

History and Modes of Operation

Cross-country networks of news and correspondence are no nineteenth century invention. Louis XI established a news service as early as 1464. In the same period the Counts of Thurn and Taxis were in charge of news transmission within the Hapsburg Emperor's court. The first commercial news agency opened in Venice in 1536 and news correspondence gained in stature within business families such as the Fuggers, the Medicis and, later, the Rothschilds with their European-wide operations. During the French Revolution the need for international news increased dramatically and cross-regional newspapers started to hire foreign correspondents. In Germany, for example, this was the case for the *Augsburger Allgemeine Zeitung* and generally speaking the London *Times* served as a role model in this respect. The Great Revolution had opened an "information expectation gap" which some papers readily promised to fill with "true facts" (Quote from the *Neueste Weltkunde* in Requate 1995, 122–3).

Admittedly, logistical problems hindered the fulfillment of this promise. Thus, the London *Times*, which eventually broke the news to the world of the death of Napoleon on May 5, 1821 in St. Helena, was unable to do so earlier than July 4, almost two months after the event (Höhne 1977, 21). The modern agency system in effect required a technical kick-start which eventually came in the form of electric telegraphy. Nevertheless, the first modern news agency, *Havas,* opened in 1835 two years prior to this path-breaking invention. In 1865 the news of President Lincoln's assassination took 12 days to reach London, but in 1881 the English Press by means of the transatlantic cable was able to report the assassination of President Garfield within 24 hours. This development of almost instant communication proved to have political consequences. Thus, after the death of Garfield the English government ordered a period of State mourning, which had not been the case after Lincoln's death (Read 1999, 97).

The global news system which emerged on the basis of an expanding telegraph network fundamentally differed from its historic ancestors. One of its effects was that the political power of the press increased exponentially. From now on it was even possible to influence wars, as illustrated by the example of the *Emser*

Depesche or Ems Dispatch of 1870. This account by a member of the King's staff of an informal meeting between the King of Prussia and the French ambassador had been cleverly edited by Bismark to suggest that the French ambassador had insulted the King and that the King had insulted the ambassador. The Prussian chancellor then orchestrated its publication by the Berlin agency, *Wolff*, and the upshot as a result of its republication throughout Germany and in a series of mistranslations in France was public outcry, mobilization and the German-Franco war of 1870–1.

Now too news—and specifically international news—became a commercial commodity. The news agencies that were founded in the middle of the century sold this new form of merchandise in steadily increasing quantities and at increasing speeds of transmission. In addition, the customers of the new news agencies were, in contrast to the systems of the Fugger or Rothschild families, almost exclusively newspapers. Thus, the agencies soon became what they still are today: the most important link between events, journalists and a growing global audience.

The Cartel Treaties

In 1859 the three largest European agencies signed their first cooperation treaty which divided the world into three geographically defined news monopolies. At this point the *Associated Press* was already linked to the cartel by the means of bilateral treaties before joining the cartel formally in 1893—there are varying accounts of the actual date (Siebold 1984, 55; Evans 2010, 213; Cooper 1942, 16). The cartel treaties had to be renegotiated from time to time and this led to differences in the ways in which they stipulated the "principal of business exclusivity" (Siebold 1984, 54). This principle, however, meant that no agency was allowed to produce or sell news outside its own territory but had to obtain news outside its territory from its partner agencies. The territorial divide of the news world proved to be astonishingly congruent with the individual influence zones of the nation states whose governments authorized, delivered and supervised the agencies.

Reuters controlled the British colonies and the Far East and shared the Netherlands, Turkey, Egypt and Belgium with *Havas*. The latter's territories stretched over France and the French colonies into Italy and Spain and comprised the entire South American continent by the end of the nineteenth century. The Prussian agency *Wolff* held sway over Scandinavia, South Eastern and Eastern Europe including Russia, or more specifically Moscow and St. Petersburg. The *Associated Press* retained responsibility for the United States and Mexico and parts of Canada, Cuba and the Philippines were successively added to its sphere of influence.

Objectivity, Politics and the Selection of News

The selection of news by the individual agencies can be compared synchronically along the territorial influence zones. A number of questions emerge from this approach. What news reports were produced, how and by whom within a particular sphere of influence before being passed on to the partner agencies? To what extent were proposed news items the result of the political context of competing nation states and how close was the connection between the value of news and the interests of national audiences? Which steps marked development from what was originally a local incident to a national or even global event?

   In addition, the process of production, transmission and publication of news can also be historicized diachronically. From the middle of the nineteenth century up to the interwar period the organizational modes, political dependencies and self definitions of news agencies changed fundamentally. *Wolff* remained under the strong influence of the Prussian Home Office far into the twentieth century. Dieter Basse speaks correctly of a "state controlled agency in private hands" (Basse 1991, 5, 35). *Havas* successively gained independence after being obliged in the Second Empire to present every news item to the relevant ministry prior to publication. *Reuters*, whose expansion ran parallel to the expanding British Empire, developed in the opposite way. As it was prohibited to sell news within Britain—an exclusive right of the *Press Association* (Scott 1968)—the enterprise gradually became a key institution of the Empire without ever being a state agency.

   The *Associated Press* which resulted from the cooperation between six New York newspapers was the most independent agency and worked as a cooperative with any profits being distributed among partners according to their investment percentage. Commercial considerations were nonetheless an important motivation in the quest for maximum political independence. In the early cartel period the *Associated Press* was something of a junior partner to the European agencies but it gained in importance over the years. This was mainly because the different newspaper associations comprising the *Associated Press* were able to protect themselves efficiently from local competition, not to mention the impact on the agency of the steadily increasing US-American market. The AP was the only US-agency that offered a fast, continuous and reliable international news service. It had an exclusive contract with *Western Union* that was able to monopolize the American telegraph system for a long time. Above all AP had at its disposal exclusive access to the transatlantic cable that had operated since 1866 and through the cartel treaties with *Havas*, *Reuters* and *Wolff* it had exclusive access to news from all over the world.

   New journalistic methods, that became predominant in North America by the end of the nineteenth century contributed further to the special position of the *Associated Press* within the cartel. The "New Journalism," which marked the AP dispatches, is what we know as investigative journalism and with it came new standards of objective and accurate news reporting.

   Despite these far reaching differences in political connections and journalistic ethos, all four agencies claimed unreservedly that they delivered objective messages completely free of political, commercial, or professional considerations. Given this business maxim, it is useful to investigate to what extent the news presented by the agencies as reliable facts was actually influenced by changing contexts and requirements. The agency system provides insights into evolving patterns of journalistic *objectivity*, *accuracy* and *factuality* and consequently helps in understanding the historical development of journalistic standards.

Objectivity, Accuracy and Factuality

The basis of the objectivity that the agencies claimed grew out of a practical business perspective rather than having a scientific or epistemological basis. Like Michael Schudson, who has traced the historic development of the idea of objectivity in journalism (Schudson 1990), Lorraine Daston and Peter Galison date the origins of modern idea of objectivity to the second half of the nineteenth century and plead for an investigation of objectivity in terms of concrete practices rather than in terms of an abstract concept (Daston and Galison 2007, 56). The agencies had to develop techniques within their daily business that not only encouraged newspapers to trust the incoming messages but ensured that the messages did not turn out to be erroneous (Siebold 1984, 77).

It is interesting to question the extent to which a specific historical context influenced the production, transmission and communication of news reports that were continuously being presented as neutral facts. What role did the simultaneous competition and interdependence of worldwide agents play when, each in its own way, was constantly dependent on its national governments? The connection between the politically and economically motivated selection of news items circulating between agencies on the one hand and their linguistic constitution on the other deserves special attention.

The commercial and economic aspects of the claim of objectivity are obvious. As businesses the agencies were required to make profits (Silberstein-Loeb 2009; Hochfelder 2000). *Havas* financed its correspondence at first by publicity. The agency sold its news to newspapers in exchange for advertising space in them. The more advertising the papers published, the more news they obtained.

However, the newspapers were not only customers but also dangerous competitors as they could subscribe to the services of the various agencies or hire their own foreign correspondents and so produce news independently. As a result, the subscription to an agency required a written statement that the newspaper would not seek to obtain news from other agencies. The *Associated Press* pushed this idea to an extreme in that it sold its services to no more than one local newspaper per city. The cooperation between the four big agencies was designed to eliminate potential competition by covering the entire planet, something which they alone were able to achieve.

This system became problematic in that it posed the threat of increasing uniformity in the news world. It meant that a growing number of newspapers were working to an increasing extent with the same basic material. This made it more and more difficult for them to distinguish themselves from their competitors. Thus, in the middle of the 1890s not only the London *Times*, but also the French daily,

Le Matin, decided no longer to subscribe to the *Reuters* service, but instead to rely on a newcomer agency, *Dalziel*. *Le Matin* justified this decision to its customers by claiming that its identity was that of a "journal d'informations télégraphiques, universelles et varies" (a newspaper of telegraphic, universal and diversified news) (Requate 1995, 84).

New journalistic techniques also began to mould the world news order. The increasing standardization of news items required accompanying comments which transgressed the boundaries between fact and interpretation. In its correspondence, for example, *Havas* determined what was the most important message of the day without further justification and it furnished additional comments without identifying what stemmed from local reporters and what from the Paris office.

With increasing international tensions in the wake of the First World War, cooperation between the individual agencies proved to be an increasing political problem. This was especially so when, as in the case of *Wolff's* cooperation with *Havas* and *Reuters*, military alliances were transcended. Nationally oriented newspapers more and more refused to publish dispatches that came directly from foreign opponents and this started to put pressure on particular agencies. In Germany, for example, nationalist publicists opposed indoctrination by what was labelled a "liar press" with the slogan "Away from Havas and Reuters" (Anton 1916; Rotheit 1915).

The First World War marked a decisive caesura in the agency system. After 1918 the international agencies progressively freed themselves from state control and at the same time queried the principle of their self-imposed restriction to geographically defined zones. In this process reference to the objectivity of news turned out to be the decisive argument. After all, an agency's reputation and its capacity to survive as a business rested upon this assertion of objectivity. It proved to be increasingly complicated to publish news items which could not be verified independently under the agency's name.

Parallel to this evolving political and economical framework, the increasing speed of transmission of news from all over the world generated a hitherto unknown proximity, relatedness and participation in events that were taking place as far away as the other side of the planet. Consequently, new commercially and culturally motivated definitions and selection criteria for what was newsworthy began to emerge. This can be illustrated by a message of instruction from the *Associated Press* to its reporters at the beginning of the twentieth century.

> Political news is only desired if it is of great general interest or of specific significance for the United States. Fire damages and bankruptcies are only to be cabled from one Million Dollars upwards or in case of considerable loss of human life. Ship wrecks, especially of American ships are to be cabled as soon as possible. [...] Assassinations, if not committed by American citizens, are only to be cabled in case of extraordinary interest. Lawsuits are only to be reported if they are pursued against state officials or against Americans.

> [...] News not required: Robberies, abnormal birth, kidnappings, seductions, suicides, scandals, child murders, promotions of less important people etc. [...] Always keep in mind that events of local significance are not necessarily of oversea interest (Quoted in Hansen 1914, 81–2).

The careful on-the-spot selection as well as the extraordinary importance of national relevance is especially noteworthy in this instruction. This suggests the importance of interrogating the different forms of linguistic and audience-specific constitutions of objectivity which nationally constituted networks observed within the global cooperative framework that they had established. Below are instructions issued by *Associated Press* in 1854, six years after it was founded and almost half a century before the instructions given above:

> In regard to the character of the news required, we would say that we want everything that is IMPORTANT and everything that would be of General Interest in the City, State, or the country at large. In preparing dispatches for transmission, it is desirable always to bear in mind that we want only the *material facts* in regard to any matter or event, and those facts in the fewest words possible compatible with a clear understanding of correspondent's meaning. All expressions of opinion upon any matters; all political, religious, and social biases; and especially all *personal feelings* on any subject on the part of the Reporter, must be kept out of his dispatches [...] (Schwarzlose 1989, 180. Italics in the original).

Techniques for Establishing Objectivity

The question arises as to the techniques that were intended to generate and guarantee objectivity. What examples could be consulted for such an investigation of the formative years of global news agencies? First of all examples of erroneous reporting are of interest. This is because they describe moments of failure that highlight the fragility of the objectivity claim and the cooperation that it depended on. In many cases erroneous reports also illustrate also the fragility of the transmission mechanisms. Continuous false reports damaged the reputation of agencies which as businesses were only able to survive if the newspapers and their readers had no doubts about the veracity of the dispatches they were receiving. In 1854, for example, *Havas* prematurely reported the fall of Sebastopol in the Crimean War. An unknown tartar had reported this supposed news to Omar Pasha, a General of the Ottoman army, from whom it found its way to the Parisian headquarters of *Havas*. When the agency had to acknowledge its mistake several days later it was the object of mockery on the part of newspaper readers and of vexation on the part of its customers. Up to the end of the century erroneous reports continued to be referred to as "tartar news." (Höhne 1997, 64).

The problem, however, transcended the existence or the trustworthiness of eye witnesses. It also involved the professional practice of journalists with regard to incoming news. In the nineteenth century this practice varied among the various nations. In comparison with other countries, Germany especially took a relatively long time for independent investigation to become the journalistic standard. When in 1887 false rumors about the premature death of General Moltke circulated, the staff of the *Wolff* agency did not even consider sending someone to his home in Berlin. Instead they consulted the official authorities for confirmation. As this was only given some time later, the gossip factory was further fuelled by the delay (Höhne 1997, 64).

Beside the reliability of eye witnesses and evolving journalistic practices, technical equipment played a major role in the definition and establishment of objectivity standards. Telegraph lines, where they existed, were often busy or interrupted. News did not come through at all or arrived mutilated so that it was not always possible to tell what part if any of a dispatch was missing. Most of all, the actual transmission of dispatches was not in the hands of journalists, but was the responsibility of the staff of the telegraph companies. Especially the communication of foreign language news proved to be a continuous source of mistakes.

Of particular interest are the emerging standards of objectivity that were directly related to specially elaborated forms of language. It was very expensive to send news over the telegraph and for all of the agencies this proved to be their biggest single budget item. As a result they developed special language codes in order to reduce the costs of news transmissions. Instead of the four words "the Emperor of Austria," they cabled the word "Austriaemperor" and thus reduced the costs by 75 per cent. Take this 31 word long German message: "Gegenüber dem Dementi der offiziösen Blätter bestätigen die, National- und Vossische Zeitung' das von der ‚Neuen Freien Presse' gemeldete Gerücht, daß eine Begegnung der Kaiser von Österreich und Russland stattfinden werde" ("Contrary to the denial of the semi-official newspapers, the *National Zeitung* and the *Vossische Zeitung* confirmed the rumor reported in an article of the *Neue Freie Presse* that the meeting of the Emperors of Austria and Russia will take place"). This was transformed into: "Gegenüberoffiziosdementi bestätigen Nationalvoß Neupreßmeldung Ostkaiserbegegnungsgerücht" (Quoted in Höhne 1977, 58). With only five words it became a far cheaper news item than in its original form. However, this dispatch required a politically informed recipient in order for it to be interpreted correctly and to be re-transcribed into its original form. When a *Reuters* correspondent tried to report the horsemanship abilities of the Prince of Wales, his very short message, "princes skill," was converted into "prince (i)s kill(ed)" and thus announced the sudden and untimely death of the heir to the throne (Höhne 1977, 59). These examples are of more than anecdotal interest because they reveal fundamental problems that the agencies had to resolve. Because of the existence of the cartel treaties any solutions needed mutual consent. This was indispensable not so much for the sake of journalistic ethos, but rather in order to avoid putting the commercial basis of the enterprises at risk.

Beside affordable reliability, the agencies had to ensure the neutrality of news items if the items were to be useful for every newspaper. Once more, new language techniques were brought into play. *Wolff's Telegraphisches Bureau* labeled news items that stemmed directly from governmental sources with the letters AC (Allgemeine Correspondenz) which allowed it to send these messages over the wire for free. The agencies agreed in 1909 to equip all news coming directly from governmental sources with the word "tractatus" (Mathien and Conso 1997, 59). Even earlier, during the Boer War (1899–1902), *Havas* systematically added to its communiqués of South-African news originating from *Reuters* that it was "De source anglaise" (Read 1991, 59).

Progressive changes in standards of objectivity manifested themselves in other language and practices in reporting news items. After an earthquake on the US-American west coast had destroyed many telegraph lines, the *Associated Press* began to cable news from the disaster region headed by the word "flash." This indicated not only the urgency, but also the dramatic character of the news item. Almost immediately correspondents from the other agencies began to title their dispatches as "flashes" in order to be allowed priority access to cable. As a result a mere two weeks later the agencies had to negotiate rules for the flash dispatches of international news (Gramling 1940, 159–60). The question then became: when and for whom had a message priority? What practical criteria could be defined that would indicate the veracity and value of a dispatch for all parties who might be interested in it?

Reuters was the first agency systematically to put its name on top of each of its dispatches (Höhne 1977, 71). By thus indicating its source, the reader was to be assured of the accuracy of the incoming news. At the same time the indication of sources enabled the publication of apparently biased news as the agencies could pass on the responsibility for inaccuracy to others. Preserving the anonymity of the reporter or the importance of reporters signing their dispatches had become an intensely debated issue by the end of the nineteenth century.

In article 197, the Versailles treaty of 1919 prohibited defeated Germany to transmit news into the world by radio. While this did not prevent *Wolff* from continuing its transmissions, the agency did decide to publish only news that was "less politically biased." By the "skillfull selection of news" and the "transmission of pure facts" *Wolff* hoped to overcome the restrictions placed on its activities by the Allied countries (Wilke 1991, 162, 166).

The Decline of the News Cartels

These examples provide insight into the modes of operation of the international agencies and contribute to our analysis of modern standards of journalistic objectivity and the tools that helped to establish them. In fact, the complex demands of accuracy, neutrality and factuality contributed to the end of the news cartel. In 1934 the cartel treaties were not renewed, despite having been continuously modified over the years. This put a definitive end to the territorial self restrictions that they had imposed on themselves.

The development that led to this outcome had started after the First World War and well beyond Europe. In Argentina more and more newspapers had become dissatisfied with the *Havas* service as it contained progressively less news from Germany. As a result they were increasingly unable to fulfill the requirements of German immigrant readers and continuously lost customers. In Japan it proved to be decreasingly acceptable that all incoming and outgoing dispatches were formulated and transmitted by *Reuters* correspondents. Here too, newspapers started to open up the market to other providers.

As early as 1909 a type of counter cartel had been formed that consisted of the recently founded *United Press*, the French agency *Fournier*, the German agency *Hirsch* and several Australian and Japanese agencies. In the same year the new *Hearst* service began—in opposition to common practice—to serve newspapers that did not belong to the *Hearst* group (Höhne 1977, 100–1). International competition began to put pressure on the cartel of the big four and the takeover or creation of their own agencies by totalitarian regimes meant another nail in the cartel's coffin. In 1925, for example, the Soviets founded the *TASS* agency and on January 1st, 1934 the Nazis transformed *Wolff* into the *Deutsches Nachrichtenbüro* (DNB).

Even with their cartel treaties *Havas, Reuters, Wolff,* and the *Associated Press* could no longer cope with this growing competition. Their mutual dependency made it increasingly difficult to meet the requirements of all potential customers. And because they were locked into their cartel-based relationships, they were seen as no longer able to guarantee that the daily selection of news mirrored the most important events throughout the world and not the obligations of business partnerships.

Any investigation of the agency system from the middle of the nineteenth century to the interwar period, as I have shown, clearly needs to consider the changing international media landscape and methods of news production. It must address linguistic practices within the emerging profession of journalism, the evolution of news in a worldwide distribution system and, last but not least, the political forces that continuously and massively influenced these processes. Ultimately, it was the *Associated Press*, the Presidents and General Managers of which were all journalists—as opposed to the European agencies—that successfully brought the cartel system to an end. Thus, the history of the formation of global news agencies is the story of a growing awareness that the provision of objective news was a much trickier problem than Charles Havas, Bernhard Wolff and Julius Reuters had ever imagined.

References

Anton, Reinhold. 1916. *Die Lügenpresse: Noch eine Gegenüberstellung deutscher und feindlicher Nachrichten, u.a. der W.T.B-, Reuter-, Havas- und P.T.A-Telegramme über den Weltkrieg 1914/16*. Leipzig: Zehrfeld.
Basse, Dieter. 1991. *Wolff's Telegraphisches Bureau 1849 bis 1933: Agenturpublizistik zwischen Politik und Wirtschaft*. Munich: Saur.
Blondheim, Menahem. 1994. *News over the Wires: The Telegraph and the Flow of Public Information in America, 1844–1897*. Cambridge: Harvard University Press.

Boyd-Barrett, Oliver and Palmer, Michael B. 1981. *Le trafic des nouvelles: Les agences mondiales d'information*. Paris: Moreau.

Cooper, Kent. 1942. *Barriers Down: The Story of the News Agency Epoch*. New York: Farrar & Rinehart.

Cooper, Kent. 1959. *Kent Cooper and the Associated Press: An Autobiography*. New York: Random House.

Daston, Lorraine and Galison, Peter. 2007. *Objektivität*. Frankfurt/M.: Suhrkamp.

Desmond, Robert W. 1978. *The Information Process: World News Reporting to the 20th Century*. Iowa City: University of Iowa Press.

Evans, Heidi J. 2010. "'The Path to Freedom?' Transocean and German Wireless Telegraphy, 1914–1922." In *Global Communications. Telecommunication and Global Flow of Information in the Late 19th and Early 20th Century*, edited by Roland Wenzlhuemer, 209–33, Köln: Leibnitz Institut.

Frédérix, Pierre. 1959. *Un siècle de chasse aux nouvelles: De l'agence d'informations Havas à l'Agence France Presse, 1835–1957*. Paris: Flammarion.

Gramling, Oliver. 1940. *AP: The Story of News*. New York: Farrar and Rinnehart.

Hansen, N. 1914. "Depeschenbureaus und internationales Nachrichtenwesen." *Weltwirtschaftliches Archiv*, 3: 78–96.

Headrick, Daniel R. 1991. *The Invisible Weapon: Telecommunications and International Politics, 1851–1945*. New York, Oxford: Oxford University Press.

Hochfelder, David. 2000. "A Comparison of the Postal and Telegraph Movement in Great Britain and the United States, 1866–1900." *Enterprise & Society*, 1: 739–61.

Höhne, Hansjoachim. 1977. *Report über Nachrichtenagenturen: Die Geschichte der Nachricht und ihrer Verbreiter*, vol. 2. Baden-Baden: Nomos.

Lefebure, Antoine. 1992. *Havas: Les arcanes du pouvoir*. Paris: Grasset.

Mathien, Michel and Conso, Catherine. 1997. *Les Agences de presse internationals*. Paris: PUF.

Meier, Werner and Schanne, Michael. 1980. *Nachrichtenagenturen im internationalen System*. Zurich: Publizistisches Seminar der Universität.

Meier, Werner and Michael Schanne. 1979. "Nachrichtenagenturen und globales Schichtungssystem. Eine Forschungsperspektive." *Publizistik*, 24: 213–22.

Palmer, Michael B. 2002. "Crieurs et diffuseurs de journaux: perspectives françaises et britanniques, 1860–1900." In *La distribution et la diffusion de la presse, du XVIIIe siècle au IIIe millénaire*, edited by Gilles Feyel. Paris: Editions Panthéon-Assas.

Palmer, Michael B. 1983. *Des petits journaux aux grandes agences: Naissance du journalisme moderne (1863–1914)*. Paris: Aubier.

Palmer, Michael B. 1976. "L'agence Haves, Reuters et Bismarck: L'échec de la triple alliance télégraphique (1887–1889)." *Revue d'histoire diplomatique*, 90: 321–57.

Read, Donald. 1999. *The Power of News: The History of Reuters*, 2nd edn. Oxford: Oxford University Press.

Requate, Jörg. 1995. *Journalismus als Beruf: Entstehung und Entwicklung des Journalistenberufs im 19. Jahrhundert. Deutschland im internationalen Vergleich*. Göttingen: Vandenhoeck & Rupprecht.

Rotheit, Rudolf. 1915. *Die Friedensbedingungen der deutschen Presse: Los von Reuter und Havas*. Berlin: Puttkammer & Mühlbrecht.

Schudson, Michael. 1990. *Origins of the Ideal of Objectivity in the Professions: Studies in the History of American Journalism and American Law, 1830–1940*. New York. London: Garland.

Schwarzlose, Richard. 1979. *The American Wire Service. A Study of Their Development as a Social Institution*. New York: Arno Press.

Schwarzlose, Richard. 1989. *The Nation's Newsbrokers*, vol. 1: "The Formative Years, from Pre-telegraph to 1865." Evanston, IL: Northwestern University Press.

Schwarzlose, Richard. 1990. *The Nations's Newsbrokers*, vol. 2: "The Rush to Institution from 1865 to 1920." Evanston, IL: Northwestern University Press.

Scott, George. 1968. *Reporter Anonymous: The Story of the Press Association*. London: Hutchinson.

Siebold, Thomas. 1984. "Zur Geschichte und Struktur der Weltnachrichtenordnung." In *Medienmacht im Nord-Süd-Konflikt: Die neue internationale Informationsordnung*, edited by Reiner Steinweg and Jörg Becker. Frankfurt/M.: Suhrkamp.

Silberstein-Loeb, Jonathan. 2009. "The Structure of the News Market in Britain, 1870–1914." *Business History Review*, 83: 759–88.

Stevenson, Robert L. and Donald L. Shaw (eds). 1984. *Foreign News and the New World Information Order*. Iowa City: Iowa State University Press.

Storey, Graham. 1951. *Reuters: The Story of a Century of News-gathering*. London: Parrish.

UNESCO (ed.) 1953. *News Agencies: Their Structure and Operation*. Paris: UNESCO.

Weaver, David H. and Wilhoit, G. Cleveland. 1984. "Foreign News in the Western Agencies." In *Foreign News and the New World Information Order*, edited by Stevenson, Robert L. and Shaw, Donald L. Iowa City: Iowa State University Press, pp. 153–85.

Wilke, Jürgen (ed.) 1991. *Telegraphenbüros und Nachrichtenagenturen in Deutschland: Untersuchungen zu ihrer Geschichte bis 1949*, Munich: Saur.

Wilke, Jürgen (ed.) 2000. *Von der Agentur zur Redaktion: Wie Nachrichten gemacht, bewertet und verwendet warden*. Cologne: Böhlau.

Wilke, Jürgen and Rosenberger, Bernhard. 1991. *Die Nachrichtenmacher: Zu Strukturen und Arbeitsweisen von Nachrichtenagenturen am Beispiel von AP und dpa*. Cologne: Böhlau.

Wilke, Jürgen. 1984. *Nachrichtenauswahl und Medienrealität in vier Jahrhunderten: Eine Modellstudie zur Verbindung von historischer und empirischer Publizistikwissenschaft*. Berlin: de Gruyter.

Winseck, Dwayne R. and Pike, Robert M. 2007. *Communication and Empire: Media, Markets, and Globalization, 1860–1930*. Durham, NC; London: Duke University Press.

Chapter 3

"In the Pursuit of Colonial Intelligence":[1] The Archive and Identity in the Australian Colonies in the Nineteenth Century

Heather Gaunt

Introduction

This chapter contributes to research in the field of information history by exploring aspects of the Australian experience in the late-nineteenth century and early-twentieth century, particularly about the transit of information between the northern hemisphere and Australia. It addresses contemporary perceptions about the importance of documentary information and printed materials in the consolidation of cultural identity in the British settler colonies of Australia by focusing on a specific area of information: official documents and scientific and literary publications. It also touches on Australian interest in acquiring international material of this sort, the conduits for supply, and the key repositories for this material once it arrived. In the general context of the push from various influential individuals, societies, and "information" organizations, such as public libraries, to share and utilize information for the purpose of nation building and the consolidation of local identity, the chapter focuses on the work to establish a local "archive" of historical documents for Australia.

In the period known as the Belle Époque in Europe, Australians eagerly sought international intelligence in the form of official documents and scientific and literary publications from local as well as international sources. Key conduits for supply were established, such as Boards for International Exchanges and libraries. The major repositories for the information, including newly established public libraries in the colonial capitals, were cognizant of the prime resource value of this material. One aspect of the pursuit of information in the last decades of the nineteenth century and the first decades of

---

1   James Bonwick, "The writing of colonial history," *Sydney Morning Herald*, 25 May 1895, p. 6.

the twentieth century was the new significance assigned to documents relating to the historical development of the British colonies in Australia. This chapter focuses on this aspect of information in particular, addressing the specific interest of Australian individuals, institutions, and colonial governments in the printed sources needed by the newly evolving empirical and "scientific" historical method. This provides a wider context of evolving Australian perceptions of the value of written history for an emerging national entity and the impact of the absence on Australian soil of local records from the early decades of white settlement.

"The girdle around the globe"

In a climate of evolving national identity in the decades immediately before and after the Federation of the six separate self-governing colonies into the new Australian nation in 1901, Australians eagerly sought information in many forms that would reduce the "tyranny of distance" (in historian Geoffrey Blainey's influential phrase) of the southern continent from the "mother country" (Blainey 1966). Intelligence from outside of the colonies was essential for the colonists to maintain their official and cultural connections with Britain. To understand the importance of information to the young colonies of Australia, it is useful to remember the extent of the geographical distance that sets the Australian continent apart from Britain—a distance of some 12,000 nautical miles for the sea route through the Suez Canal. Even in the late 1860s, it was still possible to have a delay of two months between mail deliveries by sea, still the principal means of communication in this period. However, the final decades of the nineteenth century saw dramatic improvements in the speed of communication particularly after 1871 with the laying of the telegraphic cable from Britain via Madras, Penang, to Singapore, and Darwin in the north of Australia. From Darwin, the Overland Telegraph was laid to Adelaide, which then connected all the southern colonies. The enormity of these developments to social and intellectual life were highlighted in the press. This example from 1871 entitled, "The girdle around the globe," from a regional Australian newspaper, is worth quoting at length:

> The electric wires will ... annihilate that sense of distance, which now operates to the disadvantage both of settlers in these colonies and of their fellow countrymen at home. Emigration will be divested ... of the many difficulties and drawbacks, which now surround it. To people in England who read at the breakfast-tables the intelligence of any event of importance which happened a day or two previously in Melbourne, Sydney, Adelaide, Brisbane, or Hobart Town, each of those places will appear to be brought as near to London as some of the principal cities of Europe. To colonists ... daily appraised of every occurrence of moment in the Old World, expatriation will lose most of its inconveniences. We shall feel that we are within speaking distance of the great centres of political and intellectual activity, and have been brought within the circumference of that

activity. But one of the greatest benefits which we anticipate will arise from this intimate union with the mother country, will be the breaking down of that feeling of provincialism which manifests itself in all outlying and isolated communities ... We shall feel a daily interest in the larger concerns of larger communities, governed by greater men, and as in New York the debates in the British Parliament seem to excite quite as much attention and discussion as the proceedings of Congress, so we do not doubt that a similar condition of public feeling will arise here. The Imperial sentiment will gain in strength ... [and] there will be a more acute perception and a more abiding consciousness of the fact that we are citizens of a great empire ...[2]

Physical conveyance of information also rapidly improved in the last decades of the nineteenth century. Iron-hulled steam-powered ships brought increasing speed and comfort to the voyage between Britain and Australia, offering an alternative to the Clipper sailing ships that had undertaken the majority of the voyages until the 1870s, while the opening of the Suez Canal in 1869 provided a shorter and quicker alternative to the traditional route around the Cape of Good Hope. The colonies of Australia now received new printed material such as newspapers, books, and journals more rapidly than before these developments and this helped reduce their perceptions of social and intellectual isolation. Access to journals, scientific papers, and books in the colonies stimulated participation in debates that reached beyond the physical locality of the colonies to an international intellectual sphere (Ballantyne 2002).

Intellectual societies, such as the Royal Societies that were established in a number of the colonies from the mid-nineteenth century played an important role in maintaining libraries for members, thus providing access to printed resources on international affairs and scholarship. Great pride was taken in these collections and the international flavor of their ongoing enrichment through donations from a wide range of local and overseas scientific and cultural societies and government organizations. For example The Royal Society of Tasmania, in the small island colony at the southern-most point of Australia and so geographically the most distant colony from Europe, described the form of intellectual sociability represented by publication exchange programs in one of its reports in 1871 as "our usual friendly intercourse with kindred societies in various parts of the world."[3] Some 2,500 kilometers to the north, the Royal Society of Queensland was quick to consolidate an international publication exchange program after it was established in 1884. In 1886 it reported "additions to the library, received in exchange for the publications of the society" from the Imperial Geographical Society of Russia, the Anthropological Institute of Vienna, the Conchological Society of Great Britain, the Linnean Society of New South Wales, the Asiatic Society of Bengal, and the Manchester Literary

2   "The girdle around the globe," *Border Watch* (Mount Gambier, South Australia), 7 June 1871, p. 4.

3   "Royal Society Tasmania," *Mercury* (Hobart), 1 February 1872, 2.

and Philosophical Society, among others. Great value was placed on material that enriched understanding of the Australian colonial environment as well as providing information about other cultures. The "munificent contribution" of "the whole of the geological publication of the Indian Government," for example, was considered to be of "extreme value to Queensland geologists in consequence of the close relations between the fauna and flora of the two countries in past times."[4]

Similarly, the Royal Society in Melbourne, formed in the flourishing colony of Victoria in 1854, rapidly and regularly received publications from a great variety of international sources, including the Meteorological Society of Mauritius, the Smithsonian Institution in the USA, the Royal Academy of Belgium, and scientific academies in Cherbourg and Leipzig, amongst many others. The Melbourne Society expanded its exchange program in the 1870s, making a greater effort to supply its own publications in exchange for those received. In 1872 it noted "a most liberal response" to "invitations to exchange publications," such that "scientific journals and publications of the highest value [had] since been regularly received from every part of Europe, America, India, etc."[5] In the absence of governmental financial support, the Society was limited in the quantity of its own publications it could dispatch. For many of the colonial institutions and organizations involved in exchanges, insufficient financial resources appears to have been a common reason for the disparity of the quantity of material exchanged out of the colonies compared to the quantity received by them. This situation was addressed in the 1880s through the establishment of government funded Boards of International Exchanges discussed below.

The major public libraries which had been created in the various colonial capitals from the mid-nineteenth century onward, also enthusiastically enriched their reference collections with donations from international scientific and government bodies. The public libraries saw that they had a "national" role in providing quality printed resources that would promote the social and financial progress of Australia in a period in which a discrete nation did not yet exist. As "universities of the people," these colonial public libraries were important in holding printed material which would otherwise be extremely difficult to access for the great majority of the colonial population who sought this type of information but lacked the financial resources to purchase it privately. Exchange of publications was crucial for the young colonial libraries, both for establishing collections and for affirming their participation in trans-colonial and international networks of information exchange. This can be seen in the case of the Sydney Free Public Library which was created in 1869 in the first and largest of Australia's colonial cities.[6] The trustees hoped that the new institution would become a "great monument of national intelligence." They saw the importance of establishing a reference collection as the library's core. This was considered to include the "transactions of the chief learned Societies" as

---

4   "'Scientific and useful Royal Society," *Queenslander* (Brisbane), 15 May 1886, p. 786.

5   "Royal Society of Victoria," *Argus* (Melbourne) 12 March 1872, pp. 6–7.

6   The Sydney Free Public Library was renamed the Public Library of New South Wales in 1895.

well as the "local history of the Australian colonies." They considered that "the collection of books by donation and by international exchange" was crucial "for in this manner the wealth of the library may be largely increased."[7]

The Melbourne Public Library, established in 1856 in the capital of the colony of Victoria, was from its inception active in national and international publication exchanges. International donations increased in quantity as international mail became more frequent and less costly as the century progressed. In 1872, for example, the library received some 600 publications through donation or exchange, including 73 from Europe, 17 from America, 1 from Africa, and 3 from Asia.[8]

The receipt of international publications was equally important to the principal libraries in the more isolated and less-populated colonies in establishing their status as central points of information for local residents and scholars. The island colony of Tasmania, for example, opened its public library in the capital, Hobart, in 1870, and thereupon actively sought to acquire international publications. Only two years after its establishment the Library's *Annual Report* noted "a munificent gift of Books, Maps, Plans, and Charts from Her Majesty's Government" in Britain, as well as donations from the governments of Tasmania, Fiji, and New Zealand, and from the Royal Geographical Society of England, the Royal Historical and Archaeological Association of Ireland, the Royal Society of Art and Sciences in Mauritius, the Ossianic Society of Ireland, and the political gentlemen's club in London, the "Cobden Club," amongst many others. Donations were also eagerly sought from the British Museum library and regional British libraries. The American Consul in Tasmania facilitated donations from the United States Government, requesting "such works, published under its auspices, likely to be of value for reference, of interest scientifically or historically, or of curiosity, either from Antiquarian or other sources." Tasmanian Public Library Trustees were reassured that "such recognitions" were "the best proofs which can be adduced of the status which the Library has attained" (Trustees of the Tasmanian Public Library 1873, 2).

Similarly, the Public Library of Queensland, established in 1896 some decades after most of the other major colonial libraries, eagerly sought donations from local, national and international sources. By 1898 the Library had acquired a total of 15,000 volumes in its collection—only 2,000 volumes of which had been purchased. The majority of the collection was formed from the "liberal response to applications" made by the library to "various learned societies for grants of their publications," ranging from the Huguenot Society and the East India Association, to the Royal College of Physicians and the Institute of Civil Engineers. The international application for donations had been more fruitful than local requests. The Honorary Secretary of the library regretted that "a more liberal response had not been made by the public of Brisbane."[9]

---

7 Editorial, *Sydney Morning Herald* (Sydney), 3 June 1870, p. 4.

8 Editorial, *Argus*, 3 December 1872, p. 5.

9 "Brisbane Public Library. Large and increasing collection," *The Queenslander*, 19 November 1898, p. 979.

Boards of International Exchanges

The creation of a number of government-funded boards of exchange in the different colonies in the late-nineteenth century and early-twentieth century facilitated national and international exchanges. These boards of exchange were created in response to two international exchange conventions signed in Brussels under the aegis of the Government of Belgium in March 1886, namely "Convention A for the International Exchange of Official Documents, Scientific and Literary Publications," and "Convention B for the Immediate Exchange of Official Journals, Public Parliamentary Annals and Documents." These international conventions sought to extend the existing international exchange system through formal exchange arrangements between signatory governments (Kidder 1890, 106; also Metz 2000). Exchanges were arranged from bureau to bureau, each of which typically operated under the auspices of the national libraries in each participating nation or colony.[10] The bureaus served as "intermediaries between the learned bodies and literary and scientific societies, etc. ... of the contracting states," and governments provided funding necessary for packaging and international transport (Kidder 1890, 112). Although none of the Australian colonies was signatories to the Brussels agreements, they participated voluntarily in exchanges in the decades following the implementation of the conventions.

The United States Government was one of the most active governments in adopting the international exchange agreement, along with Belgium. In the year ending June 1888, the Smithsonian Institution, on behalf of the Government of the United States of America, undertook official exchanges with 54 other governments, including those of England, Spain, Belgium, Brazil, Chile, and the Australian colonial governments of New South Wales, Victoria, Queensland, South Australia, and Tasmania. However, the USA was generally far more productive in distributing publications than other governments involved in the exchanges. In the year ending June 30, 1888 the Smithsonian sent out "about twenty times as many packages to other Governments as it receives from them," and this was a situation that had been "on the increase for several years past" (Langley 1890, 26–33, 107). Only 20 percent of governments in the exchange system supplied the USA with their publications. Australian colonial governments were amongst those that did not send any publications to the USA in these early years of the formalized exchange system, although they benefited from receipts. In 1888 Tasmania received five cases of publications from the Smithsonian; Victoria received nine cases; Queensland received eight cases; South Australia received seven cases; and New South Wales received ten. Two decades later, in 1908, the Smithsonian sought to address the imbalance of exchanges, receiving funding to

---

10 For a description of the history of the International Exchanges from the perspective of the Smithsonian Institution, see Langley 1890, pp. 26–33, and Kidder 1890, pp. 103–16, which includes a full history of the international exchanges and translation of the original Brussels conventions documentation.

more actively seek "returns from foreign countries for the exchanges sent to them," and was consequently successful in receiving more material from international sources, including Australia (Walcott 1909, 53). In 1908 Tasmania received 1,159 packages of publications from the Smithsonian, and sent two packages in return; Victoria received 2,827 and sent 510; Queensland received 1,374 packages and sent 182; South Australia received 1,469 packages and sent 80; and New South Wales received 2,732 packages and responded with 477 packages (Walcott 1909, 56–7).[11] Although Australia was now a nation, Boards of International Exchange, in what had become Australian states created from the former colonies, continued to operate their own exchanges rather than transmuting to a national system.

The Boards of Exchange were appointed to "act on behalf of the Government in the matter of International Exchanges of Literary and Scientific Works, Official Publications, etc.," in those colonies where they were established (Government of New South Wales 1890, 97). The Boards were composed of public servants from diverse professional backgrounds that typically involved information as a core business. The Board for International Exchanges that was established in New South Wales in the late 1880s, for example, included in its membership over the next decade Robert Cooper Walker, first Librarian at the Public Library in Sydney (1869–93) who acted as Chairman from 1890–96 (Richardson 1976).[12] He was followed as Chairman in 1897 by the then Principal Librarian of the Public Library of New South Wales, H. C. L. Anderson and membership included the New South Wales Superintendent of Technical Education, Robert Newton Morris (Mitchell 1974),[13] the Parliamentary Librarian, Frank Walsh, and the director of the Botanic Gardens and government botanist Joseph Henry Maiden, who was also a long-term council member of many of Sydney's learned societies (Lyons 1986).[14] The Smithsonian Institution commended the New South Wales Board of Exchanges as "very efficient" (Winlock 1894, 48).

The Queensland Government's Board for International Exchanges was established in Brisbane in 1901. It aimed at "forming a large central bureau for the distribution of all official and other publications that may concern other States of the Commonwealth than Queensland, and other countries than Australia."[15] Members of the Brisbane Board included the Government Printer, the Librarian of the Public Library of Queensland, and the undersecretaries of the Home Department and the Chief Secretary's Department.[16] The Board noted that "... much valuable knowledge in administration, in industrial pursuits, in education,

---

11   Note that figures are now quoted as "packages" rather than "cases" as was the case in 1888.

12   "Public Service Gazette," *Sydney Morning Herald*, 12 September 1896, p. 5.

13   "The Public Service," *Sydney Morning Herald*, 14 May 1898, p. 13.

14   "Public Service Gazette," see note 12.

15   "International publications," *Brisbane Courier*, 5 November 1903, p. 4.

16   "Honorary positions for Government officials," *Brisbane Courier*, 14 February 1901, p. 4

in economics, art, science, and literature may be interchanged with all parts of the civilized world, and this means of reciprocity will enable Queensland institutions to learn what equal or higher efficiency in all such matters is achieved elsewhere."[17] Within two years the Brisbane Board was exchanging publications with the Smithsonian Institution and with exchange bureaus in Belgium, Honolulu, Manila, New Zealand, and the British colony of Natal, as well as the New South Wales Board of International Exchanges and the Public Library in Melbourne (which had undertaken international exchanges in the colony of Victoria in the absence of a formal Board of International Exchanges) (Winlock 1894, 50). The Brisbane Board was proud to highlight the social and political utility of exchanges, citing as examples providing a complete set of Queensland Statutes to the government of the Transvaal Colony "which was no doubt required for the drafting of bills for the new Transvaal government," in the aftermath of the Anglo-Boer War.[18]

The Public Library in Perth, in the colony of Western Australia, also participated actively in the international exchange scheme. Only a decade after it was established, the Library was proud to note in its 1899 Annual Report that it had dispatched "the various official publications of the colony to the learned societies and public libraries throughout the world." It had distributed 137 publications to 44 societies and institutions "from whom many exchanges of value have been received."[19] A decade on, in 1909, the Chief Librarian reported receipt of 248 parcels from other governments, national and international, and the distribution of 1,330 volumes amongst 256 societies and institutions "in connection with the international exchange of official publications."[20]

Information about the Past: The New Scientific History in the Colonies

This chapter now addresses a discrete aspect of information acquisition in the Australian colonies, specifically the emergence of a new interest in accessing material for the production of written histories about the British colonial enterprise in Australasia. In the period of the Belle Époque, redemptive, scientific, and nationalizing history was increasingly the currency of historical consciousness across the western world. Because history as a professional discipline evolved in the nineteenth century in a context of increasing nationalism and bureaucracy, it was in Australia, as elsewhere, especially concerned with identity and documents (Griffiths 1996, 210–11). History was no longer a matter of rhetoric and persuasion, but rather an empirical science of "detachment and documentation" (Dalziell 2009, 102–3). In a climate of active exchange of international information and the keen awareness

17   "International publications," see note 15.
18   "Exchanging publications," *Brisbane Courier*, 20 March 1907, p. 9.
19   "Victoria Public Library. Annual Report," *West Australian* (Perth), 5 December 1899, p. 2.
20   "The Public Library. Annual Report," *West Australian*, 23 October 1909, p. 4.

of the value of research and documentation for enhancing national pride and social and economic development, access to information about Australian history became especially important to many Australians in the last decades of the nineteenth century.

Much impetus to write Australian histories emerged from forms of history being produced in Britain, information coming to the colonies from newspaper reports and from the new historical publications themselves. Colonial historian, James Bonwick, noted that "in writing colonial history we have some good models for our guidance in England," pointing to the "honest recorders" like Sharon Turner with her *History of England* (1814–29) and *History of the Anglo-Saxons* (1799–1805). The Historical Manuscripts Commission, established through a Royal Commission in Britain in 1869 became an inspiration for Australian efforts to unearth locally-specific historical sources. The British Historical Manuscripts Commission was charged with recording the existence and whereabouts of manuscripts of value for the study of history, covering in scope the whole of the United Kingdom, including sources in other countries. The work of the Commission, disseminated through published reports in parliamentary papers and calendars, was regularly covered by Australian newspapers. The Tasmanian *Mercury* observed in 1870, for example, that in the "short period of [the Commission's] existence most valuable information has been brought to light," and found "great credit ... due to the present [British] Government for the interest it has taken in originating the Commission and furthering the objects which it has in view."[21] Nearly 20 years later, the *Sydney Morning Herald* lauded the progress of the project, observing that the Commission was "almost prodigal in its bounty to students." It provided a "collective preservation" of the past for which the "English public may congratulate itself."[22]

The progressive release of the English *Dictionary of National Biography*, published in 63 volumes 1885–1900, was also enthusiastically reported in the Australian press. Australians were particularly interested when biographies of figures associated with Australia were included. Absence of Australian content in the volumes, as they were being published was also noted. This was the case for the fourth volume published in 1886 which "unlike its predecessors [contained] no name connected with the history of these colonies."[23] In 1888, the *Dictionary of National Biography* up to the letter "D' was published. It contained an entry for Captain James Cook (the popularly conceived "discoverer" of Australia). A reviewer for the Melbourne *Argus* complained that "one would have been glad of a somewhat more detailed account of his voyages and their geographical results ... Three and a half pages hardly seems an adequate memorial of England's greatest navigator."[24]

21   "Discoveries by the Historical Manuscripts Commission," *Mercury*, 13 January 1870, p. 3.

22   "Historical manuscripts," *Sydney Morning Herald*, 19 April 1889, p. 8.

23   "Dictionary of National Biography," *Argus*, 5 June 1886, p. 4.

24   "The Dictionary of National Biography," *Argus*, 5 April 1888, p. 11.

The year Cook appeared in the *Dictionary* was a significant one for Australia, in that it marked the centenary of the landing of Captain Arthur Phillip and the First Fleet at Port Jackson (present-day Sydney) in New South Wales on January 26, 1788. This anniversary was marked by public celebrations in almost all the Australian colonial capital cities. The *Sydney Morning Herald* found tangible evidence during this centenary celebration of "Australian sentiment [that] has grown with the growth of the colony" [of New South Wales].[25] The centenary of white settlement inspired a "flood of celebratory volumes" published in Australia. They included the *Picturesque Atlas of Australia* (1886–88), histories of each of the colonies, and biographical publications about "eminent colonists' (Arnold 2001, 282–91). Awareness of the changing nature of Australian colonial self-regard, where individuals could identify themselves through a number of contexts such as British, imperial, colonial (Tasmanian, Victorian), and Australian, prompted a "looking back" in order to "look forward." It is also important to remember that Australia suffered a relatively unflattering image in the mid-nineteenth century. This was the combined effect of the penal origins of a number of the colonies, the continent-wide dispossession of indigenous Australians, and the inevitable lack of sophistication of a young society. This colored both Australian self-regard, and the perception of Australia of other countries.

The new historians of late-nineteenth century Australia believed that properly researched histories based on documented facts would help to improve Australians' self-image, promote Australian affairs in international contexts, and substantiate Britain's activities in the colonies within a global and unifying imperial history. The search for historical identity in Australia in this period was frequently evoked in a metaphor of the movement from obscurity to clarity: "bringing to light," and "snatching from out of the darkness" the history of the colonies. This common phraseology, found in much writing about history through the nineteenth century, had a certain moral subtext in the Australian colonies as they were emerging from a penal past and were transforming themselves into a single national entity. In the year of the Federation of Australian colonies in 1901, the Adelaide *Advertiser* newspaper observed that the "chief hope of a really energetic nation is in the present and in the future, but it should not forget its past, and everything which tends to preserve the lights and shades of its real life makes for the general benefit."[26] Historian, James Bonwick, expressed this same idea when he wrote in 1895 that

> Australia is but making its history. The settlement thereof began in gloomy trials, it struggled against difficulties never again to be encountered, it emerged from deep shadows into brightsome sunshine, and its future will, doubtless, be associated with the real progression of our race, and the brotherhood of man (Bonwick 1895, 6).

---

25   "Editorial," *Sydney Morning Herald*, 28 January 1888, p. 12.
26   "Records of the past," *Advertiser* (Adelaide), 21 February 1901, p. 4.

The anonymous Queensland reviewer of Bonwick's *The first twenty years of Australia* (1882) concluded: "[Australia's] is a curious history, a history of savagery and immorality, and yet a history which illustrates the ascendancy of the principle of authority as sufficient to reduce lawlessness into order."[27]

"To Collect Materials from all Quarters"

Historical societies were established in the colonies in this period as a direct outcome of the new interest in historical studies and to support the writing of Australian history and the search for original documentation.[28] The Historical Society of Australasia, created in Melbourne in 1885, promoted the production of "authentic and genuine history," rather than "legend" and "trifles." In his opening speech at the inaugural meeting, the President, David Blair, drew attention to what he perceived as the difference in historical consciousness between English people "who were intensely historical in their predilections" and "the average Australian [who] cared nothing for the history of the cosmos around him, and was perfectly indifferent to the bygone history of this splendid land":

> To show the necessity of some such association as this, in order to collect the historical memorials of Australia, and place them before the world in a proper form, [Blair] quoted the inch and half of information that was all that could be found in Chambers's Encyclopedia relative to the discovery of the continent, and more than one of the few statements made were erroneous ... In the absence of authentic and genuine history, legend invariably rushed in and took the vacant place; and it was remarkable how many legends had been already palmed off as forming part of the true history of Australia.

In conclusion, Blair stated that the "great object" of the newly formed Historical Society would be "to collect materials from all quarters for a complete and authentic history of the British Empire in the south."[29]

Major multi-volume historical publication projects with a focus on the utilization of original documents emerged at this time. One of the most significant was the *Historical Records of New South Wales*, issued by the Government Printer in Sydney, 1892 through 1901. In his "Preface" to the second book in the series, published in 1892, Alexander Britton stated that the object of the publication was to afford "the fullest information obtainable concerning the foundation, progress, and government of the mother colony of Australia" (Britton 1892, v). Unlike

27 "The reviewer. First twenty years of Australia," *Queenslander*, 9 September 1882, p. 330.
28 "Editorial," *Sydney Morning Herald*, 28 January 1888, p. 12.
29 "The Historical Society of Australasia," *Argus*, 9 May 1885, p. 10.

the first book in the series, the second (and the subsequent volumes that were being planned) were to consist predominantly of the historical papers themselves "presented without comment, and without any attempt to explain the story they tell ... The Records are given here as they were found, and they speak for themselves" (Britton 1892, xviii). A later, equally significant project that assumed truly national dimensions was the *Historical Records of Australia*, published between 1914 and 1925 in 33 volumes under the auspices of the Australian Commonwealth Government.

There was, however, a major problem for Australian historians dedicated to writing the new scientific histories based on original sources rather than "hearsay." Historians had found that there were major gaps in the local records. It quickly became evident that governmental record-keeping in the Australian colonies had been fickle, political, and haphazard, particularly in the first 30 years of settlement (Curthoys 2006, 364). Local records systems were officially maintained in only two of the Australian colonies, in the Office of the Registrar-General in the colony of Victoria (Russell and Farrugia 2003) and in the Chief Secretary's Department in Tasmania. But the records in these local depositories were typically incomplete.[30] Surveyor-General of Tasmania and amateur historian, James Erskine Calder, wrote in 1873 that

> ... through negligence, that is little less than criminal, the history of the infant
> times of some of the colonies we live amongst, though so recently established,
> seems ... fast passing out of knowledge ... Through the too frequent decay of
> old official documents, very little has been preserved of the story of the career
> of Tasmania in the first fourteen or fifteen years of its existence (Calder 1873).

In other colonies, such as New South Wales, there had been no attempt to maintain records systematically. Britton noted in the preface to the *Historical Records of New South Wales* (1892), that "unfortunately manuscripts of great interest and importance, which are known to have existed, cannot now be found" (Britton 1892, x). Of Dispatches sent to England by the Governors of the colony of New South Wales and original Dispatches received from the Home authorities prior to 1800, there could be found "no trace," while "the orders, proclamations, and other official papers showing how authority was exercised in the early days are found only in fragments—in fact they can scarcely be said to exist" (Britton 1892, xii).

The historians recognized that missing records were absent not only through historical mischance and official negligence, but also through active destruction. Britton conjectured that the New South Wales governmental papers from the early decades of the colony had most probably been destroyed, along with other "public records of corresponding dates, for which the Governors were not responsible" (Britton 1892, xi). In New South Wales and Tasmania in particular a desire to

---

30 "Our record office. Nothing like it in the other states," *Mercury* (Hobart), 3 February 1905, p. 3.

obscure embarrassing information about the penal past was identified as a factor in this record destruction. Robert McNab, editor of *Historical Records of New Zealand* (1908), who utilized Australian record sources extensively, frankly stated that "the destruction of old records … is attributed to persons in the Government employ who were interested in the suppression of facts contained in them, losing no opportunity of getting rid of the incriminating documents housed in the Government pigeon holes by carting them away as old lumber and burning them."[31] The "suppression of the information contained in these old records," to use McNab's words, was a consequence of the efforts of individuals with the means to do so to destroy evidence of their relationship to convicts or their own convict past.

Unlike their contemporaries in Britain, the major colonial public libraries did not have an active role in preserving local government or other historical papers in the nineteenth century and there were no formal Government Archive offices until the twentieth century (Golder 1994; also Piggot 1994). Any archiving of the records of Australian government officials or of other colonial institutions, such as missionary societies, typically occurred outside of the Australian colonies in the English headquarters of these institutions. Thus, when the interest emerged in writing history based on archival records, the materials had to be sought in multiple sites around globe, not only from official sources in England in the Public Record Office and Colonial Office but also from other colonial states, Europe, and America. The Public Record Office in London held the most substantial proportion of Australian colonial records. In his presentation entitled, "Writing of Colonial History," given at the Royal Colonial Institute in London on March 26, 1895, James Bonwick observed that "London is, with all deficiencies [sic], the best searching-ground for contemplating colonial historians. To every Australian colonist London will ever be the Mecca of his wandering, if only for association with early history" (Bonwick 1895, 6).

The work of individuals undertaken around the globe was crucial in reconstructing on-shore archival sources for Australian history. In searching out historical records around the world, historians set up intellectual linkages characteristic of this period, connecting not only the colony to the metropolis, but also to other nodes in the imperial web. Francis Peter Labilliere was among the first historians to examine records relating to Australian history in the Public Record Office in London. He was "set upon hunting for … dust-covered documents that might throw light on the early history of Australia."[32] Labilliere was a member of the English Royal Colonial Institute, which had been founded to "promote in England a better knowledge of the colonies and of India." He was convinced that properly documented histories would aid that mission (Penny 1974). In his Introduction to *Early History of the Colony of Victoria* (1878), Labilliere drew

---

31 "An Australasian historian," *Mercury* (Hobart), 20 June 1914, p. 4.

32 "Review. The early history of the Colony of Victoria," *Sydney Morning Herald*, 28 September 1878, p. 7.

attention to the way in which American historians utilized historical documents from international sources. He noted that "Some years ago a literary society in New York sent over to this country and obtained copies of all documents relating to the history of that State, which have since been published. The same has been done for other portions of the American union" (Labilliere 1878, xvii). Labilliere replicated this practice in his book about the colony of Victoria, two thirds of which comprised transcribed historical documents. He hoped that this method of international scholarship and research would serve as an example for local historians. "Not until the forgotten materials for the history of each of the colonies are collected, as I have endeavoured to collect those of Victoria," he observed, "can a comprehensive history of Australia be written" (Labilliere 1878, xvii).

Historians such as Labilliere and Bonwick alerted the colonial governments to the richness of the resources for Australian history held off-shore and encouraged them to support transcription projects that were related to such official historical publishing ventures as the *Historical Records of New South Wales*. Bonwick had first published a plea for retrieving important historical records for Australia in 1883 in the preface to his history of the Port Phillip Settlement in the colony of Victoria. He had been delighted with the "immense repository" he had found when he went to the Public Record Office in London, particularly the "vast amount of correspondence between our own ruler, and with people not only of our own race, but with those of various foreign nations" (Bonwick 1902, 238). Bonwick urged that "faithful copies of such interesting documents should be in the public libraries of colonial capitals" (Bonwick 1883, iv). He was successful in obtaining government support for his transcription projects in London and subsequently undertook copying for the Queensland government in 1883, for the South Australian government in 1885, for the Victorian government in 1886, the Tasmanian government from 1887 to 1894, and for the government of New South Wales from 1887 until 1902 when he ceased his work due to ill health (Featherstone 1969).

In Victoria and New South Wales the transcripts went directly to the colonial public libraries, where key individuals such as Principal Librarian of the Public Library of New South Wales, H.C.L. Anderson, contributed to the project by making suggestions about material to be copied to fill specific gaps in the local record. In Tasmania, Bonwick's transcriptions remained in Government possession in the Chief Secretary's department, rather than in the Tasmanian Public Library. Historian and Trustee of the Tasmanian Public Library, James Backhouse Walker, however, acted as the government's official correspondent with Bonwick, offering advice and support for Bonwick's arduous and time-consuming work. Walker gave Bonwick specific instructions on types of documents needed, particular logs and charts that he should look for, and to copy photographically Tasmanian charts, plans, and views.[33] The Royal Society of Tasmania applied to the Government for custody of Bonwick's

---

33   Walker, James Backhouse. 1852–1899. Papers. Special and Rare Materials Collection, W9/C2/3. University of Tasmania Library.

transcriptions but the request was denied on the grounds that some items were "of a private nature and ought not to be made public" in the climate of ongoing sensitivity to aspects of the colony's penal past (Roe 2001, 229–32).

New Zealand also experienced increasing self-consciousness about national identity in the period under discussion. A number of New Zealand historians undertook research and writing activities similar to their Australian counterparts. Politician and historian, Robert McNab, for example, developed a keen interest in New Zealand historical documents in the late 1890s and travelled extensively from 1904 to 1907 in order to transcribe documentary sources that he then utilized in his histories. In his first published historical work on New Zealand, *Murihiku and the Southern Islands* (1907), McNab described the "extensive" search for "necessary information," that took him to Australia to the Colonial Secretary's Office in Hobart and the Public Library in Sydney, to historical societies and public libraries along the East Coast of the United States, and to the Public Record Office and British Museum in London (McNab 1907, vii–viii). His correspondence with archivists in Paris and Madrid resulted in additional transcriptions and information. McNab was encouraged by New Zealand Prime Minister, Richard Seddon, to edit and publish his entire transcripts of historical documents. These appeared as the two-volume *Historical records of New Zealand* (1908, 1914) (Traue 2010).

Australasian governments were stimulated to fund historical document transcription and publication projects not only by the work of local historians but also by the example of successful copying projects undertaken by other British colonial governments. Cape Colony, for example, had arranged in the early 1890s for copies to be made from the Public Record Office in London of all Governors' Despatches from 1796 through 1826, together with confidential letters and other documents.[34] The success of the Public Archives of Canada (founded 1872) and of copying projects undertaken by that body were cited when the Australian government became actively involved in filling in gaps in the local archive in the first decade of the twentieth century and sought to establish the first national archival institution.[35] When the librarian, Frank Murcott Bladen, editor of the *Historical Records of New South Wales* from 1896 to 1899, made a report for the Australian Commonwealth Government on international archives with a view to establishing a central national archive in Australia, he compared Australia's position to that of Canada and Cape Colony. There he had found "one universal regret, viz., that no records of the genesis of their nationhood had been preserved" (Bladen 1903).[36] He urged the Australian government to take steps to ensure the future safety of Australian records of the past as well as those of the present.

In addition to government funded copying projects, there was growing opinion that original documents that were central to the formation of the nation should

---

34 "Cape News," *Mercury*, 9 September 1895, p. 4.
35 "The Commonwealth," *Argus*, 16 July 1903, p. 7.
36 Quoted in "Australia's historical records," *Sydney Morning Herald*, 17 July 1903, p. 3.

be returned to Australian shores. When Frederick Watson was appointed the first official historian to the Commonwealth in 1912, a *Sydney Morning Herald* correspondent hoped that "in the name of the people of Australia, [Watson would] help to bring back to these shores those records of our past which have found their way either to general collections in England, or into the possession of individuals."[37] "Their proper resting-place," the correspondent concluded, "is in the land to which they relate, or from which they were taken."[38] However, attempts by the Australian state libraries to repatriate original documents from Britain in the period from 1901 to the 1930s were typically not successful (Powell 2005).

In the face of the negative response from the British Public Record Office, Australian efforts re-focused instead on effective and comprehensive duplication. This process was facilitated in the 1940s by the inauguration of the Australian Joint Copying Project. Filming began in the Public Record Office in Britain in 1948 and within ten years all the records from the Colonial Office dealing with the founding and early years of all the Australian colonies, as well as other important documents such as the papers of explorers James Cook and Matthew Flinders, had been copied to microfilm.

Indigenous History in the Australian Archive: "Silence on all Matters Aboriginal"

In pursuing an "ideal" history of British Australia, grounded in documents rather than "hearsay," the historical experience of Australia's First Peoples before and after colonization was largely ignored. In the 1940s, Kenneth Binns, chief librarian of the Commonwealth National Library, wondered "what future historians and anthropologists will think of our neglect to preserve adequate records of our own Australian aborigines" (Binns 1948, 24). The generalized failure to document and write Aboriginal history was described by influential anthropologist, William Stanner, as a "silence on all matters aboriginal," in his critique of Australian history in the influential 1968 Boyer Lecture series *After the Dreaming*. Stanner characterized this long-standing attitude as a "cult of disremembering'" that could not be explained by "absent mindedness" (Stanner 1969, 24–5).

Certainly some pamphlets and books published in the nineteenth century addressed specific topics such as Aboriginal tribal language that typically were based on free colonists' and missionaries' personal experiences with indigenous people. Examples of this are the publications of Lancelot Edward Threlkeld, missionary and Congregational minister in the colony of New South Wales from the 1830s to 1850s, such as *An Australian Grammar* (1834), which are now regarded as landmarks in Aboriginal studies (Gunson 1967). Public libraries in the colonies held many of these books and pamphlets on Aboriginal languages by

---

37   "A Federal historian," *Sydney Morning Herald*, 28 August 1912, p. 16.
38   "A Federal historian," p. 16.

different authors. For an article published in the Melbourne *Argus* newspaper in 1882, entitled "A plea for the Australian languages," Melbourne Public Librarian, Thomas Francis Bride, produced a list of 27 published volumes held by his library that dealt directly or indirectly with Aboriginal languages. Bride also included in the newspaper article a list of 23 manuscripts in private hands that had been used by the civil servant, Robert Brough Smyth, for his landmark publication, *The Aborigines of Victoria* (1878) (Hoare 1976).[39] Nevertheless, the relative poverty and dearth of sources on indigenous history was acknowledged at this time. Professor John Elkington, at the inaugural meeting of the Historical Society of Australasia in 1885, for example, considered that "the society should endeavour … to pay greater attention than had hitherto done to the study of the customs and habits of the aborigines of the Australian continent."[40]

When Frederick Watson was appointed as the first Federal Historian in 1912, the *Sydney Morning Herald*, reporting on Watson's appointment, observed that "Australia's history under white man's rule" was of short enough duration for its records to be "rendered easily accessible."[41] Curiously, the correspondent considered that Australia did not have the "complications of race and colour" that had been experienced in the development of other colonies, such as South Africa and Canada. In this skewed view of history neither the "deep time" of Australia's past before white colonization nor the highly divisive and painful history of relations between the colonizers and colonized since settlement, was considered significant. As an additional complicating factor, the strengthening perception that historical documents legitimated a nation's past, present, and future simultaneously diminished the perceived value of unwritten forms historical evidence, specifically oral histories. As Mark McKenna has noted in his award-winning study, *Looking for Blackfellas' Point: An Australian History of Place*, "the history of the settlers' forgetting of Aboriginal Australians was far more complex and subtle than the simple locking of a cupboard or the obscuring of a window … Narratives that acknowledged frontier violence, for example, coexisted with historical narratives that erased the frontier from settler memory" (McKenna 2002, 63). McKenna writes that this was especially true in the nineteenth century prior to Federation. Further exploration of the relationships between availability of documentary sources, prevalent attitudes to race, and keenness to disremember particular aspects of the past in relation to indigenous history and historical records, is far beyond the capacity of this chapter, but is certainly a fruitful area for future research.

---

39   "A plea for the Australian languages," *Argus* (Melbourne), 18 November 1882, p. 6. Brough Smyth published the list of sources on pp. 1–2 of *The Aborigines of Victoria* (1878).

40   "The Historical Society of Australasia," *Argus*, 9 May 1885, p. 10.

41   "A Federal historian," p. 16.

Conclusion

The "pursuit of colonial intelligence," in James Bonwick's phrase, assumed growing importance for Australians as the nineteenth century moved into the twentieth century (Bonwick 1895, 6). As this chapter has shown, it became apparent to many people in the colonies that applying the new empirical historical method to Australian history was a logical and powerful way for the new nation to write itself into the wider histories of the British Empire, and indeed the world. Historical information was central to identity formation. It acted to stabilize and direct an emerging national consciousness. It also offered a vehicle to project a formalized Australian identity to a global audience that was rapidly becoming more intimately linked through information exchange of all types. By contributing to the formation of an historical archival resource for Australia, and promoting it in support of foundational and celebratory narratives of possession and progress, early historians, historical and scientific societies, along with information institutions such as public libraries contributed to an "agreed meaning" about the Australian past and to a developing common worldview for the young nation.

References

Arnold, John. 2001. "Reference and Non-fiction Publishing." In *A History of the Book in Australia 1891–1945: A National Culture in a Colonised Market*, edited by Martyn Lyons and John Arnold. St Lucia, Queensland: University of Queensland Press, 282–91.

Ballantyne, Tony. 2002. *Orientalism and Race: Aryanism in the British Empire*. New York: Palgrave Macmillan.

Binns, Kenneth. 1948. "The Commonwealth National Library." In *Across the Years: The Lure of Early Australian Books*, edited by Charles Barrett. Melbourne: Seward.

Bladen, Francis M. 1903. *Archives: Report on European Archives*. Melbourne: Government of the Commonwealth of Australia.

Blainey, Geoffrey. 1966. *The Tyranny of Distance: How Distance Shaped Australia's History*. Melbourne: Sun Books.

Bonwick, James. 1883. *Port Phillip Settlement*. London: Sampson Low, Marston, Searle, and Rivington.

Bonwick, James. 1895. "The Writing of Colonial History." *Sydney Morning Herald*, May 25, 6. http://trove.nla.gov.au/ndp/del/page/1366691.

Bonwick, James. 1902. *An Octogenarian's Reminiscences*. London: J. Nichols.

Britton, Alexander. 1892. *Historical Records of New South Wales*, vol. 1. Sydney: Government Printer.

Brough Smyth, Robert. 1878. *The Aborigines of Victoria: With Notes Relating to the Habits of the Natives of other Parts of Australia and Tasmania Compiled from Various Sources for the Government of Victoria*. Melbourne: Government Printer.

Calder, James Erskine. 1873. "Tasmania. The First Years of Settlement. To the Editor of the Mercury." *Mercury* (Hobart), May 16, 3. http://trove.nla.gov.au/ndp/del/page/787972.

Curthoys, Ann. 2006. "The History of Killing and the Killing of History." In *Archive Stories: Facts, Fictions, and the Writing of History*, edited by A.M. Burton. Durham, NC: Duke University Press. 351–74.

Dalziell, Tanya. 2009. "No Place for a Book? Fiction in Australia to 1890." In *The Cambridge History of Australian Literature*, edited by Peter Pierce. Port Melbourne, Victoria: Cambridge University Press.

Dawson, Robert. 1830. *The Present State of Australia: A Description of the Country, its Advantages and Prospects, with Reference to Emigration: And a Particular Account of the Manners, Customs, and Condition of its Aboriginal Inhabitants*. London: Smith, Elder, and Co.

Featherstone, Guy. 1969. "Bonwick, James (1817–1906)." *Australian Dictionary of Biography*. http://adb.anu.edu.au/biography/bonwick-james-3022/text4429.

Golder, Hilary. 1994. *Documenting a Nation: Australian Archives—the First Fifty Years*. Canberra: Australian Government Public Service.

Government of New South Wales. 1890. *Consolidated Fund Appropriation Act, XXXII, 20 December 1890*. http://www.legislation.nsw.gov.au/sessionalview/sessional/act/1890-32a.pdf.

Griffiths, Tom. 1996. *Hunters and Collectors: The Antiquarian Imagination in Australia*. Cambridge: Melbourne: Cambridge University Press.

Gunson, Niel. 1967. "Threlkeld, Lancelot Edward (1788–1859)." *Australian Dictionary of Biography*. http://adb.anu.edu.au/biography/threlkeld-lancelot-edward-2734.

Hoare, Michael. 1976. "Smyth, Robert Brough (1830–1889)." *Australian Dictionary of Biography*. http://adb.anu.edu.au/biography/smyth-robert-brough-4621/text7609.

Kidder, J.H. 1890. "Report upon International Exchanges under the Direction of the Smithsonian Institution, for the Year Ending June 30, 1888." In *Annual Report of the Board of Regents of the Smithsonian Institution ... to July 1888*. Washington: Government Printing Office, 103–16.

Labilliere, Francis Peter. 1878. *Early History of the Colony of Victoria: From its Discovery to its Establishment as a Self-governing Province of the British Empire*. London: Sampson Low, Marston, Searle, and Rivington.

Langley, S.P. 1890. "Report of the Secretary: Exchange System of the Institution." In *Annual Report of the Board of Regents of the Smithsonian Institution ... for the Year Ending June 30 1888*. Washington: Government Printing Office, 26–36.

Lyons, Mark. 1986. "Maiden, Joseph Henry (1859–1925)." *Australian Dictionary of Biography*. http://adb.anu.edu.au/biography/maiden-joseph-henry-7463/text12999.

McKenna, Mark. 2002. *Looking for Blackfellas' Point: An Australian History of Place*. Sydney: University of New South Wales Press.

McNab, Robert. 1907. *Murihiku and the Southern Islands*. Invercargill, New Zealand: William Smith.

Metz, Johannes. 2000. "International Exchange of Official Publications." *International Journal of Special Libraries*, 34 (2): 80–89.

Mitchell, Bruce. 1974. "Morris, Robert Newton (1844–1931)." *Australian Dictionary of Biography*. http://adb.anu.edu.au/biography/morris-robert-newton-4253/text6873.

Penny, B.R. 1974. "Labilliere, Francis Peter (1840–1895)." *Australian Dictionary of Biography*. http://www.adb.online.anu.edu.au/biogs/A050055b.htm?hilite=labilliere.

Piggott, Michael. 1994. "Beginnings." In *The Records Continuum: Ian Maclean and Australian Archives First Fifty Years*, edited by S.M. McKemmish and M. Piggott, 1–17.

Powell, Graeme. 2005. "The Quest for the Nation's Title Deeds, 1901–1990." *Australian Library Journal*, 54 (1): 55–65.

Richardson, G.D. 1976. "Walker, Robert Cooper (1833–1897)." *Australian Dictionary of Biography*. http://adb.anu.edu.au/biography/walker-robert-cooper-4788/text7973.

Roe, Michael. 2001. *The State of Tasmania: Identity at Federation Time*. Hobart: Tasmanian Historical Research Association.

Russell, Edward and Farrugia, Charles. 2003. *A Matter of Record: A History of the Public Record Office Victoria*. North Melbourne, Victoria: Public Records Office.

Stanner, William E.H. 1969. *After the Dreaming: Black and White Australians – an Anthropologist's View* [The Boyer Lectures, 1968]. Sydney: Australian Broadcasting Commission.

Threlkeld, L.E. 1834. *An Australian Grammar ... of the Language, as Spoken by the Aborigines ... of Hunter's River*. Sydney: Society for Promoting Christian Knowledge in conjunction with the Government of New South Wales.

Traue, J.E. 2010. "McNab, Robert." *Dictionary of New Zealand Biography*. http://www.teara.govt.nz/en/biographies/3m30/1.

Trustees of the Tasmanian Public Library. 1873. *Tasmanian Public Library Report for the Year 1872, 1873*. Hobart, Tasmanian Archives and Heritage Office, MCC 16/24/1/1.

Walcott, Charles. 1909. "Report on the International Exchanges." In *Annual Report of the Board of Regents of the Smithsonian Institution ... for the Year ending June 30 1908*. Washington: Government Printing Office, 53–61.

Winlock, W.C. 1894. "Report of the Curator of Exchanges." In *Annual Report of the Board of Regents of the Smithsonian Institution ... to July, 1893*. Washington: Government Printing Office, 45–54.

Chapter 4

Divided Space—Divided Science? Closing and Transcending Scientific Boundaries in Central Europe between 1860 and 1900

Jan Surman

The development of science and scholarship under imperial circumstances has attracted the growing attention of historians of science. Because of the nature of language and political–cultural entanglements, "an empire" as opposed to "a nation" offers a privileged field of inquiry as to how power—be it cultural or political—influences scientific production and communication (Buklijas and Lafferton 2007; Ash and Surman 2012). The turn of the nineteenth century— whose enormous cultural productivity has attracted much attention—adds another perspective. With manifold language conflicts and changes of language of instruction at Central European universities, this period is characterized by a reevaluation of cultural achievements and the stabilization of communicational and educational infrastructure in Slavic cultures. This is particularly true for the post 1867 Habsburg Monarchy, which at the time was a multicultural entity in which several languages were privileged culturally. The German and Russian empires at the same time were pursuing a policy of centralization and acculturation. In the Romanov Empire, for example, the privileges enjoyed by Polish, German and Ukrainian languages in higher education were renounced. The establishment of the German Empire was similarly linked to a strengthened Germanization policy that affected the position of languages other than German in education and in academia broadly defined. The period after 1860 also brought a reconfiguration of Slavic cultural geography in which Cracow and L'viv, based in the Habsburg Monarchy, replaced the previously dominant Poznań (in the German Empire) and Warsaw and Kiev (in the Russian Empire) where scientific institutions began to face partial linguistic restrictions. Nonetheless, the infrastructural basis developed throughout the century resulted in an intensification of discussion over how to improve scholarship in the Slavic languages, while previously the question of privilege had played a much more prominent role.

The most serious changes in scientific infrastructure took place between 1860–1890 so that the period called the *belle époque* was characterized by more or less stable boundaries as defined by citizenship and language. The primary places of scholarly production—universities—were not only scientific, but also cultural strongholds, a role which they reassumed after a period of being essentially schools for civil servants (Havránek 1990). The Academies of Sciences in Cracow, Prague and L'viv (the last, the Ukrainian Shevchenko Scientific Society, did not have the official status of an academy of science but had a similar structure and standing) were established to replace less productive and more specifically local scientific societies. Through concentration on humanistic disciplines, on the popularization of science, and on cultural development through scholarly achievements, these institutions and the networks built around them, replaced more trans-culturally oriented and often bilingual local institutions. In Bohemia, for instance, not only was the Charles University divided into Czech and German components, but the Technical Academy and two distinct academies of sciences were also established: the Czech Academy of Science and the Arts (Česká akademie věd a umění císaře Františka Josefa I., est. 1890) and the German Association for the Fostering of German Science, Arts and Literature in Bohemia (Die Gesellschaft zur Förderung Deutscher Wissenschaft, Kunst und Literatur in Böhmen, est. 1891). They replaced the bilingual Bohemian Royal Society (Königliche böhmische Gesellschaft der Wissenschaften/Královská česká společnost nauk, est. 1784), which remained active but with almost no influence.

In Galicia, Polish was the language of instruction in Cracow and L'viv, causing problems for Ruthenians. Ruthenian is the ethnic name for Galician that Greek-Catholics used until the early twentieth century when the group was merged into the Ukrainian nation. For historical accuracy, I use Ruthenian to denote Galician Ukrainians as compared to Ukrainians in the Russian Empire. Thus the Cracow Academy of Sciences and Arts, committed to Polish, had almost no contacts with its Ruthenian counterpart, the Shevchenko Society. It became normal for scholars to prefer to publish in the language of their parent institutions and not predominantly in German as had been the case previously. While the different political and social circumstances involved were often seen as detrimental for the development of a "national" scholarship, nevertheless the institutional infrastructural developments successfully emulated the one-language communities of nation-states, with all the adverse consequences this had had for multicultural and multilingual regions. To be precise, some of these language communities also faced political boundaries that physically divided the internal communication networks. This was the case with the Ukrainian and Polish communities which were divided between the empires. However communication structures were developed that transgressed the political borders, not only in terms of publishing and membership in scholarly organizations, but also through joint institutional endeavors such as trans-imperial conferences (Cabaj 2007).

The changes outlined above allow one to speak of institutionalized Slavic scientific communities. At the same time, as language communities, they also succeeded in establishing boundaries inhibiting the transfer of scientific information. To be sure, the earlier supremacy of the transnational German language in Central Europe was itself not without problems, especially as it was linked to a political domination that was increasingly perceived as oppressive. However, as I argue in this chapter, the existence of the cultural boundaries that derived from the process of nationalization was also regarded as problematic. As the nineteenth century drew to a close, its gradually internationalizing scientific community saw national isolation as a threat to cultural and scientific productivity. International participation increasingly assumed the function of a marker of degrees of culture and civilization. Representing a nation in international forums was thus regarded as particularly important.

In the following, I discuss two examples of how the nature and effect of the existence of political and linguistic borders were debated at the turn of the century in academic scholarship and analyse the practices that were employed to overcome isolation at a time of increasing nationalization. In the first case, I will discuss the practice of publishing in languages other than the local or institutional language, something which began to emerge as a result of the increasing emphasis on an international ideal of scholarship. The second example, the recruitment systems of university instructors, shows a similar transition from local expertise to experience abroad as being the most important condition for tenure. In both cases, success in consciously transcending local borders was accompanied by an increasing acceptance of the importance of the international ethos of science. These developments suggest how science and scholarship were understood in the Slavic communities around 1900 but they also hint at differences of perception about them as compared to discussions in mainstream German or French language communities.

Remaking Scientific Communication

An innovation towards the end the nineteenth century was that institutions in Central Europe devoted initially to national scholarship now undertook a new function for another public. While popularizing knowledge was seen as a primary institutional goal of these institutions, in the period under discussion they undertook a supplementary role by seeking to participate in the internationalizing scholarship of the time. This duality of local and global orientations was evident throughout the century, but about 1890 the "global" orientation in terms of scientific communication began to predominate.

Scholarly publications were commonly regarded as having an intercultural function in nineteenth century Central Europe, a region in which there was a hierarchy of local (that is "national") and imperial administrative or scholarly languages. In Bohemia and Moravia, for example, the use of the local Czech language for scholarship grew slowly in the first half of the century. German predominated until at least the 1860s. While specialized journals in Polish appeared early in the century, many Polish scholars continued later in the century to publish in German, French and Latin as well, depending on their cultural background. An intensive growth of professional journals and other publications in the late 1860s, slowly intensified the use of local languages as the main medium for the transmission of scientific knowledge. The first specialized professional journals in Ukrainian, for example, only appeared in the 1890s, primarily due to the lack of a previous scholarly infrastructure. Earlier, as in the case of Polish, articles on professional matters were often published in German, French or Latin. In the first half of the nineteenth century there were also a few Slavic general journals issued by the still nascent scientific societies and museums in one or other of the local language. Books, especially those directed to a broader public, were mostly issued in the local languages (see Ostrowska 1973; Hlaváčková 2003; Janko and Štrbáňová 1988). One cannot, however, simply describe the phenomenon of use of German or French in this period as a kind of internationalism, because it represented a pre-national phenomenon of multilingualism in which each language had its social function and changing languages was not seen as a transgression from the local to the international (cf. Rindler-Schjerve 2003).

The relationship between scientific "international" and "national" languages in Central Europe became increasingly problematic, however, towards the end of the century, since the idea of universal science was conflated with the ideology of scientific betterment of society that began to gain prominence in the Slavic regions especially among nationalist organizations. Moreover, the leading scholarly language, German, was also the imperial administrative language making linguistic discussion a political issue. This was especially so because the imperial administrative language was emphasized as the language of higher education, the primary aim of which was to prepare civil servants for state service (Stachel 1999). The discussion and use of languages was thus tied to political changes, most importantly following increasing centralization in the Russian and German empires in the 1860s.

In the Hapsburg Empire political change after 1860 proceeded along different lines and this was accompanied by an increasing recognition of local languages. In Galicia after a period of German-language dominance in higher education, a change of language at the universities and the creation of a number of "national" academies were permitted. In the 1860s, Polish was gradually allowed as language of instruction at universities in Cracow and L'viv with some recognition of Ruthenian. In the 1890s Ruthenian was given additional privileges, although these were limited by Polish dominance in the province. In Bohemia, Czechs acquired a technical academy in 1869 and university in 1882.

While these developments were occurring in the Austro-Hungarian Empire, the two other Central European empires were reducing privileges for the use of local languages in education and in academia. In Germany this took the form of limitations on the use of Polish in schools and the imposition of other measures known as German *Kulturkampf.* In the Russian Empire the Polish language university in Warsaw and the German language university in Dorpat/Tartu were turned into Russian language imperial institutions in 1870 and 1893 respectively. With Valuyev's circular (1863) and Ems Ukaz (1876), the use of Ukrainian was banned in print and teaching and for artistic performance.

From the "imperial" side, the linguistic move was frequently inscribed in a political argument that merged the idea of a cross-cultural supra-language with claims of cultural superiority. In a memorandum to establish a university in

Bukovina (now on the borderland between Ukraine and Romania), for example, the Austrian philosopher, Tobias Wildauer, argued that with the change of language of instruction at the Galician universities into Polish, they "have lost their universal significance and taken on the character of camp establishments ... the whole of the wide Eastern stretch of the Empire lacks a universally accessible site for the fostering of science" (Anonymous 1900, XXIII). The first rector of the Chernivtsi University in Bukovina and the driving force behind its establishment, Constantin Tomaszczuk, agreed with this point of view. He observed that "German science has the claim of universality. And only because German education has universal standing do the non-German sons of Bukovina strive for a German university" (Turczynski 2008, 215). The claim that language change would restrict civilization and inhibit cultural development in the East can be frequently found in German-language literature, including official governmental publications (Anonymous 1853; Dumreicher 1873).

The claim of a higher value for "Kultursprachen" (cultural languages) or "Weltsprachen" (world languages), as the same concept was named in different contexts, was not limited to the imperial languages but also occurred in what one might call micro-imperial language as in the case of Polish versus Ruthenian. In an apology for Polish as a scientific language in the period after 1860, the well-known jurist, Antoni Helcel, wrote that Ruthenian had no developed scientific literature and that Polish served as the cultural language for the educated strata of this mostly "rural population" (Helcel 1860, 38). "Ruthenians themselves," he said, "do not treat Polish as a foreign language, but rather as a literary language of their dialect, better able to express the higher subjects of their public affairs." He concluded that the "best means to fulfil [their] moral needs" would be to use Polish in higher education in Galicia (Helcel 1860, 39).

The tension between the imperial and national attitudes to science is well represented in a discussion that took place in 1902 between the Swiss anatomist, Rudolf Fick, and the Czech-Bohemian biologist, Emanuel Rádl. In an article on scientific "Babylonian confusion," Fick criticized scientific journals for commissioning reviews of works published in languages other than German, English, French and Italian (Fick 1902; Hermann 2003, 446–50). In his eyes, such endeavours encouraged writing in national languages that could only ultimately hurt the scientific community. Reviews could not represent an article's depth and so this practice did a disservice to general science [allgemeine Wissenschaft]. In a reaction to this article, Rádl criticized Fick's shortsighted attitude to science in not considering how national character influenced scientific production. He pointed out that scholarship had a social role in that disciplines concerned with a society should be accessible to it. He suggested that omitting the reviews whose inclusion Frick was criticising would result in the creation of non-disciplinary journals that published only reviews and abstracts. Rádl also underscored the fact that the most important scientific articles initially appearing in the local language tended to be republished in translation into an international language. While recognizing the importance of international contacts for scholars, Rádl argued that: "the cultivation of science should answer two aims: the advancement of the nation [des Volkes] and development of general science" (Rádl 1902, 29). Thus for the Bohemian biologist, exclusive commitment to the "international" languages was not acceptable. Fick, however, saw them as necessary for the development of the scientific community. Fick's argument, although apparently apolitical, is reminiscent of imperialist claims of the importance of a "world language" for the propagation of civilization. Rádl's argument on the other hand is reminiscent of arguments which were frequently used in deliberately nationalistic propaganda.

In the 1890s, an important transition in the media landscape took place when the academies of sciences in Cracow and Prague began regularly publishing journals entitled *Bulletins Internationaux* (in 1889 and 1895 respectively) in addition to their regular journals in the local languages.[1] This almost simultaneous occurrence not only signaled a change of understanding in the nature of scientific communication, but also represented the self-positioning of Slavic cultures on the scholarly map of Europe. The *Bulletins*, at first with extensive abstracts, then with original articles in German, French, Latin and English, comprised substantial annual volumes of 500 to 1000 pages in length. Their purpose was to make their "national" scholarship not only more widely available but also recognizable as such because the publications were institutionally embedded in national academies. The *Bulletins* also included materials that had previously been published in local languages. The *Bulletins* were sent to a wide range of organizations internationally with the view that this would potentially secure a broad readership. In 1894, for example, Cracow's *Bulletin* had a monthly circulation of 590 copies (Anonymous 1894, 182).

The introduction in Central Europe of systematic publications in foreign languages in the 1890s points to an important transformation in the public role science was considered to play. Following the 1848 revolutions, the idea of the publication of a "nationalized" scholarship to serve the development of the national culture was predominant and was paired with the philosophical idea of so-called organic work, an ideology that claimed that the mass education of the lower classes would turn them into a modern nation. While the practice-orientation of this scholarship served political purposes, at the same time it was confronted with the rapid growth of international media that began to reflect each nation's cultural achievements. Specialized articles by Slavic scholars up to the 1890s were quite frequently published in specialized journals in foreign languages because most Slavic scholars were bilingual.

The change to more monolingual gymnasia education after this period meant that scholars in the Hapsburg Empire were not necessarily as fluent in German as they had been in the past and thus had fewer opportunities to publish in a language other than

---

1   Three years after the foundation of the Cracow Academy, in 1876, German language abstract-collections were published, entitled *Bibliographische Berichte über die Publikationen der Akademie der Wissenschaften in Krakau* (later *Literarische Mitteilungen und Bibliographische Berichte über die Publikationen der Akademie der Wissenschaften in Krakau*); it ceased to exist in 1880, however (Stachowska 1973).

the language of their primary education. Even given the large number of international scholarships for young scholars and the foreign experience that the universities began to require of academic staff, bilingualism seemed to be on the decline. In the previous gymnasia education the boundary between the "mother tongue" and the "foreign tongue" was hard to establish—although the issue was always a welcome argument for nationalists. Now educational reform stabilized the distinction and gave practical substance to what had been previously politicized claims for one or the other language. Previously the linguistic communities were divided into socially defined levels of everyday language and specialized languages (Czech and Polish versus German, but also Polish and Russian versus Ruthenian). In the 1890s, these distinctions were conflated into two concepts that have continued in use until today, that is "native language" and "foreign language." In the Hapsburg Empire the trend toward publishing in mother tongues had been so strong in the 1860–1890 period that scholars not paying sufficient attention to the local environment and its media were frequently denounced. Scholars from the German and Russian empire, however, where educational privileges for the "local" languages had been rescinded and their use discouraged, presented a completely opposite case.

It was in Paris that a temporary solution was found to what was seen as the decreasing international reception of the work of Slavic biological scientists. In 1887, the journal, *Archives Slaves de Biologie*, edited by Maurice Mendelssohn, Charles Richet and Henry de Varigny, was established. It had the explicit aim of bridging linguistic boundaries to allow Slavic scholars to participate more directly in the development of universal science: "In a word, a whole fraction of the scientific world is, so to say, standing on the sideline; divided from the rest of the world by an almost insuperable barrier: ignorance of language." This constituted "a double misfortune." "On one hand, Slavic scholars see their works, which are often excellent and have required long efforts, ignored and unknown; on the other hand occidental scholars cannot really follow the progress of science, because science is universal; it has neither a motherland nor language; it is done by Slavs as by Westerners (les Occidentaux) (Mendelssohn and Richet 1886, IV, III).

The aim of the journal was to provide a special medium for communication. Articles were accepted in the various Slavic languages and either translated into French or provided with one to two page summaries in French. For its editors, this was the main difference of the *Archives Slaves de Biologie* from other European journals that would publish Slavic articles only if they were "written correctly and clearly in a language, which is not their [that is, the Slavic authors'] mother tongue" (Mendelssohn and Richet 1886, IV). Such a requirement very much reduced the publication possibilities for Slavic scholars and for their work to reach an international public. Without giving reasons for its termination, the *Archives* ceased to exist after four large volumes. Prominent scholars writing originally in Russian, Czech or Polish had contributed original articles to the journal, so it is probable that the journal ceased not because there was a problem in the supply of articles but in the demand for them—or the work load for the editors was too high.

While the argument of transcending boundaries was common in the discussions about what language to use for publication, the argument of "ignorance of language" as discussed by Mendelssohn and Richet can be read in two ways. It could be that it was indeed the Slavic scholars who were considered linguistically ignorant. However the French editors of *Archives Slaves de Biologie* readily acknowledged that many of these scholars actually read French and/or German but practical reasons hindered them publishing in languages other than their mother tongue. However the "ignorance of language" argument of Richet and Mendelssohn can also be read as pointing up a deficiency of the "occidental" scholars: they could not read publications in language other than the common western languages of scholarship. While recognized as a common phenomenon, this deficiency of knowledge of languages was described as unbridgeable (Mendelssohn and Richet IV, V). The *Archives Slaves de Biologie* was thus positioned as a mediator between communities which nevertheless would not or could not change their habits. But unlike the publications of the national academies of sciences, the *Archives Slaves de Biologie* was not concerned with issues of nationalist representation but only with those of international communication.

The "honorable initiative" (Raciboski 1888, 669) of the French scholars was one of the ignition points of a discussion carried out in the Warsaw journal, *Wszechświat* (The Universe). The biologist, Marian Raciboski, wrote in an 1888 article that Polish biologists, and thus Polish biology, had almost completely vanished from the perception of "Western" scholars, and was certainly less visible than Hungarian or Russian biology. He blamed nationalization for this and especially the tendency not to publish in foreign languages, "a report or a summary of [Polish biological] works in one of the three commonly accepted languages [that is, English, German or French] that would make them immediately known to the general scientific world" (Raciboski 1888). Appealing to the natural scientist to represent Polish scholarship abroad, he mentioned in the same breath the *Archives Slaves* and a commission of the Cracow Medical Society [Towarzystwo Lekarskie Krakowskie]. The latter was established to prepare resumes of articles published in Polish for German journals. In the subsequent discussion, the existence of similar bodies for other sciences such as for mathematics, to represent Polish science in foreign languages was also raised. The general tone was that such endeavors should be institutionalized and that it should be compulsory to publish in both local and international language (for example Gosiewski 1888; Wrześniowski 1888).

Thomas Masaryk voiced a similar idea in Prague in a series of articles entitled, "How to Augment our Scientific Literature?" [Jak zvelebovati naši literaturu naukovou?]. The Moravian philosopher mentioned international journals as one of the vital points for the defense of Czech rights (Masaryk 1885, 274–5). He proposed that an interdisciplinary German language "*Böhmisches Revue*" should be published by the Czech academy of sciences. "Our scientific work, like everything we do, is in large part, even if indirectly, a political work" (Masaryk 1885, 275). The Vienna and Brno educated scholar was, however, primarily concerned with questions of establishing a literary canon in Czech in order to fill the gap that he considered a significant challenge for national scholarship. His main points were thus concerned with yet another version of internationality—translations of classic and modern scholarly literature into Czech and more intensive teaching of foreign languages and literatures at schools and at the university. The demand for the popularization of Czech scholarship abroad was more vigorously proposed a few years later by August Seydler. In his eyes it was the duty of the academy of sciences to make Czech scholarship available in other languages—"in our region, German is best" (Seydler 1890).

Masaryk and Seydler were certainly not the only ones advocating the internalization of Czech scholarship but theirs were the first articles to appear immediately after the division of the Charles University into Czech and German halves in 1882 and before the creation of the Czech Academy of Sciences in 1890. They reveal how much weight was put on the intersection between commitment to national progress but also to its being embedded in international trends. Both scholars also pleaded for German to be the vehicle for international publications, a surprising choice at the height of national conflicts in Bohemia when German was certainly not everybody's ideal.

Mobility and Monolingual Institutions

Changes of language at scholarly institutions, however, not only created national boundaries, they also opened distinct spaces that transcended imperial boundaries. Nineteenth century Slavic universities, which created trans-imperial enclaves within the imperial networks, could be described, following Foucault, as "national heterotopias." The process of language change at the universities created an environment that not only was primarily concerned with cultural spaces, both within the monarchy (the Czech case) and extending beyond it (as in the case of Polish and Ukrainian cultures), but one in which their academic staff was paradoxically increasingly international. For Foucault, heterotopias are (utopian) spaces confronting an hegemonic paradigm. This term seems appropriate here instead of something like "nationalistic internationalization" to describe a situation in which the process of nationalization results in a paradoxical opening towards scholarly heterogeneity. A short account of the development of academic structures in the Habsburg Monarchy at the end of nineteenth century will clarify this point.

The university as one of the primary places of scholarship took a sharp "national turn" when Polish and Czech became languages of instruction. Some small exceptions were permitted—professors of German language and literature at both universities in Galicia continued to lecture in German, although not in Prague where complete linguistic separation was legally binding from 1882 onwards. In L'viv, lectures in Ruthenian were also allowed but were strongly limited because of the Polish supremacy. Language replacement strengthened the role of national linguistic communities. Beginning in the 1870s, however, Galician universities were explicitly advised to search for candidates abroad as well as locally. An early and influential appeal to young scholars willing to habilitate in Galicia for the sake of Polish science was issued in 1861 by rector Józef Dietl. His widely circulated essay, apart from referring to the past glory and bright future of Polish science, also mentioned the necessity to travel abroad to other universities as well, "because one cannot finish learning at one university, but only begin" (Dietl 1861, 1).

This not only institutionalized a practice which had been widespread in the Germanophone Habsburg universities since 1848, but also limited the personal exchange between the universities in the empire that had different languages of instruction. Although a number of scholars educated in Vienna still taught in Cracow, L'viv or Prague, they had to speak the appropriate national language, a requirement which seriously limited the pool of possible recruitments. At the time, the definitions of national "own" and "other" were also clarified, thus influencing the conditions of entry into the academic world. All three Slavic Habsburg universities (Cracow, L'viv and Prague) attracted a large number of young scholars to habilitate and thus to become the first candidates for professorships. In fact the Philosophical Faculty of the Czech University in Prague was the second largest in the Austrian part of the Monarchy from 1900 onwards and the Medical Faculty the third largest in 1910 (having been the fourth largest in 1900). Universities in Cracow and L'viv were similarly high on the list for aspiring younger scholars.

The number of instructors in the Galician universities who had acquired their doctoral degree in an institution abroad hovered around 45 percent in 1910 as compared with the 10 percent in the Germanophone universities in the Monarchy (Surman 2012, 371). Galician scholars continued also to complete their habilitation process at Universities in Graz, Vienna, Innsbruck, and Chernivtsi. Even so, in 1910 a quarter of instructors in Galicia were still being appointed from the Russian and German empires. The result was the creation in the Galician universities of a widespread "mixture" of different styles of research. For example the Warsaw-L'viv school of analytical philosophy was started by Kazimierz Twardowski (habilitation in Vienna), Waclaw Sierpinski (degree in Warsaw) and Zygmunt Janiszewski (degree in Paris). One of the biggest successes of chemistry in Galicia, the liquefaction of oxygen, was accomplished by a modification of Cailletet's cryogenic apparatus. The professor of physics in Cracow, Zygmunt Wróblewski had bought this in France, where he had studied for his PhD. The pump had been created by the Viennese scholar Natterer and was used by the chemist Karol Olszewski in Cracow (PhD in Heidelberg). The two scholars collaborated in Cracow, making Jagiellonian University's laboratory a world center for the study of the liquefaction of gases. The campaigns for Darwinism, which was introduced belatedly in conservative Galicia in comparison to the Russian Empire, was also undertaken by the non-Galician scholar, Benedykt Dybowski, who was appointed professor of zoology in L'viv in 1883. He had been educated in Dorpat, Breslau and Berlin, taught for a short period in Warsaw and was then banished for political conspiracy to Siberia. His Darwinist comrade-in-arms, Józef Nusbaum-Hilarowicz, father of the so called L'viv school of zoology, was educated in Warsaw and Odessa. In fact, there is hardly a discipline in Galicia which was not heavily influenced by international or trans-imperial mobility.

Ironically, at the time when what the language of education was finally secured, the ideology of language purity changed as well. In the 1870s, appointing a scholar without fluent knowledge of Polish was possible only against the will of the faculties. Throughout the 1870s, the chairs of geography in Cracow and L'viv were unoccupied because no competent Polish-speaking scholars could be found. However, at this same time, searching for a pharmacologist, the Philosophical Faculty in Cracow invited Zygmunt Radziejewski and Wilhelm Zülzer from Berlin to apply although both of them spoke only basic Polish. The successful appointee, Edward Sas-Korczyński from St. Lazar hospital in Cracow, had to spend two years abroad before taking up the position to ensure that he was up to date with modern developments (Surman 2012, 295; Wahholz 1931). Among the proposed appointees in Cracow in the 1890s, one finds among others chemist Julius Braun, dermatologist Ernst Finger and agriculturalist Leopold Adametz. The first two spoke basic Polish but both declined the call. Adametz, who spoke no Polish at all, was appointed professor of animal husbandry in Cracow with an obligation to learn Polish within two years. Two German-Empire educated Bohemian physicians were appointed to the medical faculties, although they spoke no Polish at the time of their habilitation. Previously this would have led to the failure of their applications because language fluency had to be proved (Surman 2012, 295).

The trend toward an opening up to non-local scholarship was also pronounced at the Czech university. Although linguistically an island in the Habsburg Monarchy, the creation of a "national" Czech university in Prague in 1882, made the issue of international scholarship important. The Czech university had to be almost created from scratch because for various reasons most scholars at the existing university in Prague did not speak Czech and when it was divided they went to what became the German Charles University. Ironically, the linguistic boundary which separated the Czech University in Prague from other universities was not entirely the result of pressures within the university itself. During the discussions about the university before 1882, Czech scholars and politicians demanded a bilingual university with parallel chairs in both Czech and German. Only in this way, the argument went, could Czech-speaking cadres be educated because the political circumstances had until then prohibited their development. The argument was not accepted and when the complete division of the university into two separate parts occurred, its effects proved to be a kind of confrontation between the sturdy nationalism of local and politically active "Czechs" and the quest for higher internationally recognised scientific standards.

Although from the beginning the Czech University appointed scholars who had spent most of their careers abroad, it did so only when a lack of qualified local scholars was evident or because of a special circumstance. One such circumstance was the influence of internationally oriented Bohemia-born Vienna star-surgeon, Eduard Albert. He was an informal consultant to the ministry of education and achieved several nominations from foreign scholars with international experience and was appointed against the will of the Prague faculties. In many cases nominations of "foreign" scholars to positions in the Czech university led to conflict because the more conservative older professors pleaded for a Czech-only policy. They claimed that only appointments of young Prague Privatdozenten as professors could lead to the development of a viable Czech scholarship. On the other hand, the younger scholars (frequently exactly those who were appointed from abroad) emphasised the importance of internationalist scholarly quality rather than Czechness (see Surman 2012, 267–89).

Eventually Prague had an overproduction of Privatdozenten who could not be nominated to other universities for linguistic reasons so it became potentially self-sustaining academically. This led to the issue of internationalizing the scholarship within it had to be approached in a new way. Jana Mandlerová has shown that from its creation 1882 the Czech University in Prague granted a greater number of international scholarships than its German counterpart (Mandlerová 1969). The scholarships included both private bursaries and Habsburg governmental and provincial scholarships. This was aimed at providing young Czech scholars with an opportunity to travel abroad for two years with the requirement of subsequent habilitation at the Czech university. Even though the political situation of Central Europe made German language a symbol of political dominance, German universities were for a long time accepted as the best places for gaining a thorough education (Dybiec 2005). Around the turn of the century, however, the domination of German universities as the destination for the holders of these scholarships, though still important, declined as more and more Czech scholars decided to travel to France (Hnilica 2005).

The tradition of taking extended travels abroad was mentioned throughout the century on many occasions by Polish and Czech scholars when discussing the customs of academia. In some cases Slavic scholars emphasised the importance of such travel in an argument to counter German accusations of the destructive potential of nationalised scholarship on education at the universities. In other cases, they used the argument to encourage Slavic scholars to engage more extensively in the international scientific community (Goll 1902, 13; Liske 1876).

Conclusions

The trends sketched above outline the changing scholarly landscape around the turn of the century. They reveal the interesting paradox that the desired nationalization failed because in a sense it was successful. International science as a scholarly topos, strongly linked with images of universality, generality, progress and civilization, influenced Slavic scholars in Central Europe as it did scholars in France or Germany. The aim of keeping in touch with an all-transcending, transnational movement, which was how science was presented, went hand in hand with a feeling that a straightforward nationalism led into a blind alley not only for individuals but also the community they represented. Scholars were frequently required to act as transmitters of knowledge to the people, but by the turn of the twentieth their position as representatives of the people before an international audience had become increasingly important. It was more important to ensure their scholarly competence and the academic standing of their institutions than their national allegiance and participation in national movements. Magdalena Micińska has made a similar point for literature, although for a later period. While for many years authors active abroad were seen as traitors for leaving their motherland, around 1918 they became national heroes exactly because they had made international careers, thus popularizing the "nation" from which they had originated (Micińska 2008, 191–2).

Language as an important feature of the "new" science without borders, suggests that the points of view of scientists, although directed at the same aim, still diverged. The existence of distinct languages in which science was performed was seen by Slavic scholars not as an obstacle to the efficient functioning of the *république des savants* but as productive because it was supposed to ensure diversity. While this desirable variety was often commented on positively by Slavs, it seems to have posed a problem for scholars of the dominant languages. They responded first by assertions of cultural superiority and then by searching for auxiliary languages to manage the communication gaps created by the different linguistic communities. Because "many Italians write only Italian, many Dutchmen only Dutch, whilst numerous Russians, Poles, Czechs, Hungarians, Scandinavians, and Spaniards employ only their national languages …, much escapes general knowledge and recognition, or is only accessible in a belated or mutilated form," wrote the Austrian physicist Leopold Pfaundler in the opening article of *International language and Science* (Pfaundler 1910, 2). Traces of such a critique can rarely be found in the pages of Slavic publications (see Zakrzewski 1906, 49, 59; Baudouin 1900, 13). Although Esperanto was for a time particularly important in discussions about the need for an auxiliary international language that could bridge communication gaps, it and several other artificial languages were never successful in gaining general recognition. The modern publishing situation continues to involve a mixture of international and national languages and so continues to present today many of the same kinds of problems as those examined in this chapter.

References

Anonymous. 1853. *Die Universitätsfrage in Oesterreich. Beleuchtet vom Standpunkte der Lehr- und Lernfreiheit* (Besonders abgedruckt aus dem Wiener Lloyd).

Anonymous. 1900. *Die Franz-Josephs Universität in Czernowitz im ersten Vierteljahrhundert ihres Bestandes*. Festschrift, herausgegeben vom Akademischen Senate. Czernowitz: Bukowinaer Vereinsdruckerei.

Anonymous. 1894. "Wykaz Stosunków Akademii z innemi instytucjami naukowemi." *Rocznik Akademji Umiejętności w Krakowie 1893/4*: 181–91.

Ash, Mitchell G. and Surman, Jan. (eds) 2012. *The Nationalization of Scientific Knowledge in the Habsburg Empire, 1848–1918*. Basingstoke: Palgrave Macmillan.

Baudouin de Courtenay, Jan. 1900. *Głos członka Akademii J. Baudouina de Courtenay w sprawie słownictwa chemii*. Kraków: Akademia Umiejetnosci.

Buklijas, Tatjana and Lafferton, Emese. 2007. "Science, Medicine and Nationalism in the Habsburg Empire from the 1840s to 1918." *Studies in History and Philosophy of Biological and Biomedical Sciences*, 38: 679–86.

Cabaj, Jarosław. 2007. *Walczyć nauką za sprawy Ojczyzny. Zjazdy ponadzaborowe polskich środowisk naukowych i zawodowych jako czynnik integracji narodowej (1864–1917)*. Siedlce: Akademia Podlaska.

Dietl, Józef. 1861. "O instytucji docentów w ogóle, a szczególnie na Uniwersytecie Jagiellońskim." *Czas*, 31 October.

Dumreicher, Armand. 1873. *Die Verwaltung der Universitäten seit dem letzten politischen Systemwechsel in Oesterreich*. Wien: Alfred Hölden.

Dybiec, Julian. 2005. "Prześladowca i nauczyciel. Niemcy w nauce i kulturze polskiej 1795–1918." In *Literatura, kulturoznawstwo, uniwersytet. Księga ofiarowana Franciszkowi Ziejce w 65. rocznicę urodzin*, edited by B. Dopart, J. Popiel and M. Stala. Kraków: Universitas, pp. 455–68.

Fick, Rudolf. 1902. "Vorschläge zur Minderung der wissenschaftlichen, 'Sprachverwirrung,'" *Anatomischer Anzeiger*, 20: 462–3.

Goll, Jaroslav. 1902. *Der Hass der Völker und die österreichischen Universitäten*. Prag: Bursík & Kohout.

Gosiewski, Władysław. 1888. [Untitled] *Wszechświat*, 7: 749.

Havránek, Jan. 1990. "Nineteenth Century Universities in Central Europe: Their Dominant Position in the Science and Humanities." In *Bildungswesen und Sozialstruktur in Mitteleuropa im 19. und 20. Jahrhundert—Education and social structure in Central Europe in the 19th and 20th Centuries*, edited by Victor Karady and Wolfgang Mitter. Köln, Wien: Böhlau, pp. 9–26.

Helcel, Antoni. 1860. *Uwagi nad kwestyą językową w szkołach i uniwersytetach Galicyi i Krakowa, osnowane na liście odręcznym Jego C.K. Apostolskiej Mości z dnia 20 października 1860 r*. Kraków: D.E. Friedlein.

Hermann, Tomáš. 2003. "Originalita vědy a problém plagiátu. Tři výstupy Emanuela Rádla k jazykové otázce ve vědě z let 1902–1911." In *Místo národnich jazyku ve výuce, vědě a vzdělání v Habsburské monarchii 1867–1918—Position of National Languages in Education, Educational System and Science of the Habsburg Monarchy 1867–1918*, edited by Harald Binder, Barbora Křivohlavá and Luboš Velek. Praha: Výzkumné centrum pro dějiny vědy, pp. 441–68.

Hlaváčková, Ludmila. 2003. "Čeština v medicíně a na pražské lékařské fakultě (1784–1918)." In *Místo národnich jazyku ve výuce, vědě a vzdělání v Habsburské monarchii 1867–1918—Position of National Languages in Education, Educational System and Science of the Habsburg Monarchy 1867–1918*, edited by Harald Binder, Barbora Křivohlavá and Luboš Velek. Praha: Výzkumné centrum pro dějiny vědy, pp. 327–44.

Hnilica, Jiři. 2005. "Kulturní a intelektuální výměna mezi Čechami a Francií 1870–1925." *Acta Universitatis Carolina-Historia Universitatis Carolinae Pragensis*, 45: 95–126.

Janko, Jan, and Soňa Štrbáňová. 1988. *Věda Purkyňovy doby*. Praha: Academia.

Liske, Xawer. 1876. *Der angebliche Niedergang der Universität Lemberg. Offenes Sendeschreiben an das Reichsrathsmitglied Herrn. Eduard Dr. Suess, Prof. an der Universität Wien*. Lemberg: Gubrynowicz & Schmidt.

Mandlerová, Jana. 1969. "K zahraničním cestám učitelů vysokých škol v českých zemích (1888–1918)." *Dějiny věd a techniky*, 4: 232–46.

Masaryk, Thomas Garrique. 1885. "Jak zvelebovati naši litaraturu náukovou? Článek třetí." *Athenaeum. Listy pro literaturu a kritiku vědeckou*, 2: 270–75.

Mendelssohn, Maurice and Charles Richet. 1886. "Avant-Propos." *Achives Slaves de Biologie*, 1: III–IX.

Micińska, Magdalena. 2008. *Inteligencja na rozdrożach 1864–1918*. Warszawa: Instytut Historii PAN, Wydawnictwo Neriton.

Ostrowska, Teresa. 1973. *Polskie czasopiśmiennictwo lekarskie w XIX wieku (1800–1900). Zarys historyczno-bibliograficzny*. Wrocław etc.: Zakład narodowy im. Ossolińskich, PAN.

Pfaundler, Leopold. 1910. "The Need for a Common Scientific Language." In *International Language and Science*: *Considerations on the Introduction of an International Language into Science*, edited by Louis Couturat et al. London: Constable & Company, pp. 1–11.

Raciboski, Maryjan. 1888. "Do naszych przyrodników." *Wszechświat*, 7: 668–70.

Rádl, Emanuel. 1902. "Bemerkungen zu den Vorschlägen von R. Fick, die wissenschaftliche Sprachverwirrung betreffend." *Anatomischer Anzeiger*, 21: 27–9.

Seydler, August. 1890. "Akademie česká a Společnost nauk." *Athenaeum. Listy pro literaturu a kritiku vědeckou*, 8: 65–9.

Stachel, Peter. 1999. "Das österreichische Bildungssystem zwischen 1749 und 1918." In *Geschichte der österreichischen Humanwissenschaften. Bd.1: Historischer Kontext, wissenschaftssoziologische Befunde und methodologische Voraussetzungen*, edited by Karl Acham. Wien: Passagen, pp. 115–46.

Stachowska, Krystyna. 1973. "Z Działalności Wydawniczej Polskiej Akademii Umiejętności. Starania o upowszechnienie za granica polskiej myśli naukowej w latach 1873–1952." *Rocznik Biblioteki Polskiej Akademii Nauk w Krakowie*, 19: 39–71.

Štrbáňová, Soňa. 2012. "Patriotism, Nationalism and Internationalism in Czech Science: Chemists in the Czech National Enlightenment." In *The Nationalization of Scientific Knowledge in the Habsburg Empire, 1848–1918*, edited by Ash, Mitchell G. and Jan Surman. Basingstoke: Palgrave Macmillan, pp. 138–56.

Surman, Jan. 2012. "Habsburg Universities 1848–1918. Biography of a Space." PhD dissertation, University of Vienna.

Turczynski, Emanuel. 2008. "Czernowitz, eine vom Bildungsbürgertum errungene Universität im Dienst staatlicher Bildungs- und Wissenschaftsförderung." In

Universitäten im östlichen Mitteleuropa: zwischen Kirche, Staat und Nation: sozialgeschichtliche und politische Entwicklungen, edited by Peter Wörster. München: Oldenbourg, pp. 209–25.

Wachholz, Leon. 1931. "Trzej interniści krakowscy u schyłku XIX wieku." *Polska Gazeta Lekarska,* 10: 1–11.

Wrześniowski, August. 1888. [Untitled]. *Wszechświat,* 7: 748–9.

Zakrzewski, Adam. 1906. *Historja i stan obecny języka międzynarodowego Esperanto.* Warszawa: M. Arct.

Chapter 5

Scholarly Networks and International Congresses: The Orientalists before the First World War

Paul Servais

Introduction

In the context of the rapid evolution of scholarly disciplines, orientalism in the 19th century proved itself to be both a major area of knowledge construction but also of European representations of the world, particularly of the Orient (Saïd 1980; Hagerdal 1997). With a multitude of contacts, from China to the Turkish empire by way of the Indies, Persia and the Arab world, Europe sustained a veritable "oriental Renaissance" to use Raymond Schwab's phrase (Schwab 1950). This was based on both concrete observations and on the discovery, publication, translation and analysis of a treasure trove of texts which had been increasing in number and diversifying since the 17th and 18th centuries. It led to the emergence of programs of teaching (especially of languages), of specialised institutions, of associations of orientalists and of scholarly journals. This is when the field of orientalism was established reflecting scientific elements, encouraging artistic expression and as Edward Said has so well shown, expressing imperial interests.

In terms of this paper, I will limit myself to the world of the European orientalists and concentrate on a special aspect of their scholarly milieu: international congresses which from 1873 to 1912 gave rhythm to their activity. Having recalled some of the important characteristics of European orientalism before the First World War, I will attempt to provide a chronological and geographic panorama of all of these congresses before concentrating on the first of them, that of Paris in 1873, which saw the beginning of a new approach. I will then briefly analyse participation in the following meetings and the thematic orientations that characterised them. Finally I will try to synthesise in a few words the nature and importance of these great developments and their components.

Scientific European Orientalism: Sources and Achievements between the 17th and 19th Centuries

The East has fascinated the West for a long time, whether it was far distant like the Indies or Cathay or even Cipangu, or near and perceived as a rival and present danger such as Turkey or the Arab world. Marco Polo's book of wonders *Le Divisement du Monde*, first published in 1298, played a fundamental role in stimulating this interest as did the link with Holy Land. European expansion of the 16th century only reinforced this fascination. Seafarers, merchants and missionaries, notably but not exclusively Jesuit, increasingly revealed these strange other worlds to the European elite.

In the eighteenth century, the fashion for chinoiserie overtook Europe, at the time Father du Halde (du Halde 1733) was synthesising what was know of China in this period. The translation of Arabian works, especially the 'Thousand and One Nights' (Galland 1704–1717) spread other images, while the numerous East India Companies—in England, the Netherlands, France but also in Sweden and Austria (Chaudury and Morineau 1999)—and the expansion of commerce with the Levantine trading posts required the education of specialists. Linguistic experts but also legal and commercial experts were trained in newly created specialised institutions such as the Dragoman School founded at the instigation of Colbert (Cent cinquantenaire … 1948). This need for specialist personnel was epitomised in the progressive development of British India, founded in the last third of the 18th century and became one of the great empires of the period with all that this involved for effective imperial administration and ease of access to interested English scholars (Lloyd 1995). After the parenthesis of the French Revolution and the Wars of the Empire, the return of peace in Europe opened up other horizons and permitted scholars to relaunch the movement for the discovery and analysis of the societies and traditions of the orient.

If one were able in the 18th century to speak of "Chinese Europe" (Etiemble 1988–89), in the 19th century, it is rather a "Sanskrit Europe" (Lardinois 2007; Filliozat 1987) that begins to make a strong appearance. But other areas of civilisation also appear in this new European scholarly field, especially the Arab world. And a gradually deepening of the time scale begins to occur with Egyptological, biblical, Semitic but also Mesopotamian and Persian studies.

The challenges seemed at last to become so important that several learned societies were born (Servais 2004): the Société Asiatique de Paris in 1822 (Finot 1922), followed rapidly by the Royal Asiatic Society of London (A brief history … 1923) and the Deutsche Morgenländische Gesellschaft of Leipzig (Preissler 1995). During the course of the century and across the whole continent other associations continued to see the light of day largely taking over developments from official institutions, notably the academies. One of the major activities of these organisations was the publication of journals, beginning with the *Journal Asiatique*, very quickly followed by the *Journal of the Royal Asiatic Society* and of the *Zeitschrift der D.M.G.* Each of these associations and each of these organs naturally had its own orientation and emphasis, both thematic and geographic. But they highlight an intensification of effort at a fundamental understanding that aimed to be utilitarian and applied. The university world, caught up in implementing the Humboldtian model of the research university, was active in this area and professorships and programs of education multiplied in relation to one or other parts of the East.

Germany and England played a leading role in this development. In France the most important work was undertaken at the Collège de France, the École Nationale des Langues Orientales Vivantes, then much later at the École Pratique des Hautes Études. But these three great national schools had correspondents throughout Europe, whether in the south or north or central and eastern Europe. In Italy, the Collegio dei Cinesi of Naples, founded in 1711, became the Real Collegio Asiatico in 1868, and then the Istituto Orientale Universitario and chairs in Sanskrit in Italy multiplied throughout the nineteenth century. The situation was both, similar and different in Spain and Portugal, where, despite the very early contacts with India and China, the Arab world attracted most of the attention. Smaller countries in northern Europe, mainly in the wake of German orientalism, however, in their turn brought their own twist to the movement. The trips to India of Rasmus Rask (1787– 1832) and N.L. Westergaard provided direct contact with the Indian subcontinent that was rare apart from English orientalists. In the Netherlands, also influenced by German scholarship and organisation, formal academic studies of the East were undertaken mostly in Leiden after 1850, despite the very early Dutch presence in Asia (Vogel 1954). The great international conferences that begin in 1873 are in some sense the outcome of this major movement of institutionalisation.

International Orientalist Congresses: A Global Approach

The first of the important congresses of orientalists was held in Paris in 1873 under the Presidency of Léon de Rosny. It was characterised by a kind of improvisation created by the novel character of the experience, but also by a kind of national ceremoniousness, which is perhaps not surprising in the wake of the Franco-Prussia war. The Congress brought together more than a thousand international participants only a small majority of whom were French. It created from the outset a formalisation of functions that involved the preparation of statutes to guide subsequent meetings and the publication of proceedings, though these took three years to appear.

The second session of the International Congress of Orientalists was held in London in September 1874. As Samuel Birch noted in his opening address, this second session "is an event of more than ordinary importance in the annals of Oriental studies." Its objective was defined as "bringing together these students of congenial pursuits to interchange thoughts, to discuss points of common interest, and to make each other's acquaintance" which seemed to him a "fortunate idea." As to the choice of London, among other important and insistent proposals for a location, this seemed to him perfectly justified by the fact that it is "distinguished for its extent as well as for its devotion to the study of the East, and connected with that East by a thousand of ties, the interest of commerce, the spread of civilisation, missionary labours, and the duties of governing Oriental Dependencies of various tongues and sites in that East" (Transactions 1874).

But it seemed that holding annual meetings could not be maintained and it was not until 1876, as always in September, that the third session was opened in St. Petersburg with an increased emphasis of studies on Russian Asia. Four sessions were devoted to this subject. It was also in this year that the proceedings of the two earlier congresses were eventually published, publication having been delayed by typographic problems.

The next Congress was held in Florence in 1878 but it was necessary to wait another three years for the fifth meeting in 1881, this time in Berlin. Two years later the orientalists met in Leiden and in 1886 in Vienna. But clearly organising these meetings in different cities around Europe was not a simple undertaking and delays increased. After Vienna in 1886 came sessions in Stockholm/Christiania in 1889 and in Geneva in 1894.

The publication of the proceedings appears always to have been something of a battle field. The first part of the Stockholm/Christiania proceedings was not published and the second part was delayed until 1893, four years after the Congress. The Comte de Landberg, Secretary General of the Congress, felt it necessary to attribute the delay to the authors themselves to the extent that "all those who have delivered papers have not afterwards sent them to the secretariat and then have often taken infinitely long to revise the proofs. Some authors have corrected so much or have written the manuscript so carelessly that is has been necessary almost to rewrite the paper ..." (Actes 1893)

A kind of regularity emerges, however, after the Paris Congress of 1897, Rome (1899), Hamburg (1902), Algiers (1905) and Copenhagen (1908). But the Congress held in Athens in 1912, the 16th, not only marks an increased interval between meetings but the end of the first period of the congresses. As a result of the First Word War the meetings were not to be resumed until 1928.

Thus while it was difficult to maintain a steady, regular speed of development for this enterprise of orientalist study, it existed for four decades and was able to mobilise all the important actors in the field. The only countries in which the congresses did not take place were the Balkan countries and Spain and, surprisingly, Belgium. This apparent lack of interest doubtless reflected the situation of oriental studies in these countries at least as this was expressed in terms of the availability of institutional and official resources. Nevertheless, Belgium's is a curious case in that Brussels had become the site of numerous international meetings, especially on the occasion of the great International Exhibitions that were organised there in last half of the nineteenth century and in the period up to the First World War. In this period Brussels had also became the location of the headquarters of an enormous range of international associations and societies.

The International Congress of Orientalists in Paris in 1873: A Model?

The First Congress of Orientalists in Paris in 1873 provided something of a model for those that were to follow. It was long—eleven days. It aimed at attracting the largest and widest international participation as possible (Congrès international 1876–79). A measure of its success was that at least 1064 individual or institutions participated. As mentioned above the French were in the majority, but only just with 578 participants, followed at a great distance by the English with 78 participants, the Americans with 39 participants and the Germans with 35 participants of whom 14 were Prussians, Japan and Portugal each contributed 32 participants. Several national groups also reached significant numbers in terms of their size or distance. In alphabetical order, 21 Belgians, 20 Chinese, 20 Egyptians, 12 Greeks, 23 Italians, 26 Poles, 18 Rumanians, 25 Russians and 11 from Switzerland were present. Other representatives from the region came from Arabia, Burma, Cochinchina, the Dutch Indies, Madagascar, Morocco, Persia, Siam and Turkey. While certain members were natives most were Europeans resident in the countries themselves. Together the participation from the region comprised 92 persons or less than 9 percent of the whole.

As far as the West was concerned, apart from the groups already mentioned, there were representatives from Austria, Brazil, Canada, the Canaries, Colombia, Denmark, Spain, Finland, Holland, Hungary, Luxembourg, Mexico, Norway, Portugal, Argentina, Salvador and Sweden. Thus it is not only all of Europe that was involved in the meeting—given the involvement of the Russian and Austro-Hungarian empires—but also a substantial representation from the New World. Institutional participation should also be emphasised: 44 learned societies, 17 public libraries and museums and 8 schools or a total of 69 organisations, 42 of which were foreign.

But it was clear that the Congress was not only a scientific and scholarly initiative but also had an official and public character. This was apparent in the compositon of the Congress's Committee of Patronage. Listed among its members were French and European political figures, among whom were a substantial number of the members of reigning families. A detailed analysis of the list of participants suggests the importance of participation not only by scholars, but also by aristocrats, ecclesiastics, administrators, military and business figures and, finally, missionaries.

The institutional presence reinforced the official character assumed by the formalities of the Congress which were highlighted in the opening rituals of the meeting. In the morning at the first "administrative" session, delegates of the national committees took their places on the platform. The list of academies and learned societies "who have sent delegates to the Congress and provided them with official authority to represent them" was then read out. Then came the solemn opening session which was held "in the great chamber of the Sorbonne." This was "decorated with bundles of flags and State furnishings put at the disposition of the Organising Committee by the Ministry of Public Works. The Band of the

Republican Guard ... played a fanfare created for the ceremony." The Session was presided over by Admiral Roze in the presence of the Ambassador of Japan and the Minister Plenipotentiary from the United States as well as members of the Institut and the French parliament. Addresses and official presentations followed each other.

The proceedings of the Congress properly speaking got underway at the beginning of the afternoon under the presidency of the Ambassador of Japan. There followed 17 work sessions related to the study of countries and regions: Japan (five sessions), China, Tartary, Indochina, Oceania, Egypt and specialist areas such as Assyriological, Semitic, Iranian, Dravidian, Sanscrit, Buddhist, Armenian and Neo-Hellenic studies. General studies of Orientalism and on different idioms completed a particularly full program.

Communications at the Congress on aspects of all of these geographical and cultural areas indicate the particularly wide range of approaches that were then being taken to oriental "realities." Several sessions dealt with an international European transcription project of Japanese texts and brought together experts from France, England, Germany, and Japan. Examples of other linguistic and literature projects were the contributions of Leon de Rosny on "The monosyllabic language of the ancient Chinese," Lucien Adam on "Ural-Altaic languages," Maxence Chalvet de Rochemoutier on "grammatical relations between Egyptian and Berber" and Charles Schoebel on "the affinities of the Dravidian and Ural-Altaic languages." But the program and the conference proceedings also reveal contributions that involve such diverse disciplines as philosophy, religious studies, ethnography, history, and archeology and more clearly scientific and social scientific approaches such as characterised medicine, agronomy, economics and politics.

In addition to working sessions and their protocols, a number of social events was included in the conference programme. During the second session of Congress, for example, a series of awards in the form of Diplomas accompanied by bronze medals was given to various publishers and printers, some French, some foreign including several from Indo-China, China and India, because the Congress wanted "to express publicly the gratitude of the Orientalists to the modest workers who have, through the art of Oriental typography, collaborated in their work." The founder of the Imprimerie Orientale de Meulan, Maruis Nicolas, was singled out for a special award as "a compositor of rare intelligence who, having acquired the knowledge of several oriental languages, had founded a provincial printing house which produced a whole series of works in the different idioms of the Orient." He was awarded a gold medal and a diploma which included a special mention of his services to Orientalism. The individuals recognised in this way or their representatives, accompanied by a fanfare by the band of the Republican Guard, received their awards at this session from the President of the Congress (Congrès international des orientalistes 1874, 56–8). Other social events included an official banquet at the Grand Hotel that closed the first day of the congress. The morning of the second day was devoted to a visit to an Exhibition organised for the Congress. Later the work sessions were broken up by a whole day which was devoted to visits to various museums and private collections.

The final day of the Congress involved the approval of the official statues, the designation of the location of the next congress (84 votes for England, 83 for Italy), the formation of a permanent committee and the formulation of various resolutions. The work of this last session was completed by a decision to publish not only the Memoirs that had been presented, but a list of the members of the Committee of Patronage.

In sum, the Paris Congress inaugurated an activity that was at the junction of the national and the international, at a confluence of pure science with a diverse range of other interests where academic and political rituals crossed, where a concern for establishing permanent communication links in the form of publications and further official meetings completed efforts for creating and strengthening a formal network of individuals devoted to the study of Orientalism.

Participation, Thematic Orientations, and Scientific Communication

The model that was created by the Paris Congress was characterised in scholarly terms by a participation that was institutional and individual, multinational and generalist, even if Japan carved out an special place for itself. Each subsequent congress was to follow each of these elements with more or less success.

However in terms of participation, the Congresses of London, St. Petersburg, Berlin, Florence and Leiden saw a notable decline in the number of participants. At Leiden in 1883 a low point occurred with barely 219 participants present. With some 420 members, the Vienna Congress (1886) showed an improvement in that participation had nearly doubled. The numbers at Stockholm/Christiana are not known. The Geneva Congress (1894) is the next point of reference and the numbers again rose to more than 500 registrants. They rose again at the Paris Congress (1897) with 720 registrants. This is the number around which the Congresses cluster at the beginning of the 20th century.

International recruitment fluctuated as a function of the overall number of participants, members of the host country always representing a more or less large majority of the members of each congress. The representatives of the public powers as members of the different Patronage Committees remained highly visible during the course of these first four decades of the Congresses. Finally a category of participants that continually increased was that of university and scientific institutions.

The thematic orientation remained fundamentally generalist but certain specific emphases were established. Ethnography made its appearance in London. Russian Asia received particular attention at St. Petersburg. African and Altaic languages appeared in Florence. Berlin was structured into the following sections: Semitic, "Indo-Germanic," African and a section combining Oriental Asia and archaeology. There were five sections in Leiden: Semitic, Aryan African, Far-Eastern and Polynesian. There was a similar structure at Vienna in 1886 but here Egypt was included with Africa and Polynesia was extended to include Malaysia.

This is also reflected in the proceedings, at least in the published part, of the 1889 Stockhom/Christiana meeting where there is a turn towards central Asia. At Geneva a restructuring of the whole arrangement was suggested. An Indian section was identified and related to Linguistics and Aryan Languages, with other sections devoted to Semitic and Islamic languages, Egypt and African languages, the Far-East, Greece and the East, and finally sections on geography and oriental ethnography. In Paris (1897) a number of more specific languages was identified and languages and archaeology were brought together as were folk lore and ethnography. What was especially novel in Rome in 1899 was the creation of a section devoted to American languages and traditions and an attempt to combine the history of religions, mythology and folk lore. Here Madagascar was again included in the section devoted to Burma, Indochina and Malaysia. The Hamburg Congress (1902) added a section on exchange relations between East and West. The Algiers Congress (1905) was enriched by a section devoted not only to African archaeology but also to Islamic art. While there were no surprises at the Copenhagen Congress (1908), a great surprise at the Athens meeting (1912) was the emphasis of placed on Greece's relations with the Orient. Thus we see the thematic evolution of the congresses revealing an extension of orientalism that is simultaneously geographical, disciplinary and typological.

The publication of the congresses' work was systematic, even if there were some setbacks such as the first part of the Stockholm proceedings which never saw the light of day and differences of treatment accorded to some sections where on occasion papers were simply summarised or merely mentioned. Together, however, the volumes of the proceedings represent a particularly impressive panorama of developing knowledge of the Orient and of western orientalist interests.

A last organisational element important at the first congress in Paris and later should be mentioned: the social program that accompanied the work sessions and formal opening and closing ceremonies of the congresses. The tradition of social gatherings, visits, and additional activities was established quickly and permanently. We see this for example in the second Congress in London with its Mansion House banquet and visit to the "British and Foreign Bible Society." An Exhibition as well as receptions took place in St. Petersburg, though mention of them in the proceedings is particularly laconic. In Florence in 1878 the involvement of the great families of the city and important literary figures was important in giving banquets, securing official interest and securing access to the Palazzo Pitti and the Palazzo Riccardi. Berlin in 1881 had a whole section devoted to festivities (Festlichkeiten). The Congress in Leiden in 1883 was no exception to the rule. There was a closing banquet, a reception at the town hall of Amsterdam, another by the authorities of the city of Leiden "with music in the garden," an excursion to the Hague and to Amsterdam and a special visit to the Amsterdam International and Colonial Exhibition. An exhibition of manuscripts, books and beautiful little objects of interest to Orientalists" was also organised, while the museums of the city and the university library were opened especially for delegates.

What we see happening here is that the number of receptions and banquets and peripheral visits and exhibitions increased from one Congress to the next offering the participants many places and occasions for informal meetings and the strengthening of contacts that this involved. These activities are an essential part of the process of network building that centred on the formal substantive meetings of the Congresses.

Conclusion

What can we draw from this brief survey of the basics of the international oriental congresses? Perhaps first and foremost they constituted exceptional meeting places. At the congresses ideas were exchanged but also more important perhaps were relationship ties that slowly but surely created what was a more or less permanent intellectual, scholarly and social disciplinary network.

These Congresses also took place as part of a general movement of internationalisation of scholarship that characterised the last third of the nineteenth and the first decades of the twentieth centuries. As periodic reunions they brought into contact different traditions and cultures of research across the immense oriental "terrain." They allowed a discipline simultaneously to "think" about its structure, shape and research methods and to establish an institutional network that could enable it to continue develop. If nothing is possible without men and women, nothing is lasting without institutions. It is this conviction that the orientalist congresses establishes and illustrates.

Finally, the Congresses reveal in an impressive way on the one hand the Eurocentrism of orientalism and on the other hand the powerful entanglement of scientific concerns with political, specifically imperialist, priorities. In this sense the congresses in substance and participation offer confirmation at least partially of the hypotheses of Edward Said who insisted in *Orientalism* (1980) that Orientalist scholarship arose from and reflected European imperialist and colonialist regimes and inevitably as a result created cultural and geographical misrepresentations of the societies it was devoted to studying. They also provide at least partially evidence to the contrary.

Proceedings of the International Congresses of Orientalists

1876–1879. *Congrès international des orientalistes. Compte-rendu de la première session*, Paris, 1873. Paris: Maisonneuve.

1876. *Transactions of the second session of the international congress of orientalists held in London in September 1874*, edited by R.K. Douglas, London: Trübner & Co.

1879–1880. *Travaux de la troisième session du congrès international des orientalistes, St.-Pétersbourg*, 1876, rééd. by W.W. Grigorieff and Baron Victor de Rosen, St.-Pétersbourg: Impr. Pantèlèjeff.

1880. *Atti del IV congresso internazionale degli orientalisti, tenuto in Firenze nel settembre 1878*. Firenze: Successori Lemonnier.

1881. *Verhandlungen des fünften internationalen Orientalisten-Congresses, gehalten zu Berlin im September 1881*. Berlin: Weidman (A. Asher).

1884–1885. *Actes du sixième congrès international des orientalistes, tenu en 1883 à Leide*. Leiden: Brill.

1889. *Berichte des VII internationalen Orientalisten Congresses gehalten in Wien im Jahre 1886*. Wien: Alfred Hölder.

1893. *Actes du huitième congrès international des orientalistes tenu en 1889 à Stockholm et à Christiania*. Leiden: Brill.

1896–1897. *Actes du dixième congrès international des orientalistes, session de Genève, 1894*. Leiden: Brill.

1898–1899. *Actes du onzième congrès international des orientalistes, Paris, 1897*. Paris: Ernest Leroux.

1901–1902. *Actes du douzième congrès international des orientalistes, Rome, 1899*. Florence: Société typographique florentine.

1904. *Verhandlungen des XIII. internationalen Orientalisten-Kongresses, Hamburg, September 1902*. Leiden: Brill.

1906–1908. *Actes du XIVe congrès international des orientalistes, Alger, 1905*. Paris: Ernest Leroux.

1909. *Actes du quinzième congrès international des orientalistes, session de Copenhague, 1908*. Copenhague: Imprimerie Graebe.

1912. *Actes du seizième congrès international des orientalistes, session d'Athènes, 1912*. Athènes: Hestia.

References

"A Brief History of the Royal Asiatic Society of Great Britain and Ireland, 1823 to 1923." 1923. In *Centenary Volume of the Royal Asiatic Society of Great Britain and Ireland, 1823 to 1923*. London: The Society, pp. 7–28.

Barthold, Vassili V. 1947. *La découverte de l'Asie: histoire de l'Orientalisme en Europe et en Russie*, translated by B. Nikitine. Paris.

Cent-cinquantenaire de l'Ecole des Langues orientales. Histoire, organisation et enseignements de l'Ecole Nationale des Langues Orientales Vivantes. 1948. Paris.

Chaudhury, Sushil and Morineau, Michel. 1999. *Merchants, Companies and Trade: Europe and Asia in the Early Modern Era*. Cambridge: Cambridge University Press.

Du Halde, Jean-Baptiste. 1735. *Description géographique, historique, chronologique, politique et physique de l'empire de la Chine et de la Tartarie chinoise*. Paris.

Etiemble, René. 1988–89. *L'Europe chinoise*. Paris: Gallimard.

Filliozat, Jean. 1987. "200 ans d'indianisme. Critique des méthodes et des resultants," *Bulletin de l'Ecole Française d'Extrême-Orient*, 76: 83–116.

Finot, Louis. 1922. "Historique de la Société Asiatique." In *Le livre du centenaire de la Société Asiatique (1822–1922)*, Paris: P. Geuthner, pp. 3–65.

Galland, Antoine. 1704–1717. *Les Mille et une nuits*. Paris.

Hägerdal, Hans. 1997. "The Orientalism Debate and the Chinese Wall: An Essay on Said and Sinology." *Itinerario. European Journal of Overseas History*, 21: 19–40.

Lardinois, Roland. 2007. *L'invention de l'Inde: entre ésotérisme et science*. Paris: CNRS.

Lloyd, Trevor O. 1996. *The British Empire, 1558–1995*. Oxford: Oxford University Press.

Pouillon, François (ed.) 2008. *Dictionnaire des Orientalistes de langue française*. Paris: Karthala et IISMM.

Preissler, Holger. 1995. "Die Anfänge der Deutschen Morgenländischen Gesellschaft." *Zeitschrift der Deutschen Morgenländischen Gesellschaft*, 145: 241–327.

Riviere Gomez, Aurora. 2000. *Orientalismo y nacionalismo espanol. Estudios arabes y hebreos en la Universidad de Madrid (1843–1868)*. Madrid: Editorial Dykinson.

Said, Edward. 1980. Orientalisme. L'Orient dans les yeux de l'Occident, Paris: Seuil.

Schwab, Raymond. 1950. *La Renaissance orientale*. Paris: Payot.

Servais, Paul. 2004. "Les réseaux orientalistes européens au XIXe siècle." In *Europe–Asie. Echanges, Ethiques, Marchés (XVIIᵉ-XXIᵉ siècles)*. Actes du colloque tenu à La Rochelle les 11 et 12 décembre 2000, sous la direction de M. Raybaud et Fr. Souty. Paris: Les Indes Savantes, 103–17.

Vogel, Jean Philippe. 1954. *The Contribution of the University of Leiden to Oriental Research*. Leiden: Brill.

Chapter 6

Organizing a Global Idiom: *Esperanto, Ido* and the World Auxiliary Language Movement before the First World War

Markus Krajewski

Introduction

How to provide the world with a standardized global system of currency exchange, or with a unique paper format? How to gather the world's knowledge onto millions of meticulously cut and sorted index cards when the world's inhabitants still continue to speak in thousands of different languages? This chapter describes and analyses the notion that a new and simplified form of globality could be established by the adoption of a standardized auxiliary language. The languages under discussion here were designed by now forgotten linguistic laymen like L.L. Zamenhof and J.M. Schleyer and advocated around 1900 by renowned European scholars like Louis Couturat and Wilhelm Ostwald. An important question is that of the media setting deployed to spread the unifying idea of a single language for use in addition to the mother tongue. By discussing the rise and fall of the auxiliary language movement using the example of Ido, a derivative of Esperanto, I will try to balance the movement's aims, promises and limited achievements with the secular problems, epistemic obstacles and internal struggles it encountered until World War I broke out and interrupted the endeavours it represented of uniting the world. The War also marked the end of its kind of internationalism for the time being.

The Global Communications Network

Just before World War I, they tried to save the world. No, not the European politicians who continued to threaten each other with military power. "They" were a handful of scientists with a strong international perspective who pursued heterogeneous projects to improve general living conditions and to contribute to progress. One of them was the 1909 Nobel-prize winner in chemistry, Wilhelm Ostwald. In seeking to achieve his notion of saving energy in all circumstances of life, he attempted to optimize many forms of relationship across the world, for example by advocating a unique paper format, a global currency for financial transactions, and, even more fundamentally, a standardized language to be used by literally everybody. An explanation of the roots and requirements of these globalization projects around 1900 together with their initiators, those peculiar persons called projectors (or "project makers"), is that they were an effect of the status of the world transport system (Krajewski 2013). This global network was emerging and expanding during the second half of the 19th century with the interaction and interdependence of the telegraph cable network, steamship lines, the railway system and other kinds of postal transmission systems. In this early process of globalization based on different kinds of media (paper formats, currencies, language itself), Wilhelm Ostwald played a major role. He was not only the Nobel Prize winner of Chemistry, he was one of the first official German Visiting Professors at Harvard. He was invited by Hugo Münsterberg to spend a semester there to conduct research and to exchange ideas with American colleagues. In 1904, during his stay, Ostwald visited the World's Fair at St. Louis. He noted after his visit that it arouses "a deep and lively global feeling of being at home when each time on such occasions, one comes together for the first time with men whose achievements are known from literature and, after the initial greetings, one finds oneself at once on common, mutually familiar ground" (Ostwald 1909, 171). Despite this euphoric global feeling of being at home, there was a significant problem that he continually experienced at the lectures and plenary discussions and in smaller conversations at the Fair.

> Standing there next to one another were men who had the most important things to say to each other, but they could not come to an understanding. Even if most scholars and practitioners today master multiple languages sufficiently well to be able to read technical papers, it is still a long and arduous journey from this point to oral communication in the foreign language. Thus, from this necessity arose anew the idea of international languages. (Ostwald 1911a, 453)

Couturat, Ostwald and the Need for a World Auxiliary Language

In order to remedy at international conferences this glaring drawback of the speechlessness of international science, in 1900 at the Paris World's Fair the mathematician and philosopher Louis Couturat of the Collège de France, together with the mathematician, Léopold Leau, founded the *Délégation pour l'adoption d'une langue auxiliaire international.* This organization had the goal of choosing the best from among the numerous global auxiliary languages that then existed and of achieving its world-wide adoption. The choice was anything but easy. Around 1900, there existed approximately 250 so-called planned or artificial languages. The Délégation made it part of their task to win renowned scholars and scientists to their point of view. Thus, a message from Couturat, who became aware of Ostwald's interest in artificial languages, made its way to the university in Leipzig where Ostwald was working at this time toward placing energy, or rather the avoidance of the wasting of energy, at the center of his philosophy of nature. He considered that having to learn multiple foreign languages was an example of such a waste of energy, which in turn could easily dealt with by adopting a common auxiliary language. A native of Riga, Ostwald characterized himself as the victim of a childhood of linguistic confusion. He spoke German as a first language, Russian in school and Latvian in day-to-day life. Moreover, he had already become familiar with the artificial language, Volapük, through one of his teachers at the university in the Estonian city of Dorpat. With this kind of experience, it is understandable that Ostwald eagerly pledged his cooperation with the Délégation and was to deploy unrelenting propaganda for the idea of a global auxiliary language in the years leading up to the First World War. His advertising activities encompassed lectures before the Bavarian regional assembly of the Association of German Engineers and encouragement for the establishment of the nearly one hundred American Esperantist-Clubs which were founded in the wake of his semester abroad at Harvard in 1905.

While it perhaps belongs to the "job description" of the projector that he constantly pursues a whole cluster of plans, a colorful palette of projects, at least one common purpose may be found among the plans of the different linguistic projectors. Central to many of them was the idea of advancing the "welfare of entire empires, indeed of the whole world" through a universal language – as Johann Gottlieb von Justi stated in 1761(Justi 1761, 275). As soon as the projector made it his goal to give his plans a larger more encompassing scope, with more than a regional impact, a language with particular characteristics proved indispensable. It must be easily learned and understood without great difficulty, for example, and optimally be identical everywhere. For this reason, the high level of attraction which the concept of a universal language exerted upon the projector is not surprising.

Interest in a general language did not arise in the course of the second half of the 19th century under the auspices or in the context of the existing but closed circles of academies and other learned societies with their long traditions as might have been expected. Initially the term used was not *universal language* but rather the peculiar and programmatic designation, *world auxiliary language*. From mid-century with the expansion of the transit and communications technological infrastructure, the aim of the world auxiliary language movement was to provide a further means of communication which could be utilized at any point in the world. The leaders of the movement attempted neither to find a highly sophisticated philosophical instrument nor to reconstruct a possible proto-language. Rather, their aim was to develop a new linguistic code to serve the communication purposes of trade, industry and science. As Louis Couturat noted, "an expansion of the national markets to the international level as well as technical progress, in short, the expansion of physical means, can can lead at the same time to the atrophy of the mental means of communication. Thus one needs a kind of 'mental bonding agent.'" The technical media of circulation around 1900 had for some time been superior in how far they could reach than nationally embedded, conventional media such as money or language. "What good is it if we can travel, write, and speak from one end of the world to the other," said Courturat, "if we cannot understand each other?" (Couturat 1904, 4).

Thus the term "world auxiliary language" represented a program for the world-wide spread of a language which would match the developing global system of communications and transportation. Moreover, the idea was that the "auxiliary language" had to be learned, maintained and developed as the only non-native language acceptable on a world-wide basis. According to Couturat, the international auxiliary language should be "the second language for every man." Couturat nevertheless named three target groups which would primarily profit from the advantages of such a language: scholars, merchants and travelers (though the distinction between the latter two is difficult to determine concretely). "The global language should […] not be a technical or aristocratic language which is only accessible to a few initiates, but rather a daily language which finds use on the railway and in the hotel just as in the learned societies and at conferences" (Couturat 1904, 5, 6). Thus it was logical for Couturat, an editor of the works of Leibniz, to align himself with one of the existing concepts of the universal languages of the 17th and 18th century (cf. Knowlson 1975). However, "after particular study, which I dedicated to the logic of Leibniz," he said, "I may perhaps suggest that such attempts [i.e. the universal language schemes of the pre-enlightenment era] are impracticable" (Couturat 1904, 12).

The suggestion that one could draw on one of the current languages as the basis for an auxiliary language such as English or French was also prohibited for political reasons: neutrality had to be preserved. To opt for one of these languages would give its language community the unfair and powerful advantage of already being able to use its mother tongue. Wilhelm Ostwald's view was blunt: "The selection of any natural language is precluded, for the simple reason that they are all no good" (Ostwald 1910, 12). With the simplification or adaptation of dead languages such as Latin, which was once able to encompass an empire that extended to the limits of the *pax romana*, the situation was similar. Such a language could certainly be spoken by scholars, as it was in Europe as late as the Renaissance, but for "every man" it would now be too difficult because of its numerous exceptions, special rules and "superfluous complications." Morevoer, the unavoidable incorporation of numerous new technical terms would present considerable difficulties for a "dead" language (Couturat 1904, 6 ff.).

But what language could even begin to be considered as being able to meet the requirements of the auxiliary language? How could one succeed "in finding a remedy," asked Friedrich Nietzsche (1878, 222), such that: "in some distant future there will be a new language, first as a commercial language, then as a language of intellectual communication overall, for everyone …? Why would linguistics have studied the laws of language for a century and assessed what is necessary, is valuable and succeeds in every individual language?" Nietzsche cannot have known that in that very year, 1878, this distant future had already begun for a village priest in Baden in the southern regions of Germany. For what is more logical than for someone to construct a language specifically for the purposes of global communication; a language which is easy to learn, simple to speak, and— to satisfy relentless progress—can be expanded without difficulty through its potential for compound formation? As Ostwald pointed out, this is what even a child or "every mob of lowly negroes" can accomplish (Ostwald 1911b, 18). The casual racism of this kind of statement helps reveal not only something about the semantics of the era, it also shows something about the extent to which the whole pseudo–pacifistic movement of the world auxiliary language was hegemonic in the way it included world conquering phantasies and especially the idea of the superiority of Western civilization.

Volapük

Directly after the foundation of the Postal Union in Bern in 1874, which in the years that followed would become the institution with the most impact on global standardization, the Baden prelate, Johann Martin Schleyer (1831–1912), developed his so-called universal alphabet. This was intended to simplify the global communication that had been brought about by the global mail system by doing away with different kinds of alphabets. Though his alphabet did not immediately have the effect he desired, Schleyer dreamed in 1879 in his parsonage in Litzelstetten at Lake Constance of the continued development of his notation into nothing less than a global language. He tried to put his dream into practice by creating a new language system based on a corrupt form of English. He called this Volapük, from vol = world and pük = speak, i.e. world language. Its goal consisted among other things of "helping the global postal system, this magnificent accomplishment of the modern era, to achieve easier operability and practical application" (Kniele 1884, 9). Schleyer's artificial language experienced a dynamic and unprecedentedly successful if brief career from the first conference for a global language in Schemmerberg in 1882 (already having 70 participants), to the 1884 Volapük Congress in Friedrichshafen, to the third World Congress five years later in Paris. However, its downfall was as rapid as its rise. Even as Schleyer composed a three-stanza Volapük anthem (Instructions: "Quietly solemn, for male voices"), other voices were being raised within his choir of supporters in criticism of it. It was too hard to learn. It was inadequate as a language of business. It was unsuitable for aesthetic forms of expression. There were also difficulties in speaking Volapük, particularly as official estimates suggested erroneously that only a few days were required to learn it. Under the guidance of a world language instructor, however, this time could allegedly be reduced to only a few hours. This promise was later widely mocked, as in a famous limerick in *The Milwaukee Sentinel*: "A charming young student of Grük/Once tried to acquire Volapük/But it sounded so bad/That her friends called her mad/And she quit it in less than a wük" (Okrent 2009, 105). A small sample from the Hymn of the Volapists, second Stanza:

Kìs alsò kanòs koblòden	Tell me, what then can increase
lelikà volà menìs?	the highest good of every man?
Kìs alsò kanòs menòden.	A single tongue will bring salvation

as pak bàl. omsà stadìs?	if it sounds in every land.
Klù tonòdosèz in val:	Therefore let the whole world know:
mènade balè pak bàl!	a single tongue from pole to pole!

Schleyer reacted naively to criticisms and calls for simplification or revision. With the authority of the inventor, he rejected any and all modifications. As fast as it attracted attention, the first clearly global language consequently disappeared, though leaving behind all manner of traces, such as for example in the nonsense poems of the much loved German poet, Christian Morgenstern (e.g., his famous "Das große Lalula").

Of the 283 Volapist societies once in existence, only four remained after 1900 (Schmidt 1998, 30 ff.). However, the bold supporters of the idea of a global language had long since set their hearts on a new conception. In some cases they simply changed the title and primary purpose of the existing Volapist society to reflect their new interest in the "successor" language to Volapük. This was Esperanto. Esperanto had been devised by the Polish ophthalmologist, Lazar Ludwig Zamenhof (1859–1917). In 1887, Zamenhof published his variant of a global auxiliary language as *Lingvo Internacia* using the pseudonym, Dr. Esperanto (which in Esperanto meant "the hopeful one").

Esperanto

Zamenhof, like Schleyer, lived at a crossroads of various languages. In Białystok for example, where he was born and later in Warsaw, where much of his education took place, Polish, Russian, German, and Yiddish collided (Korzhenkov 2009). These environments allowed him to gain insight as to how everyday linguistic confusion might be remedied by means of an *a posteriori* artificial language which reflected influences of the established European linguistic communities. For ease of learning, Esperanto leaned even more than Volapük on the vocabulary of spoken languages, above all Romance morphemes (about 75 per cent) and Germanic morphemes (about 20 per cent). In this way it betrayed the global target audiences that it had been devised primarily to serve. Because of its linguistic relationships, a comparatively clear set of rules emerged which could sometimes be gathered together on a single page (see Figure 6.1).

While Volapük, despite intense attempts at institutionalization, such as the establishment of a special academy, went from one crisis to the next and was limited in spread predominantly to grammar school teachers and mid-level employees in central Europe, the proponents of global language gathered increasingly under the banner of the Esperanto movement, a green five-pointed star symbolizing the continents in the color of hope. Shortly after the turn of the century, numerous national Esperanto associations were founded, first in the land of the Universal Postal Union, Switzerland (1902), then in Spain and Mexico (1903), in England (1904), in the USA and Bolivia (1905) and finally in Germany (1906). The international Esperanto

Die ganze Grammatik auf einer Seite

Die Grundlagen der Esperanto-Sprache

Alphabet und Aussprache. a, b, c, ĉ, d, e, f, g, ĝ, h, ĥ, i, j, ĵ, k, l, m, n, o, p, r, s, ŝ, t, u, ŭ, v, z. Alle Buchstaben werden wie im Deutschen ausgesprochen mit Ausnahme der folgenden: c = z (wie in Zunge), ĉ = tsch (wie in Peitsche), ĝ = dsch (wie in Dschungel), ĥ = ch (wie in Nacht), ĵ = franz. j (wie in Journal), s = ss (wie in Messer), ŝ = sch (wie in Schuh), v = w (wie in Wonne), z = s (wie in Nase). **Es gibt keine stummen Laute und auch keine Umlaute (ä, ö, ü). Die Aussprache der einzelnen Laute bleibt unverändert.** Zwei Vokale nebeneinander (ae) werden getrennt ausgesprochen: a-e. Einsilbige Doppelvokale (Diphtonge) werden gebildet durch Zusammenziehen der einfachen Vokale mit dem i-Laut (geschrieben j) oder dem u-Laut (geschrieben ŭ): aj, oj, uj; aŭ, eŭ. Konsonantenverbindung gibt es nicht. Die Betonung ruht **immer** auf der vorletzten Silbe.

Die Grundregeln der Esperantosprache von Dr. Zamenhof

Die Regeln erleiden keine Ausnahme.

Wortlehre.

Die Wörter entstehen durch Anfügung einer grammatischen Endung an den Stamm.

Es bezeichnet die Endung:

o das Hauptwort, patro = Vater
a das Eigenschaftswort, jara = jährlich
e das Umstandswort, bone = wohl
j die Mehrzahl, la patroj = die Väter
n den 4. Fall, mi amas la patron = ich liebe den Vater
i die Nennform, ami = lieben
as die Gegenwart, mi amas = ich liebe
is die Vergangenheit, vi amis = du liebtest
os die Zukunft, li amos = er wird lieben
us die Bedingungsform, ni amus = wir würden lieben
u die Wunsch- und Befehlsform, amu = liebt, ni amu = laßt uns lieben.

Das Geschlechtswort.

Der Artikel

ist für alle Geschlechter der Ein- und Mehrzahl **la**.

la viro = der Mann
la virino = die Frau
la infano = das Kind

Die Deklination.

la bona patro = der gute Vater
de la bona patro = des guten Vaters
al la bona patro = dem guten Vater
la bonan patron = den guten Vater

Mehrzahl:
la bonaj patroj
de la bonaj patroj
al la bonaj patroj
la bonajn patrojn

Die Fürwörter.

Die persönlichen

sind: mi ich, vi du, ihr, Sie, li er (bei Personen), ŝi sie, ĝi er, sie, es (bei Sachen), ni wir, ili sie (Mehrzahl), oni man, si rückbezüglich.

Die besitzanzeigenden

werden gebildet: durch Anhängung eines **a**: mia mein, via dein, euer, Ihr, lia sein, nia unser, ilia ihr (Mehrzahl).

Die Abwandlung der persönlichen und besitzanzeigenden Fürwörter erfolgt wie bei den Haupt- und Eigenschaftswörtern: mi, **de** mi, **al** mi, mi**n** usw., mia, **de** mia, **al** mia, mia**n** usw. Mehrzahlbildung durch Endung **j**: miaj fratoj meine Brüder. Hinweisende, fragende Fürwörter usw. siehe Wörterverzeichnis.

Alle Verhältniswörter

verlangen den 1. Fall: En la domo in dem Hause, per la manoj mit den Händen, dum la manĝado während des Essens.

Das Mittelwort.

Tätigkeitsform.

Gegenwart: La leg**anta** knabo der lesende Knabe; Vergangenheit: La leg**inta** knabo der Knabe, welcher gelesen hat; Zukunft: La leg**onta** knabo der Knabe welcher lesen wird.

Leideform.

Gegenwart: La laŭd**ata** lernanto der Schüler, welcher gelobt wird; Vergangenheit: La laŭd**ita** lernanto der Schüler, welcher gelobt worden ist; Zukunft: La laŭd**ota** lernanto der Schüler, welcher gelobt werden wird.

Die zusammengesetzten Zeiten werden mit dem einzigen Hilfszeitwort „esti" gebildet. Mi estas amata ich werde geliebt, li estus veninta er wäre gekommen.

Die Steigerung

des Eigenschaftswortes erfolgt durch **pli** und **plej** (gesprochen ple-i)

bela = schön
pli bela = schöner
plej bela = am schönsten
la granda domo = das große Haus
la **pli** granda domo = das größere Haus
la **plej** granda domo = das größte Haus

Die Frageform.

Wird der Fragesatz nicht durch ein mit **k** beginnendes Fragewort eingeleitet, so muß er mit **ĉu** (ob) beginnen: **ĉu** vi legas? Liest Du? (s. Wörterverzeichnis).

Die Zahlwörter.

Die Grundzahlen sind unveränderlich: unu (1), du (2), tri (3), kvar (4), kvin (5), ses (6), sep (7), ok (8), naŭ (9), dek (10), dekunu (11), dekdu (12), dudek (20), tridek (30), cent (100), mil naŭcent ok (1908). Ordnungszahlen werden durch die Endung **-a** gebildet: unu**a** erster, du**a** zweiter, adverbiale Ordnungszahlen durch **-e**, unu**e** erstens, due zweitens. Bruchzahlen durch **-on**: du**ono** (1/2), kvar**ono** (1/4), Vervielfältigungszahlen durch **obl**: du**obla** (doppelt) tri**obla** (dreifach). Sammelzahlen durch **op**: du**ope** zu zweien.

Wortbildung.

Die Wörter werden aus folgenden stets unveränderlichen Elementen gebildet: 1. durch Anhängung der grammatischen Endungen a, o, e, i, as, is, os, usw. an den Wortstamm, 2. durch Zusammensetzung wie im Deutschen, 3. durch Vor- und Nachsilben.

Die Stämme genommen aus dem internationalen Sprachgut, bilden den Wortschatz des Esperanto: patr', bon', am'. Erst durch Anhängung der grammatischen Endungen erhalten die Stämme ihren besonderen Charakter als Wort. pluvo der Regen. pluva regnerisch, pluvi regnen, pluvas es regnet.

Das Erlernen von Tausenden von Vokabeln erübrigt sich durch die Anwendung des verblüffend einfachen Systems von

Wortbildungssilben.

Vorsilben.

Dem Stamme vorzusetzen.

bo- durch Heirat verwandt: **bo**filo Schwiegersohn. **dis-** Trennung: **dis**ŝiri zerreißen, **dis**harmonio Mißklang. **ek-** beginnende Tätigkeit: **ek**veturi abfahren. **eks-** ehemaliger: **eks**oficiro Offizier a.D. **fi-** verächtlich, unsittlich: **fi**hundo Köter. **ge-** vereinigt beide Geschlechter: **ge**fratoj Geschwister. **mal-** Gegenteil: **mal**dika dünn. **pra-** ur: **pra**patro Urvater. **re-** Rückkehr, Wiederholung, **re**veni zurückkommen.

Nachsilben.

Dem Stamme anzuhängen.

-aĉ- verächtlich, unschön usw.: skrib**aĉ**o schlechte Schrift (Geschmier). **-ad-** dauernde Tätigkeit, promenado Spaziergang. **-aĵ-** sichtbares, greifbares Ding: trink**aĵ**o Getränk. **-an-** Mitglied, Anhänger, Einwohner: Leipzig**an**o. **-ar-** Ansammlung, Menge: vort**ar**o Wörterbuch. **-ĉj-** Koseformen männlicher Namen: Kaĉjo Karlchen. **-ebl-** Möglichkeit: trink**ebl**a trinkbar. **-ec-** Eigenschaft: hel**ec**o Helligkeit. **-eg-** vergrößert oder verstärkt: pord**eg**o Tor. **-ej-** Raum: lern**ej**o Schule. **-em-** Gewohnheit, Neigung: labor**em**a, arbeitsam. **-end-** mit der Bedeutung soll und muß: pag**end**a sumo zu zahlende Summe. **-er-** Teil: sal**er**o Salzkörnchen. **-estr-** Oberhaupt: urb**estr**o Bürgermeister. **-et-** verkleinert oder schwächt ab: bild**et**o Bildchen. **-id-** Nachkomme, Kind, Junges: bov**id**o Kalb. **-ig-** veranlassen, machen: pur**ig**i reinigen. **-iĝ-** zu etwas werden (wer oder wie): maljun**iĝ**i alt werden, san**iĝ**i gesunden. **-il-** Mittel, Werkzeug, Instrument: bor**il**o Bohrer. **-ind-** Würdigkeit: leg**ind**a lesenswert. **-ing-** Halter, in dem etwas steckt: cigar**ing**o Zigarrenspitze. **-in-** weibliches Geschlecht: knab**in**o Mädchen. **-ism-** Bezeichnung einer Lehre oder Geistesrichtung, Bewegung usw.: militar**ism**o, Militarismus, kalvin**ism**o Calvinismus, ekspresion**ism**o Expressionismus. **-ist-** Beruf: polic**ist**o Polizist, bak**ist**o Bäcker. **-nj-** Koseform weiblicher Namen: He**nj**o Hertchen. **-obl-** Vielfaches: kvar**obl**e kvin 4×5. **-on-** Bruchteil: kvar**on**o Viertel, dek**on**o Zehntel. **-op-** Sammelbegriff: tri**op**e zu dritt. **-uj-** Behälter: mon**uj**o Geldtasche; Fruchtträger: pir**uj**o Birnbaum; Land: German**uj**o Deutschland. **-ul-** Person mit hervorstechender Eigenschaft: fremd**ul**o Fremder. **-um-** dient zur Bildung von Wörtern, die in gewissen Beziehungen zur Wortwurzel stehen: naz**um**o Klemmer, plen**um**i erfüllen.

Figure 6.1 Esperanto at a glance

Source: "Den Arbeitern aller Länder eine Sprache." Zur Geschichte der internationalen Arbeiter Esperanto–Bewegung 1903–83. Dortmund 1983. [Exhibition of the Institute for German and foreign workers literature, Dortmund,1983]

Congress in Boulogne-sur-Mer in 1905 served as a prelude to a sustained series of yearly gatherings on the international level which would continue to the beginning of the First World War. The national associations also consolidated into a World Federation in 1908 (Schmidt 1998, 13, 22, 34 ff.; Proelss and Sappl 1922, 10 ff.).

According to its architect, Esperanto did not only serve as a lingua franca for daily use. In contrast to Volapük, Esperanto also offered the possibility of use for aesthetic or literary purposes. Zamenhof himself encouraged this by translating *Hamlet*, Gogol's *Revisor*, Goethe's *Iphigenie auf Tauris*, Schiller's *Räuber*, Andersen's fairy tales, as well as the Old Testament, and various poems and proverbs. Aware of Couturat's suggestion that Esperanto could help in achieving Goethe's notion of a world literature, others added the *Iliad* of Homer and the *Monadologie* of Leibniz to the canon to demonstrate, on the one hand, the capabilities, the "wonderful ductility of the artificial language" (Couturat 1907, 15), and to achieve, on the other hand, an exemplary impact on its future use. The translation of world literature into a single global auxiliary language, it was generally agreed, would contribute to the spread of these texts and additionally save the effort of rendering them into other "native" languages, which would be nothing more than a "frivolous squandering" of time and energy (Couturat 1907, 14). Couturat connected this claim to a media-theoretical argument which underscores the secondary character of an artificial language: "Global language will relate to the national literatures as photographs to the original works after which they were produced" (Couturat, 1907, 15).

Esperanto vs Ido

It is noteworthy that the history of the continued development or revision of Esperanto played out a kind of repetition-compulsion of the fate of Volapük. The first Esperanto World Congress in Boulogne-sur-Mer in 1905 resulted in a split. The adherents were divided into an orthodox faction, which demonstrated a nearly religious reverence for the original version of the language, the so-called *fundamento*. By contrast, a reform movement emerged through conspiratorial activities which involved demands for a departure from the *fundamento* and the introduction of profound changes. In 1907, following a gathering of the Délégation under the leadership of Louis Couturat, the dissidents ultimately decided on the creation of a separate autonomous language called Ido (i.e. in Esperanto: descendant, successor) as a modified derivative from Zamenhof's language. The "master" himself, however, opted for immutability. Initially much less autocratic than Schleyer, he had allowed a vote in 1894 regarding possible changes to Esperanto. On the opposing side were Louis de Beaufront, a strange figure whose real name was Louis Chevreux who had been an ardent proponent of Esperanto before changing sides, Louis Couturat, and above all a real *putschist*, Wilhelm Ostwald, who was an eager advocate for basic improvements. Ostwald willingly allowed himself to be named the mouthpiece of the reform movement. He not only chaired the subsequent meetings of the Ido committee but also produced a flood of pamphlets that were designed to publicise the language and to convince earlier followers of Esperanto to convert to the new version. Ultimately, however, Esperanto emerged strengthened from the fight between the languages, particularly as there were frequent modifications and revision of Ido that contributed little to its stability. By contrast, Esperanto blossomed to a certain extent in the interwar period, though Zamenhoff himself had died in 1917 (cf. Boulton 1960). Nevertheless, its spread remained limited and historians of Zamenhof's world auxiliary language vaguely characterize his construct today as the "means of communication of a world-wide diaspora community" (Sakaguchi 1996, 18).

The aftermath of this project can be told briefly: Ostwald and Couturat cooperated closely—at least in the beginning—in their endeavors to institutionalize *Ido* especially by establishing what was called *A World Language Office*. This was founded in 1910 following the deliberations of a linguistic monstrosity, the Initiative Committee for the Establishment of an Association for the Creation of a World Language Office ("Initiativkomitee zur Gründung eines Verbandes für die Schaffung eines Weltsprache-Amtes"). Needless to say, this World Language Office never commenced operations. The Ido Academy that was designated to fill its space was no more successful. As a result, Ostwald's efforts for the adoption of Ido as a world auxiliary language, rather than resulting in regular progress, created a state of permanent tentativeness. But only until 1915. Ostwald's preoccupation as the First World War broke out with the advantages of Internationalism came to an abrupt end and he embarked on another, though very similar linguistic project, Wede.

Ostwald's Wede and Weltdeutsch

If Ostwald's 1903 lecture before the Bavarian regional Association of German Engineers served as the prelude to his intensive engagement with international auxiliary language projects, then his speech at the end of October 1915 marked the end of his preoccupation with them. His new planned language, Wede, no longer had anything in common with the pacifism implicit in Esperanto, Volapük or Ido. If one had to ask to which nation Ostwald's new industrious world official was obligated, his so called Monistic Sunday Sermon No. 36 provided an unequivocal answer. The address reveals nothing less than Ostwald's reverse conversion from Paul to Saul:

> I propose to produce a simplified German on a scientific-technical basis for practical use in those areas [i.e. the newly occupied countries]. In this, all dispensable variations, all of the aesthetically charming richness of the language which complicates its learning so tremendously, must be removed, so that this new means of communication, for which I propose the name Weltdeutsch (World German), can be learned and used by everyone with little effort. (Ostwald 1915, 557)

What began in 1901 with the Délégation and an ambitious program to establish a single auxiliary language throughout the entire world had deteriorated under the combat conditions of World War into a now neglected political farce called "Weltdeutsch." In Ostwald's internationalist understanding before the war, "Welt" denoted absolutely everything, but now under the influence of the Schlieffen Plan as well as the apparently successful eastward expansion of the German war effort he added"-deutsch"—thus collapsing the two concepts "Welt" and "Deutsch." Thus it is the World War which first reveals openly the hegemonic impetus, previously carefully hidden behind euphoric terminology, which had always underlain Ostwald's series of world projects and indeed the entire world auxiliary language movement.

Even early critics of the movement called attention to the extent to which it was pervaded by imperial politics and its colonial influences. The "world" of a world language had never been, according to Meyer, "understood in the sense of a label for the entire inhabited planet," neither in the context of Greek and Alexander the Great nor in the case of Latin and the Roman Empire. Meyer is even more precise when he says: "One should not abandon oneself to illusions regarding the fact that the overwhelming portion of all the inhabitants of our planet hasn't the least interest in the creation of a world language" (Meyer 1893, 43). The apologists for, or apostles of, a world auxiliary language, however, seemed to take as a challenge the need to win over this "overwhelming portion" of mankind to their aims. From this point of view, the global auxiliary language movement became from the beginning a missionary's program, a kind of linguistic religious order which was to bring a message of salvation to the linguistic heathen of the entire world. However, in contrast to early Christian times, other means were now available for the delivery of the Gospel. Driven by the blessings of global transit, a pentecostal global conception emerged not only of unlimited accessibility but one in which the whole planet appeared ready for conversion by a suitable project. The positions of giving and taking, of sender and receiver, remained beyond question.

> Steam shortened distances; electricity neutralized them. From this ensued an intensive and extensive development of industrial and commercial relationships between all peoples. The civilized world, which was limited to old Europe a century ago, has incorporated new nations and entire portions of the world. The European market stretches across the entire globe. (Couturat 1904, 3)

Comments such as this raise the issue of the meaning of "world" from the perspective of the artificial language movement. Couturat characterized "world" as the progressive incorporation of other continents through the institutions of Europe and its market power. The advocates of world language without any obvious political claims to power planned to launch their rational program of a unifying language through the existing distribution and communications channels. But this meant that the global auxiliary language project was grafted conceptually onto the global transit system that was—more or less—controlled from within Europe (the British "All Red Line," the military-controlled, world-encompassing telegraph wire service may serve as evidence). And this occurred at precisely the moment in which Europe was threatened with being torn apart by wartime conflict. The apostles of world language had to recognize on the one hand that their motor and medium of transmission called the world transit system had suddenly bogged down and, in places, had come to a standstill. On the other hand, they had to acknowledge that their supposedly neutral linguistic program pursued a massive hegemonic aspiration. The whole thing has been a geo-political issue all along. The battle for world language, to quote Gustav Meyer once again, "will be connected at the deepest level to the political configuration of the Earth's surface" (1893, 40).

References

Boulton, Marjorie. 1960. *Zamenhof, Creator of Esperanto*. London: Routledge and Paul.

Couturat, Louis. 1904. *Die internationale Hilfssprache*. Berlin: Möller & Borel.

Couturat, Louis. 1907. "Eine Weltsprache oder drei? Antwort an Herrn Professor Diels." *Deutsche Revue*, 32: 1–2.

Justi, Johann Heinrich Gottlob von. 1761. *Politische und Finanzschriften über wichtige Gegenstände der Staatskunst, der Kriegswissenschaften und des Cameral- und Finanzwesens*. Koppenhagen und Leipzig: Auf Kosten der Rothenschen Buchhandlung.

Kniele, Rupert. 1884/1984. "Der erste Kongress der Weltsprachefreunde." In *Der erste Volapük-Kongreß, Friedrichshafen, August 1884. Dokumente und Kommentare*, edited by Reinhard Haupenthal. Saarbrücken: Artur E. Iltis, 15–35.

Knowlson, James. 1975. *Universal Language Schemes in England and France 1600–1800*. Toronto: University of Toronto Press.

Korzhenkov, Aleksandr. 2009. *Zamenhof The Life, Works, and Ideas of the Author of Esperanto Abridged by the author from Homarano: La vivo, verkoj kaj ideoj de d-ro L.L. Zamenhof* (kun Litova Esperanto-Asocio) Kaliningrad-Kaunas: Sezonoj, 2009 English translation and notes by Ian M. Richmond. Esperantic Studies Foundation.

Krajewski, Markus. 2014 (forthcoming). *As For the Rest. World Projects and Notions of Globality before WWI*. University of Minnesota Press: Minneapolis.

Meyer, Gustav. 1893. "Weltsprache und Weltsprachen." In *Essays und Studien zur Sprachgeschichte und Volkskunde*, 2: 23–46. Straßburg: Trübner.

Nietzsche, Friedrich. 1878. *Menschliches, Allzumenschliches I. Vol.2 of Kritische Studienausgabe*, edited by Giorgio Colli und Mazzino Montinari. Munich, Berlin: dtv/de Gruyter.

Okrent, Arika. 2009. *In the Land of Invented Languages. Esperanto Rock stars, Klingon Poets, Loglan Lovers, and the Mad Dreamers who Tried to Build a Perfect Language*. New York: Spiegel & Grau.

Ostwald, Wilhelm. 1909. *Energetische Grundlagen der Kulturwissenschaft*. Vol. XVI of *Philosophisch-soziologische Bücherei*. Leipzig: Verlag von Dr. Werner Klinkhardt.

Ostwald, Wilhelm. 1910. *Die Organisation der Welt. Vortrag gehalten im Bernoullianum zu Basel am 7. September 1910*. After a stenographic recording. Basel: Verlag des Weltsprache-Vereins Ido.

Ostwald, Wilhelm. 1911a. *Die Forderung des Tages*. 2nd revised edition. Leipzig: Akademische Verlagsgesellschaft.

Ostwald, Wilhelm. 1911b. *Sprache und Verkehr*. Leipzig: Akademische Verlagsgesellschaft.

Ostwald, Wilhelm. 1915."Weltdeutsch." In *Monistische Sonntagspredigten*, no. 36, October 31, 545–58.

Proelss, Hans and Hanns Sappl. 1922. *Die bisherigen Erfolge der Welthilfsspache Esperanto auf der ganzen Welt*. Graz: Paulus Verlag.

Schmidt, Johann. 1998. *Geschichte der Universalsprache Volapük*. Saarbrücken: Ed. Iltis.

Sakaguchi, Alicja. 1996. "Die Dichotomie: künstlich « vs. » natürlich und das historische Phänomen einer funktionierenden Plansprache." *Lingvaj problemoj kaj lingvo-planado* 20: 18–39.

Chapter 7

Beyond Babel: Esperanto, Ido and Louis Couturat's Pursuit of an International Scientific Language

Fabian de Kloe

The Belle époque witnessed several remarkable attempts to organize and store the world's knowledge. The dream of Belgian lawyers Paul Otlet (1868–1944) and Henri La Fontaine (1854–1943) to catalogue the entire universe of published knowledge with their Universal Decimal Classification System is a well-documented example of such a project. A similar initiative was an organization called *Die Brücke*, which was established in Munich in 1909 by the German Nobel Prize winning chemist Wilhelm Ostwald (1853–1932). Ostwald envisioned *Die Brücke* as a central station, an information office for information offices "where any question which may be raised with respect to any field of intellectual work whatever finds an answer either direct or else indirect, in the sense that inquirer is advised as to the place where he can obtain sufficient information" (Ostwald 1913, 6). This chapter deals with an equally ambitious attempt to make scientific knowledge accessible. But to do this, instead of organizing and centralizing knowledge in its conventional print format, Couturat proposed that a new, rational streamlined medium for its communication be developed. This was Ido, an international scientific artificial language.

Louis Couturat (1868–1914) was a French mathematician and philosopher. From 1900 onwards he became engaged in a battle at international scientific conferences and in scientific journals against the inefficiencies created by the number and diversity of national languages. In his eyes scientific knowledge fell between the cracks of incompatible vernaculars. Couturat believed that, in order to move beyond the Babel of confused voices that was beginning to characterize early twentieth century science, it was necessary to use the tools of science itself in creating Ido. He represented Ido in numerous publications and pamphlets as an improved version of the artificial language, Esperanto, arguing that it was more precise and efficient as a vehicle for the international transfer of scientific knowledge. At the same time he also represented Ido as a politically neutral language because its scientific basis would take it above politics. Its vocabulary he said was chosen according to "the principle of maximum internationality" and its grammatical structure was freed of useless rules and confusing exceptions. With these qualities Ido was, to quote Couturat, "nothing but a purified and idealized extract, a quintessence of the European languages" (Couturat 1910, 51–2).

The promotion of Ido as a global lingua franca may well be considered today as outdated and eccentric. But Couturat's internationalist claims are not. His notion of Ido's function as bridging national divisions to facilitate international cooperation fits a longstanding and familiar conception of internationalism as a political movement. The origins of this internationalist movement can be traced back to the birth of nationalism itself, though relatively little is known historically about the relationship of internationalist claims to the specific socio-cultural contexts in which they were articulated. However, one thing *is* clear about internationalism: it has usually been presented approvingly. But as historian and sociologist Perry Anderson points out: "The price of approval is indeterminacy" (Anderson 2002, 5). To compensate for this lack of determinacy, he argues that internationalism since 1750 has undergone several metamorphoses against the backdrop of successive phases of nationalism.

The following reconstruction of Couturat's linguistic internationalism seeks to flesh out a historically grounded image of internationalism. Instead of diachronically tracing various versions of internationalism against the measuring rod of successive phases of nationalism, this chapter frames Couturat's linguistic internationalism as an expression both of typically Belle époque notions of scientific objectivity and French politics. Structurally, my argument is developed in six sections. The first section introduces our protagonist and presents his claims for the necessity of an international language. The next section offers a brief account of the mechanics of Ido and its supposed scientific and political virtues. The third section points out that Couturat's linguistic internationalism was more driven by the push of individual ideological motives than by the pull of external demand. The next two sections are based on the argument that fundamental to Couturat's program were: 1) a conception of logic as a tool to generate scientific objectivity, and 2) a political conviction that was forged in reaction to the famous Dreyfus Affair. The last section concludes with a brief examination of Ido's promotion as a strategy for community building.

The Return of Babel

At the end of the 19th century there was considerable interest in the idea of an international language. Apart from promoting one's national mother tongue as a global lingua franca, there were two options: a revival of Latin or the creation of an artificial language. A serious attempt to revive Latin as an international language for commerce and science was initiated at the end of the nineteenth century by the Englishman George Henderson. The German philologist Hermann Alexander Diels, who was a member of the Berlin Academy of Science, was also an avid promoter of a revival of Latin (Waquet 2001, 265). Their attempts were rivaled by a multitude of artificial language projects. During the nineteenth century at least 38 projects were initiated (Couturat and Leau 1903). Esperanto and Volapük were two of the more successful international artificial language that emerged towards the end of the century. The Russian doctor Ludwig Lazarus Zamenhof created Esperanto in 1887. The German Catholic priest Johann Martin Schleyer created Volapük in 1897, which was mostly displaced by Esperanto in the early twentieth century (see Chapter 6 above).

Louis Couturat's involvement with the idea of an international language started in the summer of 1900 at the first International Congress of Philosophy which he and his friend Xavier Léon, philosopher and director of the philosophical journal *Revue Métaphysique et de la Morale*, had initiated and organized on the occasion of the Paris World's Fair. The Congress's aim was to promote international communication among philosophers. The congress attracted such eminent figures as Henri Bergson, Moritz Cantor, Henri Poincaré, Bertrand Russell, and Paul Tannery (Canto-Sperber 2008, 90). Noticing the hampered flow of communication between participants as a result of national language differences, Couturat became convinced of the necessity of exploring the possibilities of how an international artificial language could improve the transfer of knowledge on the occasion of such events.

In 1900 Couturat had not yet published anything on the theme of international language, but he had proven himself to be a talented mathematician and philosopher. He had joined the editorial board of the *Revue Métaphysique et de la Morale* in 1883 after obtaining degrees in philosophy and mathematics at the *Ecole Normale Supérieure* in Paris. Among his teachers were such prominent mathematicians as Poincaré, Emile Picard and Jules Tannery. In that same year he published a contribution to a heated debate between Poincaré, Russell and George le Challas on the principles of geometry. In 1895 he was appointed professor of philosophy in Toulouse, and in 1896, with the publication of his two doctoral theses, *De L'inifini Mathématique* and *De Platonicis Mythis*, he could claim to be both philosopher and mathematician. Several years later he established himself internationally as a prominent Leibniz scholar with the publication of *La Logic de Leibniz* (1901) and *Opuscules et Fragments Inédits de Leibniz* (1903). He was also an active promoter of Bertrand Russell's algebraic logic in France.

In 1901 Couturat together with the French mathematician Léopold Leau founded the Delegation for the Adoption of an International Auxiliary Language. It brought together a range of linguists, scientists, and other scholars, to decide which artificial language should be adopted as the international language. From that moment onwards his involvement with the international language began to overshadow his other interests. In fact, his passion for Ido eventually verged on the obsessive, and this contributed to the termination of an intensive and lengthy correspondence with Russell: "In the last years I have lost contact with him because he became absorbed in the question of an international language" (Russell 1998, 135).

Couturat's first book on international languages was the monumental *Histoire de la Langue Universelle* (1903) written in collaboration with Leau. The book offered an extensive overview of past attempts to create a universal language, including those of such 17th century philosophers and intellectuals as Rene Descartes, George Dalgarno, Wilhelm Gottfried Leibniz, as well as of more contemporary artificial language projects like Volapük and Esperanto. Its preface portrayed the early 20th century scientific community as rapidly internationalizing. But instead of facilitating a coherent and smooth interaction between scientists, the growing number of international scientific organisations, such as the International Bureau of Weights and Measurements created in 1875, the International Geodesic Association in 1886, and the International Association of Academies in 1899, often presented members with problems in communication owing to the diversity of their national languages. To characterize this impeded flow of international communication caused by a confusion of national tongues, the authors referred to the tower of Babel (Couturat and Leau 1903, ix).

Judging Latin's grammar to be too complex and irregular and the ideography of philosophical languages to be too complicated, "as complicated as Chinese vernacular" (Couturat 1906, 14),[1] Couturat and Leau believed that the creation of an artificial language was the best option for transcending the linguistic confusions, the Babel of many tongues, now characterising contemporary science. But this did not imply the creation of "a language that Adam could, or should, have invented" (1906, 15). Instead of creating a language from scratch, they suggested that its basic building blocks should be words drawn from existing national languages. This is what made Esperanto appealing to Couturat in that it was a product of French, Italian and some Slavic elements. But by 1905 he had become increasingly convinced of its imperfections. Besides viewing Esperanto's inclusion of Slavic elements as problematic, Couturat believed that it lacked the precision and neutrality that was required for the easy transfer of scientific knowledge.

Creating a Scientific Language: Esperanto vs Ido

Thus it was in 1906 that Couturat began to work on the creation of Ido in collaboration with the French Esperantist, the Marquis Louis de Beaufront (1855–1935). From 1900 onwards they had been corresponding extensively about matters relating to the ideal of an international language.[2] De Beaufront played a major role in the promotion of Esperanto in France. In 1898 he founded the *Société Pour la Propagation de l'Espéranto*, and a new journal called *L'Espérantiste* (Glover Forster 1982, 76). But by 1906 he too believed that Esperanto needed streamlining.

In 1907 Couturat organized a series of meetings for the representatives of the Delegation for the Adoption of an International Auxiliary Language. The aim of these meetings was to identify which would be best as an international language from the existing artificial language projects. As the delegation's president, Couturat upheld its image of disinterestedness by concealing his role as the creator of Ido. The meetings took place in Paris at the *Collège de France* in 1907. During

---

1   All quotes by Couturat in this chapter are my translation.

2   Unfortunately this correspondence no longer exists. According to Claude Gacond, curator of the Swiss *Centre de Documentation et d'Etude sur la Langue Internationale*, de Beaufront lost his entire correspondence—and his house—in a fire.

each session an international language project was presented and then discussed by a committee comprising twelve to fifteen representatives of other international language projects. Several language authors appeared before the committee. Esperanto was one of the languages discussed. Unaware of his involvement with Ido, the inventor of Esperanto, Ludovic Lazarus Zamenhof (1859–1917), sent de Beaufront as Esperanto's advocate. This was not the first act of deceit de Beaufront committed in his life. He claimed to be a descendent of the French king and to have been forced to work as a private tutor after losing his fortune. But there was never any proof that this was true (Glover Forster 1982, 75).

The committee emphasized several of Esperanto's weak points. Its use of diacritical marks (acute, grave and cedilla) was criticized as inefficient for writing and as causing too many printing problems. It had *a priori* components, that is artificially constructed components that do not exist in natural languages such as Kie (where), Kial (why), Kiam (when), Kiel (how). These were criticized as being unnecessarily complicated. Despite these criticisms, the committee selected Esperanto as the best option then available because of its relative perfection and the numerous and varied uses it had already attracted. But the committee did not recommend its adoption without modifications. These were to be devised by a permanent commission in collaboration with the Ido project, which Couturat had anonymously submitted to the members of the Committee. After 1907 the story of his efforts for the international adoption of Ido was one of a constant battle with defenders of Esperanto. When Couturat turned out to be one of Ido's most militant defenders and it became clear that he was Ido's main author, it also became clear that his apparent "scientific arbitration" in Paris in 1907 at the meetings to select an international language had been little more than a ruse to promote Ido. Not surprisingly, the Esperantists felt betrayed (Glover Forster 1982, 123).

Ido used the 26-letter Latin alphabet as opposed to the Esperanto's use of ĉ, ĝ, ĥ, ĵ, ŝ, and ŭ instead of the letters q, w, x, y. But Ido was meant as much more than a simplification of the alphabet. In his and de Beaufront's reexamination of Esperanto, Couturat applied what he thought of as a scientific methodology to make Ido more regular and more neutral than Esperanto. Ido was to be constructed according to precise and rigorous rules that left no room for uncertainty and ambivalence. The end result, what he saw as Ido's simplicity and precision, was for him nothing less than an expression of logical objectivity (Couturat 1909, 12). Couturat applied two basic logical rules in Ido's creation: the principle of univocity and the principle of reversibility. Univocity is a term in logic to describe that which is expressed in one voice. The univocity principle demanded correspondence between sign (word) and signified (idea). Every notion was to be expressed by one word and one word only. Couturat believed that all languages naturally evolved towards an increasing degree of simplicity through "a reduction of grammatical forms to their most general and essential forms," but he argued that the evolution in natural languages towards a univocal coupling of notion and word was "slow and often troubled …" (Couturat 1909, 512). An artificial language could cut this evolution short by imposing the principle of univocity uniformly.

The reversibility rule required that if a root and affix were combined, the new word must represent the combined meaning of both root and affix. The idea was that by applying this rule, every derivation was reversible. If one passed from one word to another in the same family by adding or subtracting affixes, one was able to easily arrive at any word of that same family. From labor/ar, which meant to work, one could move to labor/ist, with the suffix "-ist" as the exclusive denomination for occupation or profession. And from labor/ist one could easily move to labor/ist/o, which meant worker. According to Couturat, following this principle meant that "[…] one can proceed from any word whatsoever of a family and arrive at any other word in an absolutely unique manner, whereas if one did not observe this principle one would inevitably obtain two meanings for the same word" (Couturat 1910, 42).

Couturat's application of such logical principles was not only meant to give Ido its cutting edge as a precise and efficient tool for carrying scientific knowledge across national boundaries. He also argued that its system of logical derivation meant that the roots of existing natural languages were chosen that were understood by a maximum number of Europeans. A brochure titled, *Pour la Langue Auxiliaire Neutre*, claimed that Ido reflected "the principles of justice and neutrality" (Couturat 1909, 5). It suggested that instead of pursuing the chimerical cause of national languages, that not only scientists but all Europeans should actively pursue the cause of the international language by supporting Ido. In addition to serving the scientific community, it argued that Ido was an efficient and neutral vehicle for everyone in the conduct of their international transactions and negotiations—politicians, traders, and even ordinary travelers.

Couturat, Ido, and Early Twentieth Century Babel as Myth

In the years after the Committee of the Delegation for the Adoption of an International Auxiliary Language had made its choice of Esperanto as the best of the international languages on the condition that it be simplified, Couturat worked hard to promote Ido. Besides issuing various brochures, he continued to publish numerous articles in the *Revue de la Métaphysique et de la Morale* on the application of logic to Ido. In 1908 he established a monthly journal called *Progresso* that was published entirely in Ido. As the journal's secretary he was the main contributor of lengthy discussions about the language's technical issues and possible improvements (Lalande 1914, 676). Two years later he published an extensive international mathematical lexicon in Ido with German, English, French and Italian translations (Couturat 1910). Together with de Beaufront he also compiled large bilingual dictionaries: French-Ido, English-Ido, and German-Ido. The first was published, the second was printed but never issued, and the third was never completed. Couturat's last publication was an article on the current status of French logic for an *Encyclopédie Internationale des Sciences Philosophiques* in French, English and German (Lalande 1914, 677).

Couturat did not live to witness the publication of the French edition of this encyclopedia. On August 3, 1914, the second day of French mobilization for the First World War, he was killed in a car accident on his way from Paris to Fontainebleau. His death was a blow to the Ido movement in France. Despite constant efforts to create support by producing a waterfall of pamphlets, publications and dictionaries, he waged a lonely war not only against what he believed to be the early twentieth century confusion of many tongues, but also against the Esperantists.

But while Ido failed to gain wider support, some of his contemporaries shared his conviction that the increasing diversity of national languages at international congresses and other events was becoming problematic. Since the second half of the nineteenth century universal exhibitions, international conferences and international organizations had been increasing in number. Participants in the increasingly frequent and regular international meetings were drawn from a multitude of nationalities that offered more than a rich palette of cultures and languages. During a visit to the first universal exhibition in 1851 in London, for example, the German architect Gottfried Semper (1803–1879) was so overwhelmed by the number of languages spoken at the fair that he called the Crystal Palace a modern tower of Babel (1966[1852], 28).

In the same period an increasing number of international scientific congresses and organization emerged, replacing the older cosmopolitan ideal of the Republic of Letters by the modern idea an international scientific community. Between 1860 and 1904, 28 international organizations were established in the field of the pure sciences (Speeckaert 1957, XIII). In 1901 Gaston Darboux, secretary of the mathematics section of the French Academy of Sciences, recognizing the need for international coordination created by the increasing number of international organizations of various kinds, praised the establishment of the International Association of Academies in 1899 as a functional organ to coordinate scientific work internationally. Darboux observed that: "It is impossible not to be struck by the speed at which international organizations multiply today" (Darboux 1900, 6). In his eyes "this tendency towards association" meant that the international scientific community itself was rapidly expanding. In his view unless the enormous scientific production in a multitude of languages was "unified and coordinated" we risked a return "to the Tower of Babel" (1900, 11). Darboux, in effect, was reflecting the same concerns that underpinned Couturat's belief in the necessity of Ido.

French historian of science, Anne Rasmussen, situates Couturat's promotion of Ido at the intersection of this nineteenth century convergence of on the one hand the increasing number of international organizations and publications, and on the other the publication of science in an increasing number of national languages. Rasmussen points out that since the beginning of the century the number of national languages in which science was published had surged from ten to twenty (Rasmussen 1996, 142). Confronted with the number of national languages at congresses and in journals, scientists had to deal with difficult questions of translation. It is therefore not surprising that members of the scientific community discussed on numerous occasions the value of adopting an international language to streamline communication. A 1898 editorial in *Science* titled "International Languages," for instance, discussed an initiative by the American Philosophical Society to adopt Volapük as a possible language for international communication: "An artificial language of such kind would indeed be of great benefit for scientists, and make many publications, such as Hungarian, Bohemian, Romanian, etc. available" (1889, 24).

But the author of the editorial was also pessimistic. He concluded that the tendency among scientists to cling to their national mother tongue made a broader acceptance of an international artificial language unlikely: "There seems little hope that in this period of nationalism the majority of scientists will forego the claim that their language is the language of the most accomplished and most cultured people of the world, and that it has a right to become one of the 'world-languages'. When this period has passed, English, French and German will continue to be better means of international intercourse than any artificial language" (1989, 24).

Crucially, this view was representative of a general reluctance among turn-of-the-century scientists to support the adoption of an international artificial language. In publications and at congresses scientists occasionally proclaimed their national language to be the *true* language of science. This trend was especially common among English, French and German scientists. These languages represented the dominant vernaculars in the early twentieth century international scientific arena, so most international congresses were held in one or other of those languages (Reinbothe 2006, 11). For the majority of scientists the benefits of supporting their mother tongue—or at least the dominant French, English and German languages—outweighed the alleged communication problems they experienced at international conferences and in consulting publications. Bertrand Russell, for example, did not see the necessity of learning an additional language because in his view the majority of British scholars already knew French and German.[3] Couturat's proposal to adopt an artificial language was received with similar reserve at the second International Mathematical Congress in Paris in 1900. And even the International Association of Academies declined to support Couturat's formal request to appoint a committee to select an international language.

So Couturat's promotion of Ido was clearly not driven by a strong external demand for an artificial international language. Despite the fact that some shared his perception of a potential turn-of-the-century Babel in the international scientific community, the reality of this image is questionable in view of a general reluctance among scientists to support his diagnosis of the problem and its solution by Ido. These approaches set him apart from his scientific contemporaries. To understand what propelled his linguistic internationalism we need to take a closer look at the scientific and political convictions that underpinned his claims about Ido.

---

3   Betrand Russell to Louis Couturat December 8, 1900 (Couturat 2001, 209).

Logic as Objectivity Machine

Couturat's belief in the precision and neutrality of Ido was a product of a distinctly late nineteenth century conception of logic as a tool to construct a universal system of communication. His defence of symbolic logic placed him in the company of Charles Sanders Peirce, Alfred North Whitehead, Gottlob Frege, Bertrand Russell and others. They all operated in a branch of mathematics that made use of symbols to express logical ideas in order to create a sign system that communicated concepts exactly and clearly. Frege's *Bergriffschrift* (1879) was an early product of this field of logic that was inspired by Leibniz' logic. The book proposed a *Formel Sprache* that was meant to free logic and arithmetic from the inconsistencies and deceptions of ordinary language and human intuition.

Historians of science, Lorraine Daston and Peter Galison, frame such attempts as Frege's *Formel Sprache* as expressions of a distinct turn-of-the-century notion of scientific objectivity that they call "structural objectivity." Logicians, philosophers, mathematicians as well as physicists defended structural objectivity as overcoming the deficiencies of an older conception of objectivity that Daston and Galison call "mechanical objectivity." The arrival of the photograph at the beginning of the nineteenth century made possible a new kind of scientific image that recorded specimens exactly and without human intervention. This development went hand in hand with the emergence of the ideal of mechanical objectivity: the scientist was supposed to be a self-effacing, machine-like recorder of exact and specific detail. But by the end of the nineteenth century mechanical objectivity was increasingly questioned as it became clear that the individual traits of the scientific observer could not be eliminated from his or her observations. Structural objectivity emerged out of logic, mathematics, physics, and philosophy. Rather than relying on detailed representations of nature through images and language, scientists now sought objectivity in structures of logical relationships, law-like sequences and mathematical logic that would make "science communicable among all subjects, everywhere and always" (Daston and Galison 2007, 255).

Crucially, Couturat's defence of logic as a means to by-pass the subjectivity of the human mind and language was a typical product of this trend. He stressed that in its pure form, logical reasoning was universal since men's capacity to reason logically was timeless and of all cultures. People were not only capable of understanding one another because they formed similar ideas; they also related and constructed ideas in a similar manner. He was convinced that logical reasoning preceded sociology, psychology, and, of course, ordinary languages, which Couturat viewed as "imperfect and crude instruments" (Couturat 1905, 24).

Couturat reasoned that logic was a superior tool in the hands of a trained scientist by virtue of its pristine universalism. This made the idea of using it to communicate ideas precisely and efficiently an obvious move. During his systematic study of Leibniz' logic, Couturat found an example of how this could be done. Leibniz' plan consisted in matching language with reason in a calculus similar the logical calculus of mathematics. In *Histoire de la Langue*

Universelle Couturat argued that in contrast to other language projects, Leibniz's language offered nothing less than a "real character" what Leibniz had called a "characteristica universalis" which displayed an exact correspondence between ideas and their symbols (Couturat 1903, 60). He described it as a "complete and systematic expression of intellectualistic rationalism" (1903, 11).

Leibniz' logic-based calculus ratiocinator was especially appealing to Couturat. It governed the combination and manipulation of its ideographical symbols. It was made up of a set of rules that were meant to allow incontrovertible conclusions to be drawn mechanically from premises. Couturat believed that a logical calculus conceptualized in this manner could provide for the human mind an external and independent inference engine that was driven by a set of logical rules. The principles of univocity and irreversibility that underpinned Ido's precision and neutrality were applications of this notion of mechanical logic. They led to Couturat's claim that Ido was a scientific improvement on Esperanto by making it more consistent and efficient than Esperanto. More importantly for Couturat, these principles allowed him to promote Ido as free of national interests because it was 'independently' generated by the mechanics of logic.

Beyond Dreyfus and National Chauvinism

Couturat's relentless propaganda for Ido was not only a product of specific turn-of-the-century objectivity politics in science. It was also driven by a distinct notion of its relevance in a society that was being torn apart by the famous Dreyfus affair which held France in its grip in the 1890s and early 1900s. The scandal divided France between those who supported Dreyfus, famously referred to as the *Dreyfusards*, and those who condemned him, the conservative *anti-Dreyfusards*. Albert Dreyfus, a French artillery officer of Alsatian Jewish descent was convicted of having communicated a memorandum containing French military secrets to the German Embassy in Paris and was sentenced to life imprisonment in 1892. In 1896 it became apparent that a French army major was responsible, but the affair dragged on when Dreyfus was further accused in 1899.

Many of Couturat's famous mathematics colleagues were engaged *Dreyfusards*, including his former professor Henri Poincaré, as well as Paul Appell, Gaston Darboux, Jules Tannery, and Paul Painlevé. In 1899, Poincaré intervened in the legal proceedings at the request of Painlevé by writing a letter that was read to the court. His letter was an attempt to expose the pseudo-scientific nature of the probability theory that had been used in the graphological analysis of the memorandum that pointed at Dreyfus's guilt. He repeated his criticism in 1904 in a report that was jointly written with Appell and Darboux. This report played an important role in the eventual release and rehabilitation of Dreyfus as captain in the French army. Unlike Poincaré, Couturat was not directly involved in the affair. But his political convictions were nonetheless shaped by the event, propelling him to draw on a "mechanical" logic in the creation of Ido, a language that could help ensure that the evils of military chauvinism displayed during the affair were relegated to the past.

Couturat described the Dreyfus Affair as a "sad and shameful page in the history of the French Republic" in a letter to his friend and colleague Xavier Léon. He argued that the affair was viewed without sympathy by the world at large and added that it was only by drawing "logical" and "moral" conclusions that the affair would lead to the betterment of the French Republic.[4] A polemic between him and the politically conservative French literary critic, Ferdinand Brunetière, on Kant's essay, *Perpetual Peace* (1795) offers some answers as to what Couturat meant by the betterment of the French Republic. In reply to Brunetière's suggestion that Kant viewed war as the ultimate vehicle of societal progress, Couturat argued that nothing could be further from the truth. He pointed out that Kant argued that war was only a temporary phase in society's development towards "the state of justice and international law." Couturat further argued that this Kantian development towards a state of international peace meant that "permanent weapons must disappear entirely with time." He concluded that Kant should not be confused with "apologists of force (disciples of Bismarck and Moltke), or with the backward admirers of a military and theocratic society characteristic of the Middle Ages" (Couturat 1899). His references to Bismarck and Moltke as apologists of force and his equation of a military society with the Middle Ages reveal Couturat's contempt for militarism, which he associated with the *anti-Dreyfusards*.

More importantly, Couturat clearly positioned himself as a "true" pacifist in his defense of Kant's belief in the progress of civilization towards a peaceful world state. He believed that the betterment of the French Republic meant resisting any form of nationalism and militarism. In a letter to Bertrand Russell, Couturat singled out national chauvinism as a threat to world peace. He described it as an "insatiable thirst for conquest and domination. This jingoism, cafe-concert patriotism ..." was "clearly a danger for world peace"[5] (Couturat 2001, 152). He added that intellectuals must fight chauvinism instead of falling for the blind and brutal passions of the masses. Couturat's propagation of Ido as an internationally neutral and rational means to facilitate communication was an attempt to put into practice his own call for political engagement. Ido could contribute to advancing the principles of justice and neutrality; it provided, "the indispensable condition for any international agreement ..." (Couturat 1905, 6).

Ido as an International Community Identifier

While Couturat's linguistic internationalism was both an aspect of a distinct form of objectivity politics in science in the period of the Belle époque, and an expression

---

4   Louis Couturat to Xavier Léon. September 27, 1899. From the correspondence by Xavier Léon. Bibliothèque Victor Cousin, housed at the Bibliothèque de la Sorbonne in Paris.

5   Louis Couturat to Betrand Russell, December 24, 1899 (Couturat 2001, 152).

of a political program that was strongly influenced by the French Dreyfus affair, he continued to emphasize Ido's virtues and uses at an international level. As I have suggested, Couturat stressed that any trace of personal and national interest was eliminated from Ido by its logical grammatical structure that was a product of the scientific principles of univocity, reversibility and maximum internationality. With these qualities, it represented the exact opposite of national languages, which Couturat labeled as politically imprecise, inefficient, and associated with national chauvinism: "If anyone is the victim of national prejudice it is he who proposes his own language as the international medium, and not the one who discards every national tongue, including his own, in favor of the international language" (Couturat 1905, 6).

This contrast in Couturat's rhetoric between the virtues of an international language and the shortcomings of national languages, opens the door for me to offer a closing analysis of his linguistic internationalism as a strategy of community building. Historians and sociologists of science have done little research to investigate scientists' values as markers of cultural distinctiveness. But for historians focusing on nations and nationalism, such an approach has become conventional. Eric Hobsbawm's _Nations and Nationalism Since 1780_ (1990) is an exemplary study in this tradition. In it Hobsbawm approaches nations as uniquely modern social entities by stressing the element of artifact and social engineering which is involved in their making. As such the existence of a national identity has been constructed by "political campaigning" involving the dissemination of shared values that foster a national consciousness among the social groupings and regions of a country (Hobsbawm 1990, 12). This process of institutionalizing a shared history and a common national language, creates the notion of shared culture. Hobsbawm traces this process of national community building through the adoption of a shared language in action. He points out that during the period between 1880 and 1914 language became the principal identifier of national communities—both in the hands of nationalist agitators and in the work of government statisticians.

Couturat's attempts to install Ido are remarkable, because they represented a form of resistance to the process Hobsbawm has described. Couturat had attempted to break the association of language with nationalism by promoting Ido as a rational and neutral product of science that was meant for communication in an international community. But in doing so he copied from nationalist movements the use of a linguistic means for defining and promoting community. Instead of fostering a shared _national_ identity, Couturat's promotion of Ido was intended to foster a shared _international_ identity that was an expression of typically Belle Époque notions of science and politics. Our mother tongue is a strong reminder of our membership of a national community as well as encapsulating its historical heritage. Similarly, Ido was intended to be a reminder of science's internationalism, an _aide-memoire_ of science's illustrious past conjured up in romantic images of the exemplary cosmopolitanism of the seventeenth century Republic of Letters. And it commemorated earlier attempts to construct a universal language by learned men such as Leibniz, Descartes, and Bacon. But Ido as a signpost of science's alleged political neutrality also reveals a narrowness in Belle époque conceptions of the meaning of international community. International meant Europe and the US. Couturat claimed to have constructed a language that was "a quintessence of European languages," but he distilled a linguistic essence of Europe explicitly excluding Slavic languages, to say nothing of other language groups. Internationalism as a political movement may well be presented in positive terms. But a closer examination of its history reveals that internationalism is no less an expression of the values and interests of a distinct local community than any other political movement.

References

Anderson, Perry. 2002. "Internationalism: A Breviary." *New Left Review*, 14: 5–25.
Canto-Sperber, Monique. 2008. *Moral Disquiet and the Human Life*. Princeton: Princeton University Press.
Couturat, Louis. 1899. "Correspondance." *Le Temps*, March 27.
Couturat, Louis. 1903. *Histoire de la Langue Universelle*. Paris: Hachette.
Couturat, Louis. 1905. "Lecture inaugural, Collège de France." In *L'Oeuvre de Louis Couturat: 1868–1914: de Leibniz à Russell*. Communications présentées au colloque international consacré à l'œuvre de Louis Couturat, Paris, juin 1977. Paris: Presses de l'Ecole Normale Supérieure
Couturat, L. 1905. "An International Auxiliary Language." *Monist*, 15: 142–6.
Couturat, Louis. 1906. *Pour la Langue Internationale*. Coulommiers: Imprimerie Paul Brodard.
Couturat, Louis. 1909. *Le Choix d'une Langue Internationale*. Paris: Éditions de la Revue du Mois.
Couturat, Louis. 1909. "Rapport de la Logique et de la Linguistique." *Revue de Métaphysique et de Morale*, 17: 509–16.
Couturat, Louis. 1910. "On the Application of Logic to an International Language." In *International Language and Science: Considerations on the Introduction of an International Language into Science*, edited by L. Couturat, L.O. Jespersen, R. Lorenz, W. Ostwald, and L. Pfaundler. London: Constable and Company Limited, 42–52.
Couturat, Louis. 2001. *Correspondence sur la Philosophie, la Logique et la Politique avec Louis Couturat (1897–1913)*, edited by Anne-Françoise Schmid. Paris: Kimé, 151–4.
Darboux, Gaston. 1901. "L'Association Internationale des Académies," *Le Journal des Savants*. Année, 1901: 5–23.
Daston, Lorraine and Galison, Peter. 2007. *Objectivity*. Brooklyn, NY: Zone Books.
Eco, Umberto. 1995. *The Search for the Perfect Language*. Oxford: Blackwell Publishers.

Glover Forster, Peter. 1982. *The Esperanto Movement*. The Hague: Mouton Publishers.

Hobsbawm, Eric. 1991. *Nations and Nationalism Since 1870: Programme, Myth, Reality*. Cambridge: Canto.

Lalande, André. 1914. "L'Oeuvre de Louis Couturat," *Revue de Métaphysique et de Morale*. 22: 644–87.

Ostwald, Wilhelm. 1913. "Scientific management for scientists. 'The Bridge'. The trust idea applied to intellectual production," *Scientific American* 108: 5–6.

Rasmussen, Anne. 1996. "A la recherche d'une langue internationale de la science 1880–1914." In *Science et Langues en Europe*, edited by Roger Chartier and Pietro Corsi. Paris: École des hautes études en sciences sociales, 139–55.

Reinbothe, Roswitha. 2006. *Deutsch als internationale Wissenschaftssprache und der Boykott nach dem Ersten Weltkrieg*. Frankfurt am Main: Peter Lang.

Russell, Bertrand. 1998. *Bertrand Russell Autobiography*. London/New York: Routledge.

Semper, Gottfried. 1966[1852]. "Wissenschaft, Industrie und Kunst." In *Wissenschaft, Industrie und Kunst, und andere Schriften über Architektur, Kunsthandwerk und Kunstunterricht*, edited by Hans M. Wingler. Mainz/Berlin: Kupferberg.

Speeckaert, Georges Patrick. 1957. *The 1,978 organisations founded since the congress of Vienna: Chronological List*. Brussels: Union of International Organisations.

Waquet, Françoise. 2001. *Latin, Or, the Empire of a Sign: From the Sixteenth to Twentieth Centuries*. Translated by John Howe. London/New York: Verso.

1889. "International Languages," *Science* (editorial) 13: 24.

Chapter 8

Laboratories of Social Thought: The Transnational Advocacy Network of the Institut International pour la Diffusion des Expériences Sociales and its *Documents du Progrès* (1907–1916)

Christophe Verbruggen and Julie Carlier

Introduction: Looking for Order in the Chaos of the "nébuleuse réformatrice"

Many literary historians have been writing and thinking about the characterization of Settembrini, a figure in *Die Zauberberg* from Thomas Mann (White and White 1980; Loose 1968). Settembrini embodies the *belle époque* revival of the ideals of the Enlightenment such as humanism, women's rights, and human rights. Settembrini is also a member of the *Bund zur Organisierung des Fortschritts*. Whether a coincidence or not, an organization with a similar name really existed in Berlin, Munich and Vienna in the years prior to the First World War. The *Institut International pour la Diffusion des Experiences Sociales* (*IIDES* 1909) and its annex, the *Ligue internationale pour l'organisation du progrès* (1912), were founded by the Austrian sociologist Rudolf Broda and published journals in Paris (*Les Documents du Progrès*), Berlin (*Dokumente des Fortsschritts*) and London (*The International: A Review of the World's Progress* which was merged with *Progress* in 1909). Later editions followed in Saint Petersburg, Budapest and Madrid. The *Institut International pour la Diffusion des Experiences Sociales* and its journals constitute a transnational network that is part of what Topalov calls the "galaxy of social reform movements" that covered Europe during the period of the *belle époque* (Topalov 1999). Internationalism, pacifism, feminism and many other "isms" were an integral part of this "nébuleuse réformatrice," that connected progressive liberals, reformist socialists and Christian-democrats, all of whom shared the ambition to contribute to "social progress." Their motives were an ambiguous mixture of social optimism and underlying cultural pessimism in which optimism became an oppressive "moral duty." From the birth of socialism, the notion of social progress was, and has remained, what reconciles progressive liberals and socialists, although the concept itself has naturally been subject to changing interpretations. Alongside social-Darwinist conceptions of evolution and progress, socio-cultural evolution theories, such as those of Auguste Comte and Herbert Spencer, were *en vogue* at the end of the 19th and the beginning of the 20th centuries (Hawkins 1997; Spadafora 2002; Angenot 2003). In his classic study the *Idea of Progress*, historian John B. Bury offered as the principle definition of progress that "... civilisation has moved, is moving, and will move in a desirable direction" (Bury 1920, 2). The notions "international" and "progress" are the most important and continuous element in the scheme of intellectual cooperation we will discuss in this chapter. We study this case from the perspective of the different components of a transnational network: organizations, institutes and above all periodicals.

As a relational approach, transnational history not only deals with "the people who forge connections" (Clavin 2010, 28–9) but also with the ever-present tension between different scales and spaces, such as between the local and the transnational. This tension is also present in Keck and Sikkink's concept of transnational advocacy networks, defined as networks that "include those actors working internationally on an issue, who are bound together by shared values, a common discourse and dense exchanges of information and services" (Keck and Sikkink 2002, 89). The issues involved are understood to be universal such as human rights, women's rights, environmental problems, educational reform and social reform in general. From a long-term perspective these issues become recurring and sometimes interrelated objectives highlighted in the "history of transnational issue networks" (Saunier 2012). Interrelated objectives, movement dynamics and mobilization structures can be related to the so-called "framing" of meanings and issues in different settings (Benford and Snow 2000), for instance the framing of temperance or pacifism as related or sub-causes within the movement for women's rights.

Rudolf Broda (ca. 1880–1932)

We consider the *Institut International pour la Diffusion des Experiences Sociales* and its reviews as an excellent case for the study of what can be described as an enlarged and locally rooted transnational advocacy network, interconnecting scientific expertise and social activism. It constituted a transnational proliferation of persons and organizations that carried interest in a given social or intellectual movement into other, including local, social and cultural settings (Rodogno, Struck and Vogel 2012). The journals published by the Institute featured articles on topics ranging from social reform, sociology, feminism and pacifism to religion and literature. It became an institution that offered patronage to a worldwide network of member organizations, thereby interconnecting a wide range of social movements or "causes." One of its most interesting features as a case study is the extent to which the protagonists involved experimented with innovative schemes of international intellectual cooperation. Studying carriers of flows and interaction, the network of agents of transnational advocacy represented by the Institute, will help us to understand what Pierre-Yves Saunier describes as the "ecosystem" of "world of causes" in the nineteenth century (Saunier 2012, 31).

Rudolf Broda and his transnational network of social reformist reviews and associations have not been systematically studied in their entirety (Angenot 2003; Boussabha-Bravard 2008; Saenen 2008).[1] Broda (ca. 1880–1932) was an Austrian and Jewish liberal, sociologist, pacifist and feminist with socialist affinities. He was a professor at the *Collège Libre des Sciences Sociales* (Free College of Social Sciences) in Paris from 1907 until 1914. During the war he worked at the Minerva Institute in Zürich. In the early 1920s he was a collaborator of the International Labor Organization before moving to the USA where he worked subsequently at Harvard University and at Antioch University in Yellow Springs.[2]

We will start by charting the foundation, evolution and main characteristics of the network created by Broda through the *Institut International pour la Diffusion des Experiences Sociales* by means of a graphic approach, paying special attention to the network's direct and indirect predecessors. In a network approach it is important to differentiate between "visible" ties, consisting of formal exchanges between organizations, and "latent" ties, consisting of the informal connections created by the personal relations of network members (Diani 2011). Notwithstanding the importance of informal and loose network ties in recruitment processes and in the circulation of knowledge and reports of experiences, it was equally important for members to participate in formal (public) structures of sociability in order to establish long-term and relatively stable patterns of interactions. Subsequently, moving from a general overview we will zoom in on the movement or cause that was central to Broda's network: feminism. Through the case study of Belgian feminism and its entanglement with the network of the Institute and its reviews, we will analyze the exchange of ideas and practices between the two. This allows us to move beyond a mere map of connections and explore the question of the local impact of the transnational network. This will also allow us to discuss the importance of informal correspondence networks that coincided with the formal networks represented in printed sources.

The Foundations of the Network

An approved way of mapping the evolution of networks over time is through the multiple memberships of activist cohorts (Lemercier and Bertrand 2011; Diani 2011). Multiple memberships have already been used as an indicator for

---

1   We would like to thank both Myriam Boussahba-Bravard and Bregt Saenen for sharing their work with us and for their kind permission to quote from it.

2   "Broda, Rudolf." *Österreichisches Biographisches Lexikon 1815–1950*, Bd1, 115. Graz: H. Böhlaus Nachf, 1957.

```
1906   1907   1908   1909   1910   1911   1912   1913   1914   1915   1916   1917   1918   1919   1920
```

Die Menschheit (Wiesbaden, 1914-1930)

La voix de l'humanité (Lausanne, 1914-1922)

Die Versöhnung
(1917-1919)

Dokumente des Fortschritts. Internationale Revue (Berlin, 1907-1918)

Documents du progrès. Revue internationale (Paris, 1907-1918)

Institut International pour la diffusion des expériences sociales -
Bund für Organisierung menschlichen Fortschritts

Ligue pour la défense de l'Humanité et pour l'Organisation
de son progrès - Bund für Menschheitsinteressen und
Organisierung menschlichen Fortschritts

The International: a Review
of the World's Progress
(Londen, 1907-1909)

British Institute of Social Service

Progress. Civic, social, industrial (London, 1906-?)

Homaro. Sendependa brosuraro (Madrid, 1911-1921)

Zaprosy žizni
(St-Petersburg, 1909-1912)

Szociálpolitikai szemle (Budapest, 1912-1918)*

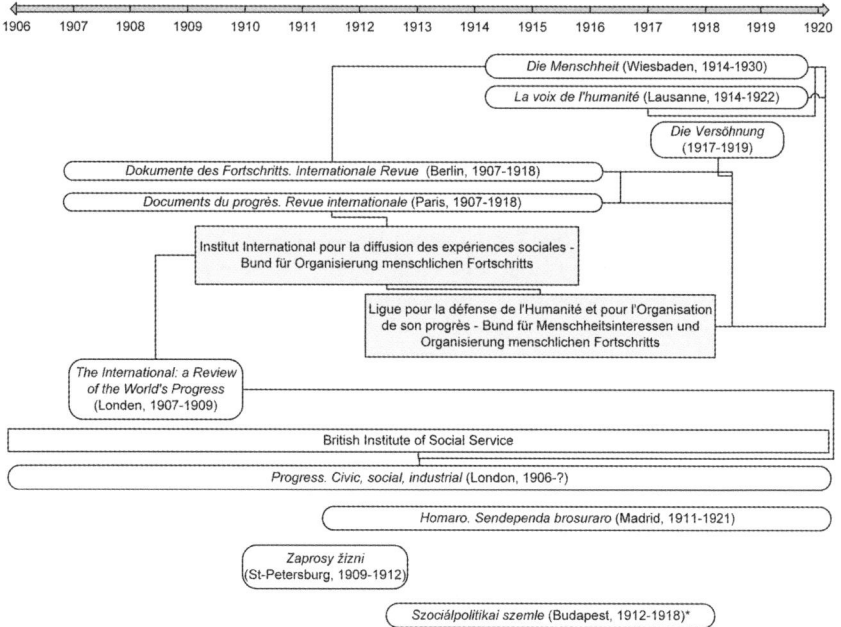

Figure 8.1   Journals related formally to the IIDES and patron institutions

organizational exchanges, for instance by Rosenthal et al. (1985), who managed
to create a genealogy of causes in 19th century New York State by focusing on
the multiple memberships of women active in social reform movements. Network
visualization improves the communication of relational data and allows for the
exploration of network properties (Brandes et al. 1999). In this chapter we will
limit ourselves to a graphic exploration of a "snapshot" in time: the network of
the *Institut International pour la Diffusion des Experiences Sociales* at its formal
creation in 1909–1910.[3] As only one membership list of this institution has been
preserved, it is impossible to adopt a detailed quantitative longitudinal perspective
in which both institutional members and private persons are included. However,
based on the editorials of the reviews and published annual reports it is possible
to create a basic longitudinal scheme of journals related to the IIDES and patron
organizations (figure 8.1).

The foundations of the network out of which the *Institut International pour
la Diffusion des Experiences Sociales* grew, were laid in 1907 when Rudolf
Broda began to publish the periodicals *Les Documents du Progrès*, *Dokumente
des Fortsschritts* and *The International: A Review of the World's Progress* as

---

3   This case study is part of TIC-Collaborative, an international collaborative project
on transnational social reform networks in the long nineteenth century.

"virtual" laboratories of social thought. Their purpose was to facilitate exchange of social experiences between the nations. He believed that not only knowledge about remote civilizations ("civilisations lointaines") could provide answers for the current ideological crisis in the west, but that the European nations could also learn from each other. Simply reporting "the facts" was not enough. Social facts had to be submitted to a "sociological analysis" in order to obtain an informed understanding of the social and intellectual evolution of humankind (Broda 1907, 3–4). As a lecturer at the *College Libre des Sciences sociales*, Broda was well situated to introduce these journals. The *College Libre* had been created by leading progressive intellectuals—authors, journalists, academics—in 1895 after the outbreak of the Dreyfus affair. It was the first French institution created specifically for higher education in social sciences. Broda also lectured at the *Ecole des Hautes Etudes Sociales* (School for Advanced Social Studies).

When Broda founded the journals, he needed firm backing from local editors. In Berlin, Hermann Beck provided such support. Beck was a member of the "bibliographic movement" in Germany and was very close to both Wilhelm Ostwald and Ferdinand Tönnies (Hapke 2005). Broda also received the support of the *Monistenbund*, founded in 1906 at the instigation of Ernst Haeckel. Naturalistic monism—a scientifically grounded pantheism—provided an important epistemological framework for numerous cultural and social movements (Weir 2012). Reconstructing the establishment of the English edition of his journal and the identification of Broda's liaison officers in London are more difficult. Broda began the English edition without any one person being in charge as editor who would also be responsible for finding new authors and subscribers. For the latter task he could fall back on the support of leading English intellectuals such as Sydney Webb, P. Snowden, W.T. Stead, Edward Pease and several other Fabians who contributed to the journal. However, none of them accepted editorial responsibilities. It was the feminist, Cicely Dean Corbett, who would become the main British agent for Broda's projects.

Myriam Boussahba-Bravard has concluded that between 1907 and 1909 the enthusiasm for Broda's internationalist project slowly diminished. She noticed a growing divergence in the content of the editions of the journals. Not every contribution was translated into French or German and vice versa. This suggested the local importance of certain topics and the intellectual prestige of particular local authors. In December 1909, Broda announced that the German and English editions had to be seen as fully-grown, separate journals. It had become clear that articles dealing with universal problems but written for a specific national audience were not necessarily of interest to readers in other countries (Broda 1909a). However, the exchange of foreign experiences remained the *raison d'être* of *Documents du progrès* and its affiliated journals and a selection of articles continued to be translated into other languages at the initiative of local editorial committees. While the national journals offered differing content in their international news or in a "review of reviews" section—a popular section in many early twentieth-century literary and scientific journals—maintenance of an international perspective and a commitment to international intellectual cooperation remained the key characteristic that bound the journals together.

This changing publication strategy offered new opportunities. When Broda founded the *Institut International pour la Diffusion des Expériences Sociales* (International Institute for the Dissemination of Social Experiences) based in Paris, his hope was that branches would also be created in other countries and that other organizations would become affiliated with it. Besides the support of the existing journals, the mission of the Institute was the organization of educational tours throughout Europe: lectures, mass meetings and public debates, "especially addresses by leading authorities from abroad, who are, from experiences gained in their own country, particularly qualified to shed light on how their country has dealt successfully with a problem which has since become acute in the country in which the lecture is held" (Broda 1909b, 7; Broda 1909c). Broda also contacted existing journals, offering them the status of a journal "officially" supported— and probably partly financed with money Broda had inherited—by the Institute. As mentioned above, the English edition, *The International: A Review of the World's Progress,* merged with the quarterly, *Progress: civic, social, industrial, educational*, the official organ of British Institute of Social service. This involved a revealing enlargement of the journal's network. The British Institute of Social service had been founded in 1905 at the instigation of the American Institute of Social Service, established in 1898. Like its foreign counterparts in Denmark, Sweden and Italy, this American institution has hardly been studied. Like the *Institut de Sociologie Solvay* in Brussels, the *Institut für Gemeinwohl* in Frankfurt and the *Musée Social* in Paris, it aimed to collect "all manner of facts bearing on human progress, so that the experience of all countries and individuals may be available for the guidance of each" (Peabody 2001, 35). The Institute also offered professional and expert counseling on social reform to philanthropic organizations.

The founding fathers of the American Institute were Josiah Strong and Willam Tolman. Strong was a protestant minister who was also the father of the Social Gospel movement that would heavily influence the "Progressive Era" and evolutionary thought in the United States (Curtis 2001). Tolman is less well known than the term he introduced into the social, political and economical sciences: "social engineering" (Östlund 2007). Strong and Tolman explained their "social mission" in an article in the *New York Times* in 1906 (Cranston 1906). They explicitly mentioned that their objective was to create analogous and interconnected organizations in Europe. In Italy, Denmark, Sweden and last but not least in Great Britain, a local branch was founded "as the result of correspondence or personal conferences" by both Strong and Tolman and in Sweden by Edward Wavrinsky. Both Strong and Tolman maintained intensive contacts with their European correspondents whom they described as "distinguished social economists, men of affairs, statesmen and publicists." They were selected to "promote its European work" and they included, for instance, the Belgian social engineer and future ILO representative Louis Varlez who contributed to the *Institut International pour la Diffusion des Experiences Sociales* journals. It is striking that there is a significant overlap between their list of correspondents and Broda's network. Moreover, Broda's Paris-based Institute seems to have been modeled after the American example. The respective mission statements are interchangeable. Whether or not they were the result of a transnational transfer and transformation, these American and European institutions were certainly entangled.

The *Institut International pour la Diffusion des Experiences Sociales* and its reviews embodied Broda's social reformist aspirations for international intellectual cooperation and the dissemination of social scientific expertise as the engine for social progress and peace. While its aims were similar to those of the American Institute as mentioned above, they were also similar to those of the *Institut International de Bibliographie* of Paul Otlet and Henri La Fontaine, with whom Broda developed friendly relations.[4] Broda's project aimed at the creation of a transnational social reformist coalition, interconnecting proponents of temperance, pacifism, feminism, educational and social reform by means of a joint "culture intellectuelle, la culture sociologique et politique" (Broda 1910a, 218; Broda 1911a). In 1912 Broda and his colleagues created a *Ligue internationale pour l'organisation du progrès* in parallel to the *Institut International pour la Diffusion des Experiences Sociales*. From then on, the *Institut* would focus on scientific research of socio-political issues, while the dissemination, lobbying and networking function was conferred on the *Ligue*, which took over the administration of the reviews and the organization of lecture tours and conferences (Saenen 2008, 31–3).

Characteristics of the Network: A Graphic Exploration

The creation of the *Institut International pour la Diffusion des Expériences Sociales* thus marked the maturing of Broda's ambitions. In the following we will present a graphic exploration of the network of the organization and its periodicals. In 1910, the Institute had 789 individual and society members from 30 countries and 181 cities. Most members came from Germany, Austria and France (see figure 8.2), the home countries of the pivotal figure, Rudolf Broda.

Within the network of the Institute we can distinguish several types of actors: journals, organizations and persons. Among the private members of the network were members of the organizations and subscribers to the journals, "active readers" who contributed to surveys and wrote letters, and finally correspondents and occasional authors (approximately 250, see: Saenen 2008). Core members, frequent contributors, and speakers such as Hermann Beck and Cicely Dean Corbett were elected members of the "*comité de patronage*." Within the category

---

4   See: correspondence of Broda to La Fontaine, 1910–1929, in: Mundaneum, Papers of Henri La Fontaine (HLF), 53–54.

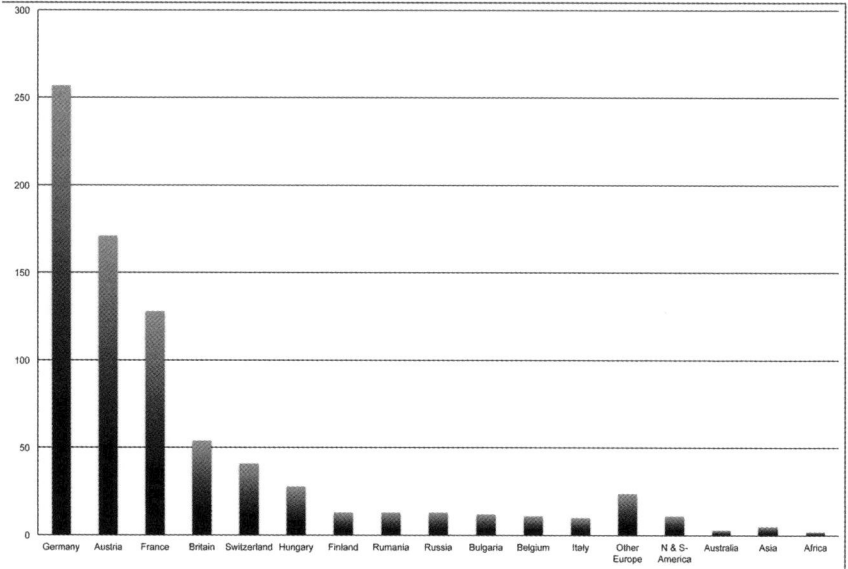

Figure 8.2   Membership by country of the IIDES (1910)

of journals and organizations, a distinction should be made between unofficial and officially recognized journals (those belonging to this category were subject to change). Between 52 and 75 cities had an official section or local branch. However it is important to note that there were no national branches. Only local branches were created. Nevertheless Broda's understanding of "internationalism" did not imply an indifference towards the nation. On the contrary, international solidarity had to start with national solidarity (Broda 1910d). Since most social reform initiatives were organized locally prior to the First World War, the dissemination of social experiences was expected to happen first and foremost at the local level. The possible—i.e. expected—activities of local branches were summarized in a report for 1913 (Broda 1914). The first of these activities was the organization of meetings with foreign "conférenciers" sent out by the Institute in Paris. A second activity for local branches was discussing issues explored in *Documents du Progrès* at evening debates. A third was an "innovation," the opening of reading rooms, and creating 'interlibrary' loans and information exchange between branches (mediated by the Paris' centre). Local authorities and politicians had to be made aware of the existence of the local branches in order to learn about foreign examples and experiences. By issuing brochures and press releases and holding public meetings, every branch was expected to influence public opinion. Very active groups were encouraged to undertake special initiatives themselves (e.g. founding popular universities, setting up agencies for

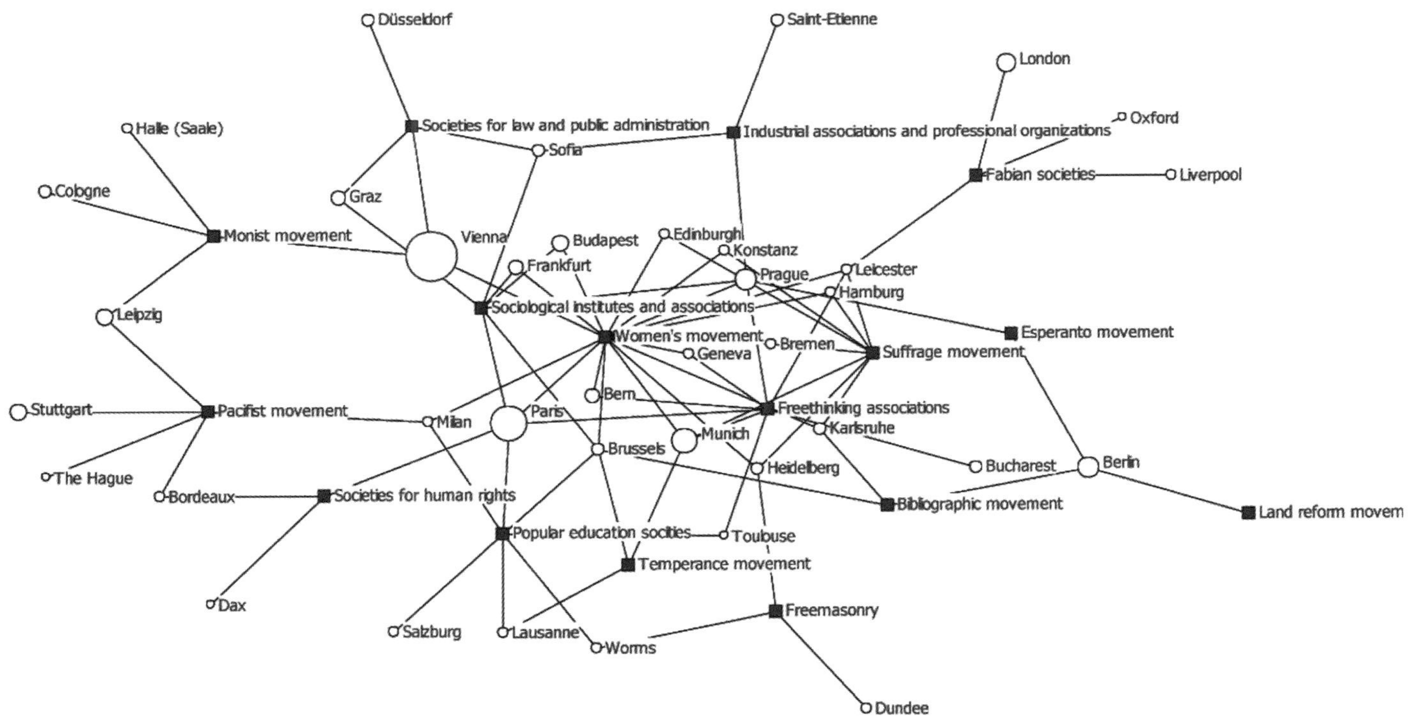

Figure 8.3 Society members of IIDES (categorized by city and cause/movement: node size = number of private and society members)

unemployment insurance, etc) and to report on their experiences in one of the journals published by the *Institut International pour la Diffusion des Experiences Sociales*. Finally a distinction was made between branches in provincial towns and groups in capital cities. The latter were expected to lobby national politicians for legislative reform.

Figure 8.3 shows the urban roots of the Institute at the end of 1910. Only cities with society members are included. Local society members were categorized according to their self-identification as actors. According to Broda, "in England the local Institute's groups were connected principally with the feminist movement (for which foreign experiences are very valuable) and the Fabian Society." For Germany, Broda mainly mentioned local branches of the temperance movement, the women's movement and freethinking societies (Broda 1911b, i–iv). Cities were thus transnationally connected by these "causes." The size of the nodes in figure 8.3 reflects the number of members (both private and society members). Most members were based in Paris and Vienna, whereas there was only one member in the Netherlands (the internationalist P.H. Eijkman—see Somsen, Chapter 12 in this book). The graphic representation of the network clearly shows the centrality of the women's and suffrage movement next to freethinking and sociological associations. Budapest is an ideal-type in this regard with the Society of Social Sciences (1901–1919) in Budapest and Feministak Egyesülete (Association of Feminists) as principle components of the network. Founded in 1904, the latter was the first feminist organization in Hungary and was affiliated to the International Woman Suffrage Alliance (Acsady 1999, 54–9 and Szapor 2004). In terms of Figure 8.3 the Esperanto movement is clearly of peripheral interest to the institute, except later in Spain where the Esperantist and freemason, Julio Mangada Rosenörn, co-founded *Homaro*, one of the official organs of the organization.

In line with the classic distinction between strong ties and strong identifications with a project, we can distinguish different ties for persons, organizations and society members. Stronger ties not only facilitate information exchange but also collective action. The identification of strong and weak ties and their respective functions can be illustrated by the journals within the network. Between 1910 and the outbreak of the First World War, there were official organs for the local branches in Madrid and St-Petersburg. They clearly had strong ties and facilitated the execution of the shared mission embodied in the *Institut International pour la Diffusion des Experiences Sociales*. However, not every journal that was connected by a strong tie to *Documents du Progrès* figures on the list of official organs. Among these "unofficial journals" were: the Belgian based *Minerva* (a journal that can be regarded as marking the beginning of the International Bureau of Education), *La Société Nouvelle*, *Isis* and several publications and journals devoted to "social economy" (Geerkens 2008; Topalov 1999); the Hungarian based *Huszadik Szazad* (Twentienth Century, directed by Oszkár Jászi); the *Revue international de sociologie* (Paris, directed by René Worms); *Revista Idealiste* (Sofia, directed by M. G. Holban); *Les Annales de la régie directe* (Genève, directed by Edgard Milhaud);

and the Indian *Modern Review* (Calcutta, directed by Ramanda Chatterjee). All of these journals exchanged advertisements and articles, constituting what can be described as a sphere of ideas within the galaxy or nebula of social reform.

The best example of an unofficial journal that belonged to the same sphere of ideas and dense information network as the official journals of the *Institut International pour la Diffusion des Experiences Sociales* is probably the Italian international journal, *Coenobium. Rivista internazionale di liberi studi* (Panzera 2007). The founder and editor-in-chief of *Coenobium* was Enrico Bignami, an influential Italian socialist politician and journalist. *Coenobium* published literary contributions and studies on religion, spirituality, theosophy, social reform and pacifism. The international section was written in French. While *Coenobium* never became the official organ of an Italian branch of the Institute, Bignami became one of the most active members of its committee of patronage or steering committee. In his turn, Broda became responsible for a section in Coenobium on the exchange of information about social experiences.

An author Broda and Bignami shared was the excommunicated modernist French priest, Marcel Hébert, then living in Brussels. He was a member of the steering committee of the Institute and one of the most productive contributors to its reviews. Hébert was not the only modernist priest within the *Coenobium-Documents du Progrès* "nebula." Paul Naudet was a well known Christian democrat priest from Bordeaux and one of the architects of the encyclical, *Rerum Novarum* on "The Rights and Duties of Capital and Labor" issued by Pope Leo XIII in 1891. Naudet was a colleague of Broda at the *Collège Libre des Sciences Sociales* in Paris and was one of the most productive authors in both *Coenobium* and *Documents Du Progrès*. He had begun his contributions with the inception of both journals in 1907. The two journals not only represented analogous and intertwining social and intellectual movements, they were also central actors within other clusters of transnational and even transatlantic intellectual sociability. *Coenobium* for instance, extensively exchanged articles with the then English-American *Hibbert Journal*, issued since 1902 by the Hibbert Trust and mainly devoted to religion, theology and philosophy.

The Exchange of Ideas in Practice: The Case of Belgian Feminism

As we have said, the actors within the transnational advocacy network of the *Institut international pour la diffusion des expériences sociales* and its reviews were bound together by shared ideas and ideals. As evident from the graphic representation of the institutional members of the Institute, the women's movement occupied a central place in this network. Many important feminists and women activists, such as the German left-wing feminist, Adele Schreiber, the Austrian pacifist and Nobel Peace Prize winner, Bertha Von Suttner, the Swedish feminist and progressive educationalist, Ellen Key, and the Italian suffragist and peace activist Alma Dolens (pseudonym of Teresita Pasini dei Bonifatti, Cooper 1985), were part of the clusters of intellectual sociability represented by *Documents du Progrès*, *Dokumente des Fortschritts* and *Progress*. The steering committee of the Institute included the British suffragist Cicely Dean Corbett. She was the sister of Margery Corbett-Ashby, a future president of the International Woman Suffrage Alliance (Crawford 2001). The Steering Committee also included the German radical feminist, Lida Gustava Heymann, and the French socialist feminist of Russian origin, Lydie de Pissargevsky, who was a colleague of Rudolf Broda at the Parisian *Collège Libre des Science Sociales*.[5] These memberships suggest the need for a critical assessment of the place of women, women's activism and feminism in this specific nebula of social reform, not only institutionally but also ideologically. It is clear that in Broda's project feminism was considered to be a pivotal part of social progress. We will study the exchange of ideas in practice, i.e. the ideological and institutional effects of advocacy within the network, through the case study of Belgian feminism and its entanglement with the scientific network of the *Institut International pour la Diffusion des Experiences Sociales* and its journals.

Feminism constituted an integral part of the Institute and of Broda's social theory. In Broda's positivist view, sociology—"the science of society"—integrating all other sciences, was the discipline par excellence that would provide the scientific route towards social progress (Broda, 1907; Broda 1908). He defined this in marked social-Darwinist terms: "In the sections of our Institute, within these associations of men and women who are determined to apply themselves consciously to the work of progress, we can awaken the ideal of the *organic development of the human species*, the ideal of human progress viewed as a *biological necessity*, as the most important development for allowing us to understand the place we occupy in nature" (Broda 1911a, 282, italics in the original).

Broda's reasoning, however, was not reductionist or biologically deterministic. He assigned a morally and socially progressive direction to evolution and presented the sociological principles of universal solidarity, democracy, and peace as the fulfillment of the biological destiny of humankind (Saenen 2008). This evolutionary progress towards social justice included women's emancipation. "From the point of view of evolutionary theory, the feminist movement, with its claims for political and social equality of the sexes and for a better education for women, also deserves our strong support. It awakens in the feminine world all the talents that have been dormant until now, for want of a proper education …" (Broda 1910c, 255).

In Broda's social reformist project, women's emancipation was not only a way of liberating the potential contribution of one half of humanity to the evolution of civilization, it also became the catalyst for all other social reforms. According to Broda women were naturally inclined towards social action and social reform because of their humanitarian sentiment. Women were the natural allies of pacifists.

---

5   See letter of Rodolphe Broda to Lalla Vandervelde, June 8, 1912. Archives of the Université Libre de Bruxelles (ULB), Fonds Université Nouvelle (UN), 1Z210.

Their so-called tender, less materialistic nature led them to detest the atrocities of war (Saenen 2008). This maternalist conception of women's caring nature was translated in particular into a relentless advocacy of women's suffrage as the key to all social and moral reform by Broda himself and by the *Institut International pour la Diffusion des Experiences Sociales* in general:

> With women's suffrage we will see all the great sentiments of the feminine spirit enter the political arena as powerful factors: sympathy for the weak, love of children, aversion to alcohol and gambling, aspiration to a harmonious life within the family and the State. Especially the pacifist idea will find strong support among women, the natural enemies of the devastations of war. … If we consider that in the everyday life of the family, it is always the woman who, as the representative of the ideas of thrift and providence, appears as the true guardian of the permanent interests of the family and the species, it comes as no surprise that she will also strive for these ideas to prevail in political life. In a way, women's suffrage is a conservative measure in the sense that it will help safeguard the social order … ; but it is also a progressive measure in the sense that it paves the way for numerous humanitarian reforms. As such it combines, in a superb synthesis, the principle of evolution and the principle of conservation. There is hardly any reform that can be considered to be as universally beneficial as women's suffrage (Broda 1910b, 7).

The Belgian case reflects this entanglement of social sciences and social reform, on the one hand, and feminism and suffragism, on the other hand. As well as individuals such as Henri La Fontaine and Paul Otlet, the Belgian members of *Institut International pour la Diffusion des Experiences Sociales* included the leading feminist organization, the *Ligue belge du droit des femmes* (Belgian League for Women's Rights 1892) and the leading women's temperance association, the *Alliance des femmes contre l'abus de l'alcool* (the Alliance of Women against the Abuse of Alcohol 1904). Another key Belgian member was Lalla Vandervelde, a socialist feminist, the wife of Henri Vandervelde and a staff member of the *Institut des Hautes Etudes* (Institute of Higher Education), the evening school of the *Université Nouvelle* (the New University) in Brussels. The *Université Nouvelle* was the Belgian counterpart of the Parisian *Collège Libre des Sciences Sociales*. Both had been established at roughly the same time— the *Université Nouvelle* in 1894 and the *Collège Libre des Sciences Sociales* in 1895—and for similar reasons. They were essentially forums for instruction on innovative social sciences by adherents of positivism who found established universities closed to their new theories. At the turn of century—before Broda entered the stage—the *Institut des Hautes Etudes* of the *Université Nouvelle* had organized several courses on the woman question that were delivered by leading Belgian and foreign feminists who were often connected to the *Ligue belge du droit des femmes* and the Parisian *Collège Libre*. Out of these entanglements of feminism and progressive social sciences emerged an embryonic version of what we would nowadays call "women's studies."

Key lecturers were Jeanne Oddo-Deflou and Marguerite Souley-Darqué. Oddo-Deflou was the leader of the radical *Groupe français d'études féministes* (French Group of Feminist Studies) and closely connected to the *Ligue belge du droit des femmes*. Oddo-Deflou and her group had "created a scholarly field devoted to the study of women—in fact, what we know now as women's studies." They promoted the concept of matriarchy as a radical feminist alternative to patriarchy (Allen 2005, 27). In 1903 the *Groupe français d'études féministes* had published the first French translation of the introduction to *Das Mutterrecht* (1861, "Mother Right" or "Matriachy") of the conservative Swiss legal scholar, Joseph Jakob Bachofen. Oddo-Deflou appropriated and subverted Bachofen's anti-feminist analysis by defining patriarchy not as the reign of enlightenment but as an era of oppression, war, and hatred. As such it was opposed to the original form of matriarchy which expressed the ideal of a nurturing, motherly state (Allen 1999, Allen 2005).

Marguérite Souley-Darqué was a French woman of letters and amateur anthropologist (Despy-Meyer and Goffin 1976). She had developed a science known as "feminology" and taught a course on it at the *Collège Libre des Sciences Sociales* (Boxer, 1982). Feminology was "that branch of sociology which studies women, their place in society, now and in the past."[6] As a pioneer of women's studies, Souley-Darqué presented feminology as the interdisciplinary philosophical, sociological, historical, and feminist analysis of the physical and moral evolution of woman.

The *Institut International pour la Diffusion des Experiences Sociales* organised lecture tours in Belgium in collaboration with both the *Université Nouvelle* and the *Ligue belge du droit des femmes*. Women's suffrage featured prominently in these series of lectures. In 1910, Cicely Dean Corbett gave an evening class at the *Université Nouvelle* on the British women's suffrage movement (Despy-Meyer and Goffin 1976), while Broda himself delivered a conference on "The results of women's suffrage in Australia and Finland" at the *Ligue belge du droit des femmes*.[7] He argued that votes for women were an instrument of progress and moral improvement (Parent 1910). In April 1913, the British suffragist and collaborator of the International Woman Suffrage Alliance, Mary Sheephanks, gave a lecture at the *Ligue belge du droit des femmes* in the name of the Institute on "The Influence of English Woman on municipal politics" (Sheepshanks 1913). And in November of that same year, Broda himself returned to the *Ligue belge du droit des femmes* with a conference on "Alcoholism conquered by women's suffrage."[8]

Conferences such as those organized by the *Institut International pour la Diffusion des Experiences Sociales* resulted in durable transnational networks of

---

6   1905. "Féminologie." *La Femme Socialiste*, December 24.

7   See invitation by the *Ligue belge du droit des femmes*, May 12, 1910, in: Mundaneum, HLF, 111.

8   L., L. 1913. "Conférence de Broda." *La Ligue* 21: 233–5.

female intellectuals and feminist activists. The series of lectures on feminist issues held in the winter of 1906–1907 at the *Université Nouvelle* is a case in point. Among the speakers we find not only Jeanne Oddo-Deflou, but also a Dutch board member of the International Woman Suffrage Alliance, Martina Kramers. Her lecture sparked the creation of the first Belgian woman suffrage association, affiliated to the International Woman Suffrage Alliance (Carlier 2010a). In May 1912, Kramers gave a second conference on women's suffrage at the *Université Nouvelle* and there the idea of forming a larger suffragist coalition, led by the *Ligue belge du droit des femmes* and joined by catholic feminists, took concrete form. By February 1913 the *Fédération belge pour le suffrage des femmes* (Belgian Federation for Women's Suffrage) was officially constituted (Carlier 2010b). Thus the birth of organized suffragism in Belgium was the result of political transfer induced by the International Woman Suffrage Alliance and mediated by the progressive scientific milieu of Brussels. The transnational advocacy network of the *Institut International pour la Diffusion des Experiences Sociales* and the *Université Nouvelle* formed the crucial intellectual setting where Belgian feminists met transnational activists such as Cecily Dean Corbett, Mary Sheepshanks and Martina Kramers and thus formed the necessary context for the creation of a Belgian women's suffrage movement.

Conclusion

In an interesting article on "l'invention de l'humanité," Marc Angenot briefly mentioned *Documents du Progrès* as a herald of modern humanitarianism where humanitarianism was defined as "having a secular faith and mandate to contribute its modest part to 'social progress.'" According to Angenot, this review was an excellent example of "a truly innovative progressism yet situated at the heart of social legitimacy" (Angenot 2003, 35). The venture was indeed one of a philanthropic imagination carried out by a transnationally operating but locally rooted intellectual elite. In keeping with contemporary moderate social democratic tendencies, the institute and its reviews embodied Broda's social reformist vision of international intellectual cooperation and dissemination of social scientific expertise as the motor for social progress and peace. The journal, *Documents du Progrès*, its pendants and supporting organization constituted a transnational advocacy network bound together by shared values, a common discourse and dense exchanges of information and services. Moreover, corresponding to its internationalist and pacifist foundations, the exchange of information and services was the actual *raison d'être* of the network. The integration of the network during and after the First World War into other "advocacy" networks deserves further research, but it is hardly surprising that Broda and his inner circle belonged to the early supporters of the creation of the League of Nations, the International Labor Organization and the Committee on Intellectual Cooperation when a new era of international intellectual cooperation that was far more institutionalized and mediated by international organizations was born.

The interaction between the transnational level and the diverse local settings is another aspect that merits further examination. The different receptions in Austria and Germany (very positive), for example, and the Netherlands (rather indifferent) might be explained by an accidental lack of individual agency and recruitment but also by different national, regional and local concerns. In this chapter we have concentrated on Belgian feminism in order to study the impact in practice of the transnational exchange of ideas within the network of the *Institut International pour la Diffusion des Experiences Sociales*. Within this "nébuleuse réformatrice," as we have seen, the women's movement and feminism occupied a central place institutionally, ideologically, and in the framing of other issues. Prominent women activists and feminist associations from all over Europe were among the key members of the Institute and among the most active contributors to its journals. In Broda's social-Darwinist project of social progress, justice and peace, women's emancipation—particularly women's suffrage—was the catalyst for all other social reforms. The transnational transfer of organized suffragism in Belgium was mediated by the progressive intellectual milieu of Brussels which was part of the transnational advocacy network of the *Institut International pour la Diffusion des Experiences Sociales*. The Belgian case demonstrates the local impact of this entanglement of social sciences, social reform and women's rights. But above all it is a fine example of the functioning of the "world of causes" at the turn of the 19th century and the early 20th century.

References

Acsady, Judith. 1999. "Remarks on the History of Hungarian Feminism." *Hungarian Studies Review*, 27: 59–64.

Allen, Ann Taylor. 1999. "Feminism, Social Science and the Meaning of Modernity: The Debate on the Origin of the Family in Europe and the United States, 1860–1914." *American Historical Review*, 104: 1085–113.

Allen, Ann Taylor. 2005. *Feminism and Motherhood in Western Europe, 1890–1970: The Maternal Dilemma*. Basingstoke: Palgrave Macmillan.

Angenot, Marc. 2003. "L'invention de l'Humanité et le sujet du Progrès." In *Le soi et l'autre: l'énonciation de l'identité dans les contextes interculturels*, edited by Pierre Ouellet. Québec: Presses de l'Université Laval, 363–80.

Benford, Robert and Snow, David. 2000. "Framing Processes and Social Movements: An Overview and Assessment." *Annual Review of Sociology*, 26: 611–39.

Boussabha-Bravard, Myriam. 2008. "The International, A Review of the World's Progress, an international venture, 1907–1909." Unpublished conference paper.

Boxer, Marilyn. 1982. "Women's Studies in France circa 1902: A Course on Feminology." *Women's Studies Quarterly International Supplement*, 1: 24–7.

Brandes, Ulrik, et al. 1999. "Explorations into the Visualization of Policy Networks." *Journal of Theoretical Politics*, 11: 75–106.

Broda, Rudolf. 1907. "Ce que nous voulons." *Documents du Progrès*, December: 3–4.

Broda, Rudolf. 1908. "L'avenir des sciences sociales." *Documents du Progrès*, February: 150–56.

Broda, Rudolf. 1909a. "A nos lecteurs." *Documents du Progrès*, December: 435–9.

Broda, Rudolf. 1909b. "The International. To our readers." *The International*, December: 7–9.

Broda, Rudolf. 1909c. "Ce que les peuples peuvent apprendre les uns des autres." *Documents du Progrès*, December: 440–45.

Broda, Rudolf. 1910a. "La Fondation de l'Institut international pour la diffusion des expériences sociales." *Documents du Progrès*, March: 216–22.

Broda, Rudolf. 1910b. "Le suffrage des femmes." *Documents du Progrès*, July: 3–10.

Broda, Rudolf. 1910c. "La théorie de l'évolution et ses applications à la conception du monde, à la morale et à la vie." *Documents du Progrès*, October: 255.

Broda, Rudolf, 1910d. "Le problème de l'internationalisme." *Documents du Progrès*, December: 467–71.

Broda, Rudolf. 1911a. "L'échange des expériences sociales." *Documents du Progrès*, March: 281–4.

Broda, Rudolf. 1911b. "Institut international pour la Diffusion des Expériences sociales. Rapport annuel." *Documents du Progrès*, December (annex): I–V.

Broda, Rudolf. 1914. "Institut international pour la Diffusion des Expériences sociales." *Documents du Progrès*, April: I–II.

Bury, John Bagnell. 1920. *The Idea of Progress. An Inquiry into its Origin and Growth*. London: MacMillan. http://archive.org/stream/ ideaofprogressin00buryuoft#page/n7/mode/2up (accessed December 15, 2012).

Carlier, Julie. 2010a. "Forgotten Transnational Connections and National Contexts. An Entangled History of the Political Transfers that Shaped Belgian Feminism, 1890–1914." *Women's History Review*, 19: 503–22.

Carlier, Julie. 2010b. "Moving Beyond Boundaries. An Entangled History of Feminism in Belgium, 1890–1914." Unpublished PhD dissertation. Ghent University.

Cranston, M.R. 1906. "What is the American Institute of Social Service; Practical Work and Scope of the Organization Conducted by Dr. Josiah Strong and Dr. W.H. Tolman Explained." *New York Times Magazine*, May 20.

Crawford, Elisabeth. 2001. *The Women's Suffrage Movement: A Reference Guide, 1866–1928*. London: Routledge.

Clavin, Patricia. 2010. "Time, Manner, Place: Writing Modern European History in Global, Transnational and International Contexts." *European History Quarterly*, 40: 624–40.

Cooper, Sandi E. 1985, "Dolens Alma." In *Biographical Dictionary of Modern Peace Leaders*, edited by H. Josephson. London: Greenwood Press, 220–21.

Curtis, Susan. 2001. *A Consuming Faith: The Social Gospel and Modern American Culture*. Columbia: University of Missouri Press.

Diani, Mario. 2011. "Social Movements and Collective Action." In *The SAGE Handbook of Social Network Analysis*, edited by J. Scott and P. Carrington. London: Sage, 223–35.

Despy-Meyer, Andreé and Goffin Pierre. 1976. *Liber memorialis de l'Institut des Hautes Etudes de Belgique*. Bruxelles: Institut des Hautes Etudes en Belgique-Université Libre de Bruxelles Service des Archives.

Freeman, Linton C. 2000. "Visualizing Social Networks." *Journal of Social Structure*: 1. http://www.cmu.edu/joss/content/articles/volume1/Freeman.html (accessed July 20, 2012).

Freeman, Linton C. 2005. "Graphic Techniques for Exploring Social Network Data." In *Models and Methods in Social Network Analysis*, edited by Peter J. Carrington, John Scott, and Stanley Wasserman. Cambridge: Cambridge University Press, 248–69.

Frickel, Scott and Gross Neil. 2005. "A General Theory of Scientific/Intellectual Movements." *American Sociological Review*, 70: 204–32.

Fuchs, Eckhardt. 2007. "Networks and the History of Education." *Paedagogica Historica*, 43: 185–97.

Geerkens, Eric. 2008. "From the Annales de la régie directe to Annals of Public and Cooperative Economics: 100 Years of Transformations in an International Economic Journal." *Annals of Public and Cooperative Economics*, 79: 417–60.

Haas, Peter M. 1992. "Epistemic Communities and International-policy Coordination." *International Organization*, 461: 1–35.

Hapke, Thomas. 2005. "Ostwald and the Bibliographic Movement." In *Wilhelm Ostwald at the Crossroads between Chemistry, Philosophy and Media Culture*, edited by Britta Görs, Nikos Psarros, and Paul Ziche. Leipzig: Leipziger Univ.–Verl, 115–34.

Hawkins, Mike. 1997. *Social Darwinism in European and American Thought, 1860–1945: Nature as Model and Nature as Threat*. Cambridge: Cambridge University Press.

Horne, Janet R. 2001. *A Social Laboratory for Modern France: The Musée Social and the Rise of the Welfare State*. Durham: Duke University Press.

Klejman, Laurence and Florence Rochefort. 1989. *L'égalité en marche. Le féminisme sous la Troisième République*. Paris: Presses de la Fondation Nationale des Sciences Politiques.

Laqua, Daniel. 2013. *The Age of Internationalism and the Belgian Hub: Peace, Progress and Prestige*. Manchester: Manchester University Press.

Lemercier, Claire and Bertrand Michel. 2011. "Introduction: où en est l'analyse de réseaux en histoire?" *Redes: revista hispana para el análisis de redes sociales*, 21: 12–23.

Loose, Gerhard. 1968. "Ludovico Settembrini und 'Soziologie der Leiden. Notes on Thomas Mann's Zauberberg." *Modern Language Notes*, 83: 422–33.

Östlund, David. 2007. "A Knower and Friend of Human Beings, Not Machines: The Business Career of the Terminology of Social Engineering 1894–1910." *Ideas in History*, 2: 43–82.

Panzera, Fabrizio and Saresella Daniela (eds) 2007. *Spiritualità e utopia: la rivista "Coenobium" (1906–1919)*, a cura di, introduzione di D. Saresella. Milano: Cisalpino.

Parent, Marie. 1910. "Le Mouvement Féministe. La Vérité sur les suffragettes." *La Ligue*, 18: 90–94.

Peabody, Francis. 2001. *The Social Museum as an Instrument of University Teaching*. Cambridge Mass.: Harvard University Press.

Rodogno, Davide, Struck Bernard, and Vogel Jacob (eds) 2012. *Shaping the Transnational Sphere. Experts, Networks, Issues (c. 1850–1930)*. New York: Berghahn Books.

Rasmussen, Anne. 1995. "L'internationale scientifique (1890–1914)." PhD dissertation, Ecole des Hautes Etudes en Sciences Sociales (EHESS), Paris.

Rosenthal, Naomi et al. 1985. "Social Movements and Network Analysis." *American Journal of Sociology*, 90: 1022–54.

Saenen, Bregt. 2008. "'Pour la diffusion des expériences socials': een onderzoek naar *Documents du Progrès* binnen de transnationale ruimte aan het begin van de twintigste eeuw." Masters dissertation, Ghent University.

Saunier, Pierre-Yves. 2012. "La secrétaire générale, l'ambassadeur et le docteur. Un conte en trois épisodes pour les historiens du 'monde des causes' à l'époque contemporaine, 1800–2000." *Monde(s). Histoire, Espaces, Relations*, 1: 29–46.

Sheepshanks, Mary. 1913. "Le Mouvement féministe en Angleterre et l'action de la femme anglaise sur la politique communale." *La Ligue* 21: 66–78, 201–12.

Spadafora, David. 2002. "Progress." In *Encyclopedia of the Enlightenment*. Oxford: Oxford University Press, 367–72.

Stone, Diane. 2002. "Introduction: Global Knowledge and Advocacy Networks." *Global Networks*, 2: 1–12.

Szapor, Judith. 2004. "Sisters or Foes: The Shifting Front Lines of the Hungarian Women's Movements, 1896–1918." In *Women's Emancipation Movements in the 19th Century. A European Perspective*, edited by Sylvia Paletschek and Bianca Pietrow-Ennker. Stanford: Stanford University Press, 189–205.

Tarrow, Sidney and della Porta Donatella. 2005. "Conclusion. Globalisation, Complex Internationalism and Transnational Contention." In *Transnational Protest and Global Activism*, edited by Sidney Tarrow and Donatella della Porta. Oxford: Oxford University Press, 227–46.

Tarrow, Sidney. 2005. *The New Transnational Activism*. Cambridge: Cambridge University Press.

Topalov, Christian (ed.) 1999. *Laboratoires du nouveau siècle: la nébuleuse réformatrice et ses réseaux en France, 1880–1914*. Paris: Ecole des Hautes Etudes en Sciences Sociales (EHESS).

Van Daele, Jasmien. 2005. "Engineering Social Peace: Networks, Ideas, and the Founding of the International Labour Organization." *International Review of Social History*, 50: 435–66.

Weir, Todd H. 2012. "The Riddles of Monism: An Introductory Essay." In *Monism. Science, Philosophy, Religion, and the History of a Worldview*, edited by Todd H. Weir. Basingstoke: Palgrave Macmillan, 1–44.

White, I.A. and White, J.J. 1980. "The Importance of F.C. Müller-Lyer's Ideas for Der Zauberberg." *Modern Language Review*, 75: 333–48.

Wils, Kaat. 2005. *De omweg van de wetenschap: het positivisme en de Belgische en Nederlandse intellectuele cultuur, 1845–1914*. Amsterdam: Amsterdam University Press.

Chapter 9

Sociology in Brussels, Organicism and the Idea of a World Society in the Period before the First World War

Wouter Van Acker

Brussels in the Belle Époque counted several dynamic organisations in which innovative research was undertaken in the field of the social sciences: the *Société d'études sociales et politiques*, the *Institut des Sciences Sociales* later redeveloped as the *Institut de Sociologie* (*Solvay*), and the *Institut des Hautes Etudes*. Among the members of these organisations were such important Belgian intellectuals as Guillaume De Greef (1842–1924), Hector Denis (1842–1913), Edmond Picard (1836–1924), and a younger generation that included Paul Hymans (1865–1941) and Emile Vandervelde (1866–1938). The Belgian socialist senator and future Nobel Peace Prize laureate, Henri La Fontaine (1854–1943), and his younger partner, the bibliographer and founding father of the notion of "documentation" Paul Otlet (1868–1944) were members of this second generation and matured intellectually within this milieu. La Fontaine and Otlet are important for their work in improving transnational cooperation in the classification, dissemination and exchange of information. In 1895 they founded the International Institute of Bibliography (IIB). Over the period of nearly fifty years of their association with it, it established a transnational network of associations, societies, institutes, and commissions that supported the development of a central bibliographic and documentary repertory in Brussels—a sort of "World Memory" as they called it (Institut International de Bibliographie 1908, 5). It propagated new methods of bibliography and documentation based on the Universal Decimal Classification. In 1907, extending the transnational collaboration that they had established through the International Institute of Bibliography, Otlet and La Fontaine founded the Central Office of International Associations, with the collaboration of Cyrille Van Overbergh (1866–1959). Van Overberg was a Christian Democrat attached to different ministries and was appointed Director-General of Higher Learning, Science and the Arts in 1900 (de Seyn 1936, 1091). The Central Office of International Associations was an umbrella organization for international organizations which did much to affirm the importance of international associations at a moment when internationalism was gaining momentum. A study of the involvement of leading internationalists such as Otlet and La Fontaine in the Brussels sociological milieu is useful in analyzing the theoretical and practical exchanges between sociology and in Brussels in the period of the Belle Époque. I argue that Otlet and La Fontaine's reflections about international organization were deeply influenced by sociological theories, in particular the theory of organicism, as these theories were being developed at the sociological societies mentioned above and with which the two men were connected. Important however were their transnational collaborative efforts in scientific communication which helped to shape the constitution of Brussels sociology as an international field. Transnational practices that led to the formation of networks and the creation of bibliographical services were considered essential for shaping the discipline of sociology. Sociological theory on the other hand helped in the creation of internationalism as a distinct scholarly field of action and reflection.

Société d'études sociales et politiques and La Fontaine and Otlet[1]

The ideas that circulated within the Brussels-based *Société d'études sociales et politiques* had an important influence on the thinking of La Fontaine and probably even more on Otlet who was fourteen years younger than La Fontaine. The *Société d'études sociales et politiques* was formally created in 1890 by the cosmopolitan, liberal politician, publicist, freemason and former Vice-President of the Belgian Chamber of Representatives Auguste Couvreur (1824–1894) as a reconstitution of the short-lived *Association internationale pour le progrès des sciences sociales* (AIPSS, 1862–1867), in which he had played a leading part. As Paul Errera (1860–1922)—lawyer, rector of the Université Libre de Bruxelles (ULB), mayor of Uccle (Bruxelles), and member of the *Société d'études sociales et politiques*— described it:

> Under the title *Société d'études sociales et politiques*, our association pursues the same goals by the same means as formerly [in the time of the AIPSS]: a scholarly goal, impartial studies and surveys and [the creation of] a centre of information. Our ambition is to assemble, as in 1862, men of thought and of action, without regard to nationality or political party. (Errera 1894, 101)

Although the majority of the leading members of the *Société d'études sociales et politiques* were liberals and freemasons, the Society, like the AIPSS, emphasized its political neutrality. Divergent social and political views were represented in the Society and this it was claimed allowed social questions to be examined from different angles "in a conscientious and profound manner" (Couvreur 1889, 2).

Like the *Association internationale pour le progrès des sciences sociales*, the *Société d'études sociales et politiques* aimed to be not only a place of debate but also a centre of information. Alphonse Rivier (1835–1898), professor of law at

---

1   In this section I follow and extend the account that W. Boyd Rayward gave in an unpublished talk (Rayward 2008). The original French text of this and subsequent quotes can be found in my doctoral dissertation (Van Acker 2011, 264–70).

the ULB and President of the *Institut de droit international*, was responsible for developing a library and a committee of the bibliographic section was responsible for developing a bibliography in the field of social and political sciences. This bibliographic section was presided over by the liberal lawyer, freemason, and president of the *Ligue de l'Enseignement* (1872–1878), Gustave Jottrand (1830–1906). Henri La Fontaine was its Vice-President. Among its members was the lawyer, socialist, and freemason, Emile Vandervelde, future President of the Second International (1900–1918), then only twenty-four years old.

The *Société d'études sociales et politiques* issued a journal, the *Revue Sociale et Politique* (1891–1895), which contained "leading articles," reports on the discussions held at the society, a section of general information about various events of national and international interest, a bibliographical review, and lists of additions to the library. Among those who contributed important articles to the journal were leading members of the society such as Couvreur, Vandervelde and a number of local and foreign social and political figures. Among the former were English Fabian and co-founder of the London School of Economics, Sidney Webb (1859–1947) and among the latter were Paul Heger (1846–1925), professor of physiology at the Free University of Brussels of which he was later Rector; the socialist sociologist, Hector Denis; and the liberal politician and future first president of the League of Nations, Paul Hymans.

Paul Otlet became involved in the *Société d'études sociales et politiques* through Henri La Fontaine, whom he assisted in the work for its bibliographic section. The men probably became acquainted through Edmond Picard who was one of the most famous Belgian lawyers and patrons of the arts of his day. La Fontaine had at one time been his secretary and Otlet upon completion of his law degree became one of his "stagiaires" or articled clerks. Both of them had worked on the *Pandectes belges*; an enormous compendium of Belgian jurisprudence edited by Picard and published in 117 volumes between 1878 and 1924 (Rayward 2010, 2). Their shared interest in bibliography, presumably established in working on this project of Picard, intensified in the environment of the *Société d'études sociales et politiques.* Here bibliography, the means by which documented social facts were recorded and communicated, was regarded as of the highest importance. Otlet attended a meeting of the Society's section on comparative law in February 1892 (Rayward 2008). In March 1892, La Fontaine set up a meeting with Otlet at the office of the Society.[2] In June, La Fontaine sent him the regulations of the bibliographic section. The bibliographic cards of the *Société d'études sociales et politiques* had pre-printed indications of what should be listed on the cards: "name of the author," "title of the article," and "analysis." In this last section members of the Society or external collaborators were asked to summarize the factual information in the article or book under review in the form of a short abstract. The information obtained in this way was completed, supplemented, and finally

---

2 Henri La Fontaine, *Letter to Paul Otlet* (11 March, 1892). Mundaneum (Mons), archives of Henri La Fontaine, Correspondance Otlet-La Fontaine, 1892.

published as full notes or book reviews in the *Revue Sociale et Politique*. This editing process made the bibliographical section immensely important and it is here that La Fontaine, Otlet, and others made major and continuous contributions to the "informativeness" of the journal.

In 1893 Otlet and La Fontaine took over the bibliographic service of the *Société d'études sociales et politiques* into what they had called the *Office International de Bibliographie Sociologique*. This had "the exclusive goal of collecting and classifying materials of all sorts related to the social sciences: legislation, statistics, social and political economy"(Otlet and Fontaine 1895). The Office was to be used as the base for coordinating the compilation and publication of two periodical bibliographies which they had been editing separately. Otlet had been a member of an editorial group producing a legal bibliography, the *Sommaires des traités et revues de droit*. La Fontaine had been compiling a sociological bibliography. They now titled the combination of the two journals, *Sommaires méthodiques des Traités et Revues de Sociologie et de Droit* (or Bibliographia Sociologica). This publication project was announced in a letter circulated by Couvreur in 1894 to members of the Society as the second and complementary publication of the *Revue Sociale et Politique*.[3] The same letter announced that from March 1894, the main office of the Society, including its bibliographic service, would be moved to the Hôtel Ravenstein. The importance of this move lay in the fact that in December 1893 the Hôtel Ravenstein had been given the new function of acting as a centre in which the offices of Belgian learned societies would be concentrated (Brion and Buyle 1987, 41). By grouping the learned societies, which at that time had no legislative basis of formal incorporation, in one place they would be able to carry more weight on a juridical level and of course would be able to develop and use the same infrastructure of services, including a central library and bibliographic office (Dubois 2002, 27–8) In his letter, Couvreur invited the society's members to visit the Hotel Ravenstein to use the library, the reading room with its collection of daily newspapers, and "the index cards of the bibliographic service, now transformed into the International Office of Bibliography" (OIB).

August Couvreur died on the 23 April, 1894. He had been the originator and driving force behind the *Société d'études sociales et politiques*. His death created great uncertainty about its future existence. In late 1894 some of its prominent members tried to ensure its survival by urging a continuation of the *Revue Sociale et Politique*, support for the library, and collaboration with the Office of Sociological Bibliography.[4] Nevertheless, at the Society's general assembly in January 1895 it

---

3   August Couvreur, Letter to the members of the *Société d'études sociales et politiques* (2 March, 1894); Mundaneum (Mons), archives of Henri La Fontaine, HLF 193 *Société d'études sociales et politiques*.

4   Letter to the members of the *Société d'études sociales et politiques*, signed by the "comité directeur" Gustave Montefiore Levi, Paul Errera, and Ad. De Vergnies (10 November, 1894); Mundaneum (Mons), archives of Henri La Fontaine, HLF 193, *Société d'études sociales et politiques*.

was decided to discontinue the connection with the OIB. The meeting "accepted the proposition of MM. La Fontaine and Otlet that they take over responsibility for the *Revue* and the bibliographic service. As a consequence, the Society will have from now on no other goal than the study of questions of a political and social order that are submitted to it."[5] In effect, the *Société d'études sociales et politiques* ceased to exist from this point, especially as its role as a focus for discussion about the social sciences was being assumed by the *Institut des Sciences Sociales* created in 1894 by Belgian industrialist Ernest Solvay (1838–1922).

La Fontaine and Otlet insisted that the progam of the *Revue sociale et politique* whose management had been confided to them was the same as it had always been, namely, "to observe facts, to let them speak, to position itself above passions and prejudices, not dictating laws, but to search for their bases, to study their necessity and legitimacy" (Otlet and Fontaine 1895). The *Revue* would give "facts, nothing but facts"—facts that were dispersed over its bibliographies, facts on "the most pertinent questions" of the time, facts from the latest publications on political, economic and social questions. To gather these facts, the OIB hoped to continue to work with an international team of sociologists with whom the Society had been collaborating or corresponding.[6] However, after the publication of four issues of the *Revue* in 1895 publication was suspended. Otlet and La Fontaine organized an international conference on bibliography for September 1895. With the assistance of La Fontaine's sister, Léonie, and using the Dewey Decimal Classification that Otlet had recently discovered, the three of them had classified 400,000 card entries in the bibliography of the Office of Sociological Bibliography. This was to be offered to the conference as a demonstration of the powers of bibliographical organization of the Decimal Classification (Rayward 1975, ch. 3). Without the organization of the *Société d'études sociales et politiques* behind them, they had none of the resources for the editorial and publishing activities needed to keep the journal going in addition to the new venture that they were undertaking.

Institut des Sciences Sociales

The *Revue Sociale et Politique* would have had a longer life if the executive committee of the *Société d'études sociales et politiques* had accepted the offer in 1894 of another sociological institute, of which Otlet and La Fontaine were members, to adopt it. The *Institut des Sciences Sociales* had been founded in 1894 by Ernest Solvay. Immensely wealthy from the manufacture of the

5   Letter to the members of the *Société d'études sociales et politiques*, signed by Auguste Beernaert, Pierre Tempels, Mauric Vauthier, Léon Dupriez and Paul Hymans (16 April, 1895); Mundaneum (Mons), archives of Henri La Fontaine, HLF 193, *Société d'études sociales et politiques*.
6   This list of collaborators is mentioned on the back cover of the *Revue*: Office International de Bibliographie sociologique, in: *Revue Sociale et Politique*, 5.3–4 (1895).

industrial chemical sodium carbonate by what is now known as the Solvay ammonia-soda process, Solvay became a major philanthropist supporting intellectual and scientific activities in Brussels (Wirtz-Cordier 1994; Despy-Myer and Devriese 1997). He had already begun to support medical research in the form of the creation of an Institute for Physiology in 1889 that was directed by his physician and friend, Paul Heger, Otlet's uncle. A building to house the Physiological Institute was begun in the Parc Léopold in 1892 and completed two years later. The *Institut des Sciences Sociales* was intended to be a sociological laboratory that followed the model of the Insitute of Physiology. If it proved to be desirable, the Institute could be linked to the *École des Sciences Politiques et Sociales*. Begun as a half-hearted approach to the study of sociology at the Université Libre in 1889 in the form of a series of free courses for graduates, when the initiative seemed to be about to fail for lack of interest, in 1897 Solvay undertook to finance in the form of the Ecole des Sciences Politiques et Sociales (Crombois 1995).

Solvay had called upon three members of the *Société d'études sociales et politiques* to lead the ISS as directors of the new Institute. They undertook to explore the Institute Solvay's own emerging sociological and economic theories. They were the sociologists Guillaume De Greef, Hector Denis, and Emile Vandervelde—the latter two were professors at the Free University of Brussels (Maheim 1892). Soon after the foundation of the *Institut des Sciences Sociales*, an approach was made to the executive committee of the *Société d'études sociales et politiques*, as mentioned above, to use the *Revue sociale et politique* as a vehicle for its own work. A meeting in early 1894 about this matter involved de Greef and Vandervelde on one side and Otlet and La Fontaine on the other. The minutes report that the executive committee had refused to give the appropriate authority to allow a successful negotiation to proceed and the idea was dropped.[7]

Because of his strong "productivist" views, Ernest Solvay had good reason not to try to make the *Société d'études sociales et politiques* the vehicle for the work of what became the *Institut des Sciences Sociales*. He had a special agenda. As in the case of *Institut de Physiology*, the *Institut des Sciences Sociales* was to study "human productivism, the enhancer of life of man and its energetic return" (Wirtz-Cordier 1988). Inspired by research in physiology and chemistry, Solvay considered each human being to be a "machine" which struggles to increase its output in order to augment its well-being. In the *Institut des Sciences Sociales*, Solvay aimed to explore this social-political ideology which he termed productivism. A society based on a productivist system would maximize the "social return" of each human group, give back the profits of work to the workers themselves and, as such, limit the capitalistic accumulation of profits (Crombois 1997, 212–13). In the first article of the *Annales de l'Institut*, Solvay outlined the productivist programme which he hoped to explore with the *Institut des Sciences Sociales*. He observed that "we

---

7   Institut des Sciences Sociales. Procès-Verbaux. Séance du 15 March 1894; MDN, HLF 121 Institut des Sciences Sociales/office International Bibliog. PV. des reunions 1894.

must—and this will be the work of the future—make the fundamental inequality of modern society disappear: *inequality from square one*" (Solvay 1894, 2). Only by "utilising scientific methods for the solution of these problems, by increasingly identifying the laws of social transformism in the frightening complexity of phenomena," Solvay claimed, "we can move towards equality without scarifying freedom, and realize progressively what will be the goal of all projects of reform: *to obtain, for everyone's well-being, the maximum return from human energy*" (Solvay 1894, 1).

The three directors of the *Institut des Sciences Sociales*, Denis, De Greef, and Vandervelde, all three of them socialists, responded in their own way to Solvay's liberal program of theoretical social accounting and productivism. Hector Denis addressed the international monetary question and issues such as Proudhon's idea of a Bank of the People (Denis 1894, 1895); De Greef wrote essays on "money, credit and the banks" and "social tranformism" (De Greef 1894, 1898); Vandervelde focused on Marx's "Capital" and on the state of affairs concerning real estate in the city and in the countryside (Vandervelde 1898, 1899). Yet, they also reserved for themselves space to study other social problems and matters of special interest to them. This eventually led to a breach with Solvay.

In 1902, ending his association with the three men, Solvay transformed the *Institut des Sciences Sociales* into the *Institut de Sociologie Solvay*. He entrusted its direction to Emile Waxweiler (1867–1916). Waxweiler was an engineer trained at Ghent University and employed since 1896 in the statistics section of the Belgian Office of Labour. He was a member of the International Institute of Statistics and as a young man he had also been a member of the *Société des études sociales et politique*. Solvay met Waxweiler in 1900 when Waxweiler was giving in courses at the *École des sciences politiques et sociales* at the ULB. Solvay provided a building in 1902 for the *Institut de Sociologie Solvay* on a hillside in the Parc Léopold not far from the Institute of Physiology that he had built earlier. Here Waxweiler created a programme of "functional sociology" that relied on biological and psychological analogies in order to explain the "social functions" of society. In 1904 Solvay had a third building erected in the Parc Léopold to accommodate the *Ecole de Commerce Solvay*, the directorship of which Solvay also entrusted to Waxweiler. The School was intended to train an elite group of business directors and instituted a new degree of "commercial engineer."

Nevertheless, in the period between 1894 and 1902, the select club of the *Institut des Sciences Sociales* gathered each week at the Hotel Ravenstein to discuss their research. In addition to Solvay and its three directors, the meetings in 1894, for example, were attended by of course Henri La Fontaine and Otlet but also by a range of prominent politicians, lawyers and others many of whom had been members of the *Société d'études sociales et politiques*.[8]

---

8   Weekly reports of its reunions in 1894; MDN, HLF 121 Institut des Sciences Sociales 1894–1895, folder P.V. des réunions 1894.

The *Institut des Sciences Sociales*, like the *Société d'études sociales et politiques*, was a relatively closed study group which relied for its bibliographical services on the *Office International de Bibliographie*. In fact on the back cover of the Institute's new journal, its *Annales*, the OIB's *Sommaires méthodiques,* its *Bibliographia Sociologica,* was advertised as a project that had grown out of its collaboration with the the *Société d'études sociales et politiques* and the *Institut des Sciences Sociales*. [9] The directors of the *Institut* had reached an agreement with Otlet and La Fontaine that the office space of the *Office International de Bibliographie* at the Hotel Ravenstein would serve as the meeting room for the *Institut des Sciences Sociales*. The *Institut* also agreed to pay the *Office* to provide it with a bibliographical services.[10] The *Office* was also commissioned to act as the secretariat of the *Institut*. It was to manage the *Institut's* correspondence, copying services and the cataloguing of documents. It undertook to organize a monthly meeting for the *Institut* and to compile its statistics. Otlet, for his part, attended most of the meetings and published two articles in the *Annales*, one on accountancy in general and one on Solvay's theory of accountancy in specific (Otlet 1896, 1898).

L'Université Nouvelle and the *Institut des Hautes Etudes*

As members of the *Institut des Sciences Sociales*, Otlet and La Fontaine were not only part of one of the most innovative sites of sociological debate in Belgium but also lived through the tumultuous events associated with attempts to institutionalize sociology in the Free University of Brussels. A beginning had been made in 1889 when Denis, De Greef, and Eugène Van der Rest (1848–1920) set up an interfaculty program in for the political and social sciences (Wils 2005, 276–8). In 1894 the Academic Board of the University, following anarchist uprisings in France, cancelled a series of lectures of the anarchist geographer Elisée Reclus whom it had appointed to a Chair. Hector Denis, who had invited his friend Reclus to give these lectures, resigned in protest as rector of the Free University. Student demonstration erupted. Several other professors, notably De Greef and Picard, also left the University in protest and together they founded a new dissident *Université Nouvelle* (Despy-Meyer 1994). Guillaume de Greef acted as rector of the new university and managed to establish an internationally oriented university programme with the social sciences as its point of focus, the

---

9   On the backside of the *Annales*, the OIB's *Sommaires méthodiques* were advertised as a project grown out of its collaboration with the *Société des études sociales et politiques* and the *Institut des Sciences Sociales*.

10   Contract between the *Institut des Sciences Sociales* and the *Office International de Bibliographie*, first version dated 15 March 1894, second final version dated 12 April 1894; MDN, HLF 121, Folder C7 Institut des Sciences Sociales/Office International Bibliog. PV des réunions 1894 Politiques.

first major academic program in this field in Belgium (Wils 2001, 314). One of the fundamental educational principles of the *Université Nouvelle* was that it aimed to offset academic specialization with a synthetic and encyclopaedist sociological perspective. In 1894 an *Institut des Hautes Études* opened its doors as part of the *Université Nouvelle* and provided a programme of evening classes and public lectures in the social sciences. Between 1899 and 1911 a Faculty of Social Sciences replaced the *Institut des Hautes Études*. When the *Université Nouvelle* was dissolved in 1918, the *Institut des Hautes Études* was transferred to the Free University of Brussels (Despy-Meyer et al. 1976, 88). The programme of the *Institut des Hautes Études* and the Faculty of Social Sciences was explicitly international in that about half of the population of the students were foreigners and it made use of the presence of many foreign professors in Brussels. The Institute was, as one of its founders Edmond Picard described it, the "crown" of the *Université Nouvelle*, open to the public and not focused on gaining grades but on providing a generalized form of education that orchestrated the multiple sciences in a series of "exciting" conferences (Picard 1897, 7). As La Fontaine expressed it: the institute emphasized the idea that "a synthetic overview, at the same time theoretical and practical, of the intellectual domain is indispensable for those who want to carry out social action in a well thought-out manner, as well as for those who want to undertake in a fully rational manner the study of a particular branch of human knowledge" (Fontaine 1900, xviii).

The *Institut des Hautes Études* at the *Université Nouvelle* may be considered, after the *Société d'études sociales et politiques* and the *Institut des Sciences Sociales*, the third sociological research centre in which Otlet and La Fontaine were involved. Unlike most of the other members of the *Institut des Sciences Sociales,* such as La Fontaine, Otlet did not immediately give a course at the *Université Nouvelle*. He did, however, eventually offer lectures on "the organization of science and of scientific work" in the academic term 1909–1910 and on "the organization of international life" in 1910–1911 (Despy-Meyer et al. 1976). He continued to give courses every year at the *Institut des Hautes Études* from 1920 until 1939 even though it had been transferred to the Free University.

Organicism at the Brussels University and at the *Institut International de Sociologie* in Paris

Through their involvement in the *Société d'études sociales et politiques*, the *Institut des Sciences Sociales*, and the *Institut des Hautes Etudes*, Otlet and La Fontaine became deeply influenced by the sociological mode of thought that was being discussed at these institutes. One particular sociological current, known as organicism, coloured sociology in the Brussels milieu. By comparing society to a biological organism, organicism brought the abstract notion "society" down to earth, turned it into something physical and visible and therefore potentially knowable, a living entity that was of the same order as animal societies and the human body.

The analogy between society and biological organisms had become widely popular in sociological thought following the publication of Herbert Spencer's *Principles of Sociology* (1874–1896). Spencer coupled a functional analysis of super-organic bodies to an evolutionary analysis of the long-term development of social forms of organization (Turner 2000, 38). Several works published almost simultaneously with or immediately following Spencer's *Principles* helped to make organicism as a sociological movement increasingly popular: Paul von Lilienfeld's multivolume work *Gedanken über die Socialwissenschaft der Zukunft* [Thoughts Concerning the Social Science of the Future] (1873–1881); Albert Schaeffle's multivolume work *Bau und Leben des sozialen Körpers* [Structure and Life of the Social Body] (1875–1878); and Alfred Espinas's *Des sociétés animales* (1877). Like a biological organism, society was considered by these theorists to be a living being that consisted of components which, despite a certain independence (at least on the level of organs), interacted in such a way that they could only survive together (Hejl 1995). Society was more than a set in which each member efficiently performed its particular but separate functions. Its members shared the same purposes because they were bonded in a higher-level inter-dependent individuality.

One of the main centres of the organicist movement and a forum for sociologists to test their own organicist ideas, was the *Institut International de Sociologie*, founded in 1893 in Paris by René Worms. On the example of other disciplines Worms hoped to give the international institutionalization of sociology a boost by founding his Institute (Sapiro 2009, 65). Sociology's international dimensions, he believed, carried it beyond the French framework. It had, he believed, sufficient authority and size to claim autonomy as a new science (Mosbah-Natanson 2005). The *Institut International de Sociologie* organized thirteen international congresses between 1893 and 1937 the results of which were published in its *Annales de l'Institut international de sociologie* (Schuerkens 1996). Membership in the Institute was by invitation and based on the eminence of the scholar involved. In this way the Institute assembled a stellar international lineup of members, including, among others, Alfred Fouillée, Gabriel Tarde, Alfred Espinas, and Léon Bourgois from France; Albert Schaeffle, Georg Simmel, and Ferdinand Tönnies from Germany; the Americans Albion W. Small and Lester Ward; and Jacques Novicow and Paul von Lilienfeld from Russia.

From its foundation, prominent and senior Belgian sociologists were elected to the institute. The first Belgian members joined in 1894 and others in the years following. La Fontaine and Otlet did not achieve membership until their work had become well known. They became associated members in 1913 and members in 1927 (Worms 1913, 1928).

Table 9.1   List of Belgian members of the *Institut International de Sociologie*[11]

Date of entry	Name and function within the *Institut International de Sociologie*	Occupation
1894	Paul Héger (1846–1925) Vice-President in 1905	Director of the the *Institut de Physiologie Solvay*
1894	Charles De Quéker (1857–)	Secretary of the City Council and the "Bourse du Travail" in Brussels
1894	Adolphe Prins (1845–1919) Vice-President in 1898	Inspector General of Prisons, founder, the *Union internationale de Droit penal*, rector of the ULB (1900–1901)
1894	Maxime Kovalewsky (1851–1916)	Russian sociologist considered to be "Belgian," as a lecturer at the *Université Nouvelle*
1895	Eugène Van der Rest (1848–1920) Vice-President in 1901	Professor of Political Economy, *Université Libre de Bruxelles*
1895	Eugène Goblet d'Alviella (1846–1925) Vice-President in 1899	Professor of the History of Religions, ULB, Liberal senator
1896	Guillaume De Greef (1842–1924) President in 1900	Professor at the *Université Nouvelle*
1897	Hector Denis (1842–1913) Vice-President in 1903 President in 1910	Co-director *Institut des Sciences Sociales;* Socialist member of Belgian parliament, Member *Conseil Supérieur du Travail*
1898	Jules Dallemagne (1840–1921)	Mining engineer and industrialist, Catholic member of parliament for Liège (1900–1919)
1913	Emile Vandervelde (1866–1938) Vice-President in 1913	Chairman, International Socialist Bureau; Socialist member of parliament
1913	Cyrille Van Overbergh (1866–1959) Vice-President in 1920	Principal private Secretary to the Minister of Science and Arts (François Schollaert)
1913	Emile Waxweiler (1867–1916) Vice-President in 1916	Director of the *École de commerce Solvay* and the *Institut de sociologie* at the Parc Léopold

---

11   Based on an analysis of the *Annales de l'Institut International de Sociologie* and reference to the *Biographie Nationale* of the *Académie royale des sciences des lettres et des beaux-arts de Belgique*.

The *Institut International de Sociologie* was a forum where diverse sociological viewpoints were expressed rather than a place where collaborative and well-delineated research was undertaken, such as was the case at the rival school of Émile Durkheim and his associates of the *Année Sociologique* (Geiger 1981, 355, Mosbah-Natanson 2005). Worms promoted organicism and even tried to impose organicism as the general theoretical view of the *Institut International de Sociologie* from his position as permanent secretary and editor of its journal, the *Revue internationale de sociology*, which he launched in 1893. At the third congress of the Institute in 1897, however, the organicist position was heavily critiqued. The polemic against it was opened by the French social psychologist, Gabriel Tarde, who campaigned against every form of biologism in sociology and who based his own sociology entirely on psychology. In a reply to Novicow's lecture on "The Organic Theory of Societies," Tarde argued that the "fruitless comparison of societies in general to living beings [...] has not only been superfluous, but also dangerous." Lilienfeld, Worms and Espinas supported Novicow. Several others including the Hungarian rabbi and philosopher of evolutionary optimism, Ludwig Stein (Haberman 1995), and the Polish Marxist, Casimir de Kelles-Krauz, attacked it (Mucchielli 1998, 272). In general, critiques of organicism said that the full theory lead to a form of authoritarianism and determinism. From then onwards, the focus on organic analogies would move to the background within and outside the *Institut International de Sociologie*. Even Worms would eventually no longer make organic analogies the main focus of his synthesis of sociology, *La philosophie des sciences sociales* (1903–1907). This is in contrast to his earlier publication, *Organisme et Société* (1896), in which he aimed to offer a complete theory of organicism (Geiger 1981, 359). Nevertheless, even after the eclipse of the theory of organicism, the key concepts that had been developed in its framework such as organ and function, morphology and physiology, the normal and the pathological, and the definition of society as a concrete, natural entity, would remain influential in sociological thought after 1900.

Organicism provided a major guideline for the research at the *Institut des Sciences Sociales* in Brussels. Considering the dominant influence of De Greef on the Institute, organicism remained a leading theme until 1902 when the *Institut des Sciences Sociales* was transformed into the *Institut de Sociology*, and when Waxweiler, who was critical of the organicist viewpoint, became its director. Because of the involvement of the members of the *Institut des Sciences Sociales* in the *Institut International de Sociologie* in Paris, it may be reasonably assumed that the sociological viewpoint of the Belgians was closely related to the organicists of the Paris Institute. This is confirmed by the publication of substantial accounts of the organicist ideas of prominent thinkers connected to the *Institut International de Sociologie* such as Paul von Lilienfeld and Albert Schaeffle in the *Annales de l'"Institut des sciences sociales*. (Schaeffle 1894; von Lilienfeld 1894–1895). But more importantly, the members of the Brussels Institute also conducted research which had a clear organicist bias. Organicist metaphors were carried through at great length in the research that Emile Vandervelde undertook in collaboration with the botanist Jean Massart on "organic parasitism and social parasitism." This was published in English and French in 1895 (Massart and Vandervelde 1895a, b). The evident socialist agenda that speaks from this research went against the inegalitarianism of much other organicist research, which assumed that elites were an essential ingredient of all societies in the same way as hierarchy was an essential feature of any complex organism (Logue 1983, 114).

Hector Denis, who had already been acquainted with a biological and physiological approach to sociology from his involvement in the *Société d'Anthropologie de Bruxelles*, also gradually followed the organicist route as a member of the *Institut des Sciences Sociales* (Wils 2001, 311). In an article in the *Annales de l'Institut des Sciences Sociales* on the "rise of the organicist conception of the economic society," Denis sought for the "morphologic laws" of the monetary system based on the organicist analogy between the circulation of riches and the circulation of the blood (Denis 1896). Political economy was at that time the most established of the social sciences and was usually part of the curriculum of the law faculties. Denis thus appropriated the earlier organic metaphors from literature in political economy—as illustrated in Figure 9.1—and gave them, within the well-developed organicist framework, what he considered a "truly" scientific basis.

Solvay, however, never fully identified with the organicist position, though he attached great importance to a biological perspective, in particular a physiological approach to society which resulted from the research that he had initiated at the Institute of Physiology.

The most important theorist of organicism at the *Institut des Sciences Sociales* was Guillaume de Greef. In "Le transformisme social," De Greef gave a brief outline of a sociological research programme for the Insitute, the results of which he published in 1895 in Paris as *Le transformisme social: essai sur le progrès et le regrès des sociétés* (de Greef 1894, 1895). Next to *Introduction à la Sociologie* (1886) and *Les Lois sociologiques* (1892), this was one of his most influential works and was translated into several languages, including Russian (Viré 1971, 264). In *Le transformisme social*, De Greef used organicist arguments to address the problem of how, in their different ways, human societies and the "great society" of which they are part arise and decline in a process of increasing organization and disorganization. Building on the doctrine of evolutionism which assumes that one living being is superior to another if it is further advanced in its "transformism," De Greef argued that the relative value of a society can only be measured by the degree of association that it had attained in its evolution towards a maximum of differentiation of the masses and a maximum of coordination of its differentiated parts. This was an argument that was attacked by Gabriel Tarde, who thought that it was impossible to value one civilization as superior to another on the basis of the biological analogies used by de Greef (de Greef 1901, xix–xxii). De Greef had a clear socialist agenda believing in a "collective contractualism" in which the difference between the working class and the capitalist class would be replaced by a representational system on many different levels of society (de Greef 1911, lxxix).

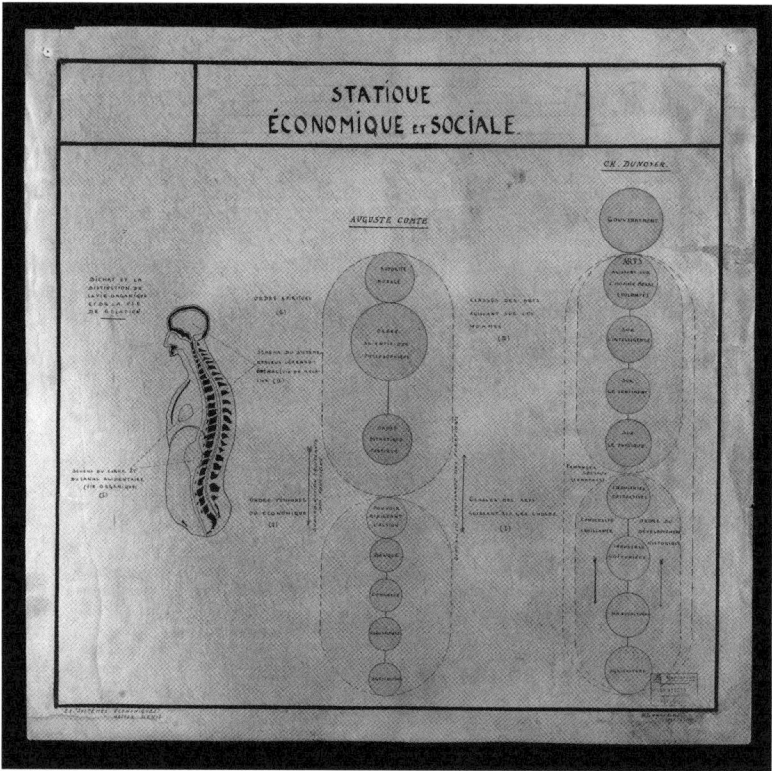

Figure 9.1   Schema of "Economic and Social Statics," based on Hector Denis's
Systèmes économiques, **copied in 1930 by George Lorphèvre for the use of the Palais Mondial**

The Organicism in the Work of Otlet and La Fontaine for International Organization

The sociological ideas and organicist theories that Otlet and La Fontaine encountered in the Brussels intellectual milieu had an important influence on their thinking and practical efforts for international organization. Their conception of international organization in terms of increasing differentiation and centralization was especially close to that of Guillaume de Greef. The internationalist dimension of de Greef's sociology is most developed in *Structure générale des sociétés* (1907) in which he interprets all nation-states as "States-of-transition"; as political borders that are transitional in the development towards a greater world society and world economy. Institutions, he suggested, were urgently needed which were capable of coordinating the process of increasing differentiation and fusion which gave rise to the world society: "An organization which is equally global

Figure 9.2   Paul Otlet, progressive extension of social structures (1912)

is therefore needed by an already de facto global life; the function exists, but the organ is still missing" (de Greef 1908, 300–301). Similarly, Otlet and La Fontaine were convinced that a gradual integration of social structures toward a state of universal association was taking place (as illustrated in Figure 9.2), and that it was now that the hitherto spontaneous organization of Humanity should become conscious and systematized" (La Fontaine and Otlet 1913, 495). As Otlet stated in his essay on the "law of amplification": "[…] the creation of an international society and the advent of the era of globalization ["ère de la mondialité"] is a phenomenon in which the community becomes increasingly self-aware of itself as the necessary outcome of its normal activity"(Otlet 1907, 157). The phrase "era of globalization" was adopted from de Greef who used it as the title of his inauguration address for the eleventh academic year of the Université Nouvelle in Brussels. de Greef had stated that a new form of cosmopolitanism had arisen, one that in contrast to the "individuo-humanitarian" cosmopolitanism of the eighteenth century recognised the existence of intermediary social groups between the individual and humanity (de Greef 1905). Following de Greef's perception that the creation of an "international organism can no longer be designated as utopia," Otlet stated that the movement towards internationalism was accelerating and amplifying the organization of social relations, and that this would lead to a new "stage in the evolution of humanity," that of "humanitarism or mondialism" (Otlet 1907, 159).

Backed up by these sociological theories, Otlet addressed the question of the intellectual unity of humanity on a more practical and explorative level. In 1907, he set up an inquiry into international organizations as a "social structure" in collaboration with Cyrille Van Overbergh. In addition to his official and administrative functions, Van Overbergh was founder in 1899 of the *Société Belge de Sociologie* and Director of the *Bureau International d'Ethnographie*, founded in 1905 (Office central des Institution internationals 1907, 327). Otlet and Van Overbergh circulated a survey form to an enormous number of international associations and as a result information was systematically obtained on the definition of their aims, history, place within the field of international activity, their organs, their activities and services, their evolution, and their publications. Analyzing this information, Otlet and Van Overbergh distinguished several types of international associations: "official associations" such as the Postal Union or what we would call intergovernmental organizations or IGOs; "free associations" such as the *Institut de droit international* or what we would call international non-governmental organizations; and "mixed associations" such as the *Congrès de navigation* or what we would call *hybrid* international non-governmental organizations (Otlet 1910, 18). The inquiry of 1907 was published as a third monograph in the "collection of sociological monographs" of *Le Mouvement Sociologique International* (the successor of *Le Mouvement Sociologique*), the journal of the *Société Belge de Sociologie*, founded by Van Overbergh (Van Overbergh 1907). The first of these monographs had been a study by Van Overbergh of *Sociologie pure,* the classic work of the American sociologist Lester Ward. The second monograph, also by Van Overbergh, summarized Guillaume de Greef's sociology.

In collaborating with Van Overbergh and in publishing several articles in *Le Mouvement Sociologique*, Otlet was part of a sociological sphere of influence that was in many respects different from the Université Nouvelle in Brussels. It is significant that Cyrille Van Overbergh was a guest professor at the *Institut Supérieur de Philosophie* of the Catholic University of Leuven, founded by the theologian and future cardinal of Belgium, Désiré Mercier, which collaborated with the International Institute of Bibliography. Under the chairmanship of Mercier the Institute of Philosophy tried to bring the sociological thought of Comte and Durkheim within a neoscholastic frame of reference thereby dissociating sociology from, and developing a well-considered counterforce against, positivism (Wils 2005, 327). The Catholic school of Mercier opposed the monopolization of the word "sociology" by the positivist camp and instead sought to develop "sociology" into a neothomist solidarist-corporative theory, acting upon the *Rerum Novarum* of Leo XIII (1891). The *Société Belge de Sociologie*, through which Van Overbergh hoped to bring together Catholic intellectuals engaged in modern sociology distributed its journal along with that of the Leuven institute until 1906 (Wils 2005, 336). However, Van Overbergh's approach to sociology was less philosophical and more progressive and eclectic than that of Mercier. He relied more on positivist arguments and was more sympathetic to the sociological approach of the *Université Nouvelle*, as his monograph on de Greef in *Le Mouvement Sociologique* illustrates. Yet, in contrast to the positivist approach of sociologists such as Denis and de Greef, who positioned themselves in the anti-Durkheimian camp by adhering to the International Institute of Sociology of Worms, Van Overbergh defended Durkheim's sociology and explicitly took Durkheim's *L'Année Sociologique* as the model for *Le Mouvement Sociologique*. Like many other Catholics, Van Overbergh especially appreciated Durkheim's sociology of religion and his theory of a corporative organization of society which, according to him, dovetailed with the encyclical of Leo XIII.

In 1907, extending the transnational collaboration that they had established through the International Institute of Bibliography, Otlet, La Fontaine and Cyrille Van Overbergh, founded the Central Office of International Associations (Office Central des Associations Internationales). This was an umbrella organization for international organizations which had their headquarters in Brussels. On the occasion of the Universal Exhibition in Brussels in 1910, the Central Office organized the first World Congress of International Associations which was attended by delegates of 132 international associations and 13 governments and in which five Nobel Prize winners took part (Speeckaert 1970, 27). As a result of this congress the Central Office of International Associations was transformed into the headquarters for a Union of International Associations (UIA). In 1913 the UIA organized a second World Congress in Brussels and Ghent which was even larger than its predecessor (Rayward 1975, ch. VIII). Before World War I, the UIA was an extremely dynamic international organism which did much to affirm the importance of international associations at a moment when internationalism gained momentum.

That Otlet, La Fontaine, and Van Overbergh could easily agree on their vision of the UIA is perhaps surprising, considering their different sociological backgrounds. Yet, the Catholic corporatism of Van Overbergh, the socialist collectivism of La Fontaine, and the positivist humanism of Otlet, shared an organic mode of thinking in terms of division of labour and functional representation that became very popular around 1900 in the Catholic as well as in the positivist milieu, and which was opposed to classic liberal individualism (Wils 2005, 324). The organicist functionalism that grounded the structure of the UIA consisted, for a Catholic such as Van Overbergh as well as for positivists such as La Fontaine and Otlet, essentially in the idea that the social bond takes precedence over the individual and that social evolution involved increasing social differentiation and integration. Following Spencer and the organicists, they considered that the more complex, differentiated and integrated an organism was, the higher its place in the evolutionary ladder of life.

The organicist functionalist viewpoint that underpinned the UIA expressed itself clearly in the network-like schema that Otlet and La Fontaine published

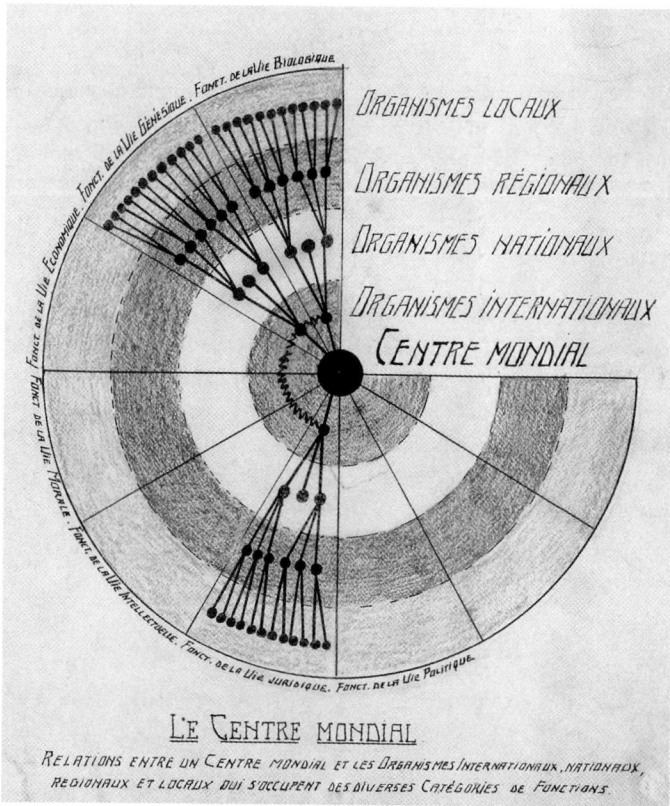

Figure 9.3   Paul Otlet, relations between a world centre and the international, national, regional, and local organisms with different categories of functions (1912)

in a brochure in 1912. The concentric rings of the schema, *Le Centre Mondial* (Figure 9.3), indicate the different scales on which the "organisms" are active— local, regional, national, or international. The wedges or sectors of the schema divide the "organisms" according to their "function," "speciality," or "branch of interest." At the centre, the different international organisms are linked into one superorganism by an international centre that was responsible for establishing the cooperation between the associations. Complementary to a static functionalist reading, the centralized network can also be read from a dynamic evolutionist point of view as an evolutionary tree of humanity that, during its evolution, has grown to a global scale. This is illustrated in an evolutionary schema of the story of the Creation (Figure 9.4), in which Otlet positions the formation of associations after the formation of communication, cities, and the States and as an aspect of a globalizing process culminating in the central world institution— the Mundaneum. Critical of the social Darwinists' point of view, and much like

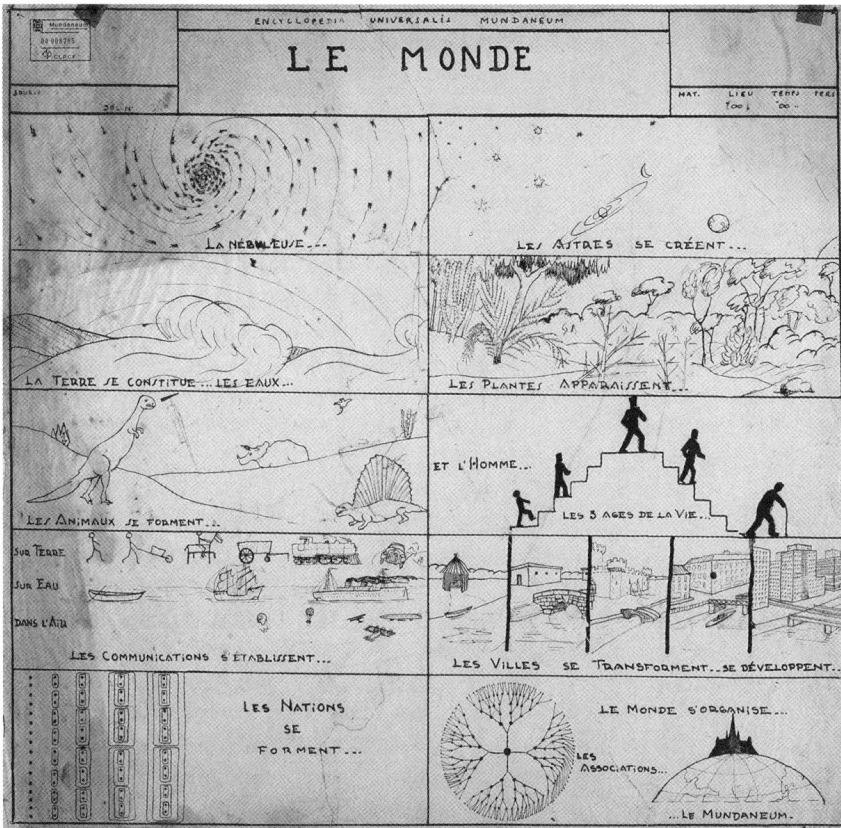

Figure 9.4   Paul Otlet, the evolution of the world's organization (n.d.)

De Greef's contractualist theories and Kropotkin's concept of "mutual aid" as an evolutionary principle, Otlet considered "association" as an evolutionary mechanism opposed to the struggle for survival (de Greef 1911, 146). Associations were "like concentrations of force" that through coordination and cooperation could attain an ever greater strength. What made these international associations grow, Otlet reasoned in a typical organicist argument, was the fact that associations implied a multiplication of force. He gave the following example borrowed from the Russian sociologist and pacifist Jacques Novicow: "One isolated person is able to raise 30 kilos; hundred persons not 3,000 but 300,000 kilos or more" (Otlet 1916, 102, 1935, 107).

Not only did Otlet borrow this example from Novicow, he also adopted Novicow's organicist definition of the concept of "association." Novicow defined an association as a "metaphysical entity without concrete reality," an "exchange of services," "a vital circulation," and "only a means" the single goal of which is

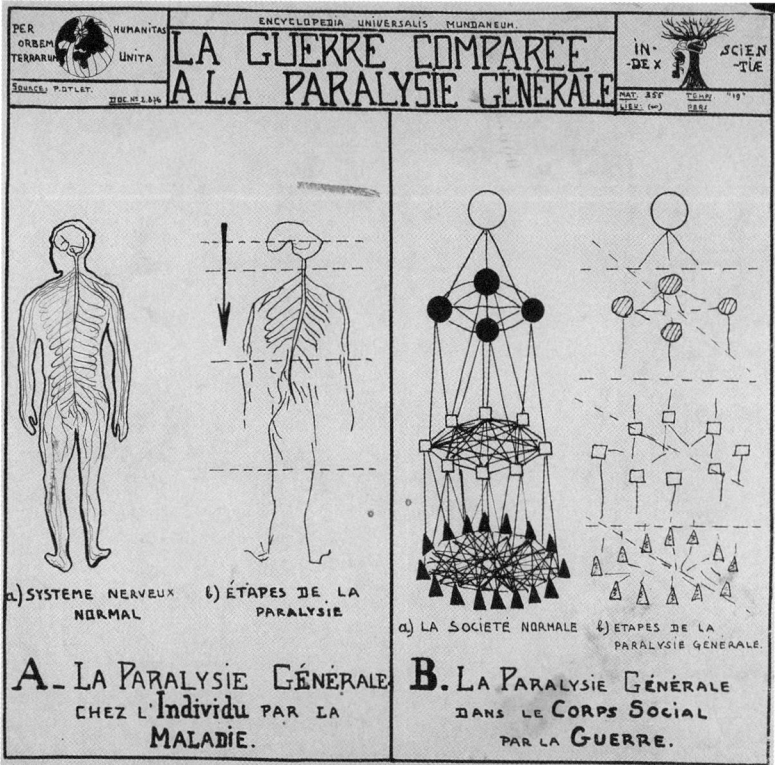

Figure 9.5   Paul Otlet, war compared to the general paralysis (1919)

"the vital intensity of the individual" (Novicow 1911, 21). Novicow believed that, because there was no limit to human association, society would evolve in the future beyond the limits of the nation-state to form a federation of states (Josephson 1985, 706). In a lecture on the "Essence of Human Association" delivered at the 1909 congress of the *Institut de Sciences Sociales* on "social solidarity," Novicow argued that "each individual, living on the earth, will reach the peak of his vital power once a general federation of the human race is formed" (Novicow 1911, 31). He suggested that such a universal federation was the normal and rational state of the human race, while non-federation or international anarchy, disorder, or disorganization was the abnormal, irrational, or pathological state. As shown in Figure 9.5, like Novicow, Otlet compared the organized and coordinated system of human association to the normal or healthy state of the human body and the dis-association of society caused by war to the illness of the human body. Quoting Novicow's organicistic vision, Otlet observed that all organisms attempt through association to attain "a maximum of vital intensity" (Otlet 1916, 102). The world society was evolving towards such a "maximum of vital intensity" or "a maximum of consciousness" through the formation of a collective mind. "Progress," Otlet concluded, "is marked by the transformation of the unconscious into the conscious" (Otlet 1935, 107).

Conclusion

At a time when the well-established university disciplines were underpinned by international organizations that regularly held international conferences, it seemed that if sociology were to be recognized as a university discipline it had to become internationally organized as well. The transnational bibliographical and networking initiatives of Otlet and La Fontaine helped the sociological laboratories in Brussels in the period of the Belle Époque in this process of international organization, although this did not necessarily help them receive academic recognition at the national level. At the same time, Otlet and La Fontaine developed out of this transnational practice a "new profession" of international organization. To use George Sarton's phrase, they were "international organisers" rather than sociologists in the sense that Spencer or de Greef understood themselves to be—that is, philosophers of the social sciences (Sarton 1913, 28). As they stated in the Union of International Association's journal, *La Vie Internationale,* the movement for international organization "seeks action; acts and in so doing differentiates itself from pure sociological explanations" (La Fontaine and Otlet 1912, 34). Nevertheless, their conceptual approach to society on a macro-sociological level drew heavily on the blend of the sociological theories of Comte and Spencer that had been made by de Greef and Denis and other more recent organicist theories with which they had become familiar. Otlet and La Fontaine's approach to "Humanity" as a historical progressive achievement of consciousness or thought was impregnated with Comtian positivism. Émile Corra (1848–1934), who founded the *Société Positiviste Internationale* in 1906, devoted a chapter of his book *L'Humanité* to the UIA which "helped to show that accompanying the grand philosophical theory of Auguste Comte, the hands of practical men, justifying and consolidating this theory, are now working for an objective constitution of Humanity […]" (Corra 1914, 59). Yet, the macro-sociological point of departure in the UIA's programme, namely that Society at large was entering into forms of association that were ever more complex, more differentiated and better coordinated, was more indebted to the organicist theories of de Greef than the theories of Comte. At the same time, Otlet and Fontaine's interpretation of the ongoing process of social integration and organization toward an ever greater state of consciousness and intelligence may also well have been imported from or inspired by the idea supported by sociologists such as Eugène De Roberty (1843–1915), Alfred Espinas, Jean Izoulet (1854–1929), Alfred Fouillée, Emile Durkheim and Gabriel Tarde, that collective consciousness is produced as a result of social interactions which make individuals act cooperatively in a group (Sorokin 1928, 433–87). Similarly, Otlet in a late lecture for the *Institut International de Sociologie* in 1930 spoke at length about the "collective intelligence" of the world society and how it needed a "collective Brain" or "world civilisation" as its substratum. Consequently, if Otlet's sociological conception is to be summarized, given these different influences, it is perhaps best expressed by what Petrim Sorokin calls in his overview of *Contemporary Sociological Theories* (1928), "psycho-organicism"(Sorokin 1928, 465).

References

Brion, R. and Buyle A. 1987. *L'Hotel Ravenstein. Siège de la Société royale belge des ingénieurs et des industriels*. Bruxelles: S.R.B.I.I.

Carr, Reg. 1977. *Anarchism in France: The Case of Octave Mirbeau*. Manchester: Manchester University Press.

Corra, Émile. 1914. *L'Humanité*. Chateaudun: Société Typographique.

Couvreur, August. 1889. *Société des Études Sociales et Politiques – son origine, son but*. Bruxelles: The Society.

Crombois, Jean-François. 1995. "Bibliographie, sociologie et coopération internationale. De l'Institut International de Bibliographie à l'Institut de Sociology Solvay." In *Cent Ans de l'Office International de Bibliographie*, edited by Andrée Despy-Meyer. Bergen: Editions Mundaneum, 215–38.

Crombois, Jean-François. 1997. "Energétisme et productivisme: La pensée morale, sociale et politique d'Ernest Solvay." In *Ernest Solvay et son temps*, edited by Andrée Despy-Meyer and Didier Devriese. Bruxelles: Archives de l'Université de Bruxelles, 209–20.

De Greef, Guillaume. 1894. "Le Transformisme social. Plan d'étudespour la Section de sociologie de l'Institut des sciences sociales." *Annales de l'Institut des Sciences Sociales*, 1 (2): 69–78.

De Greef, Guillaume. 1895. *Le transformisme social: essai sur le progrès et le regrès des sociétés*. Paris: F. Alcan.

De Greef, Guillaume. 1898. "Essais sur la monnaie, le crédit et les banques." *Annales de l'Institut des Sciences Sociales*, 4 (1): 20–57.

De Greef, Guillaume. 1901. *Le Transformisme Social. Essai sur le progrès et le regrès des sociétés. Deuxième édition, revue et augmentée*. Paris: Félix Alcan.

De Greef, Guillaume. 1905. *L'Ère de la Mondialité. Eloge d'Elie Reclus*. Bruxelles: Université Nouvelle.

De Greef, Guillaume. 1908. *La Structure Générale des Sociétés. Tome 2 Théorie des Frontières et des Classes*. Bruxelles/Paris: Ferd. Larcier/Félix Alcan.

De Greef, Guillaume. 1911. *Introduction à la sociologie. Première Partie. Éléments. Deuxième édition*. Paris: Marcel Rivière et Cie.

De Seyn, Eugène. 1936. "Cyrille Van Overbergh." *Dictionnaire biographique des Sciences, des Lettres et des Arts en Belgique. Tome Second*. Bruxelles: Éditions l'Avenir.

Denis, Hector. 1894. *Plan des recherches de sociologie économique. Extrait des "Annales de l'Institut des sciences sociales.*" Brussel: Institut des sciences sociales.

Denis, Hector. 1895. "Proudhon et les principes de la Banque d'échange." *Annales de l'Institut des Sciences Sociales*, 1 (4).

Denis, Hector. 1896. "L'école physiocratique et l'avènement de la conception organique de la société économique. La circulation du sang et la circulation des richesses." *Annales de l'Institut des Sciences Sociales*, (2): 23–36.

Despy-Meyer, Andrée. 1994. "Un laboratoire d'idées: l'Université nouvelle de Bruxelles (1894–1919)." In *Laboratoires et réseaux de diffusion des idées en Belgique (XIXe–XXe siècles)*, edited by Ginette Kurgan-van Hentenryk. Bruxelles: Université de Bruxelles, 51–4.

Despy-Meyer, Andrée and Didier Devriese. 1997. *Ernest Solvay et son temps*. Bruxelles: Archives de l'"Université Libre debruxelles.

Despy-Meyer, Andrée and Goffin Pierre, Institut des hautes études de Belgique and ULB. Service des archives. 1976. *Liber memorialis de l'Institut des hautes études de Belgique fondé en 1894*. Bruxelles: Institut des hautes études de Belgique.

Dubois, Sophie. 2002. *L'Hôtel Ravenstein. Un enjeu pour les Sociétés Savantes*, Faculté de Philosophie et Lettres. Université Libre de Bruxelles, Bruxelles.

Errera, Paul. 1894. "Auguste Couvreur." *Revue Sociale et Politique*, 4 (2): 81–102.

Fontaine, Henri La. 1900. Création d'un enseignement social international en Belgique. Paper read at Congrès International de l'Enseignement des Sciences Sociales, at Paris.

Fontaine, Henri La and Otlet, Paul, "La Vie Internationale et l'effort pour son organisation," La Vie Internationale 1 (1912): 9–34

Fontaine, Henri La and Otlet, Paul. 1913. "La Deuxième Session du Congrès Mondial." *La Vie Internationale*, 3: 489–524.

Geiger, Roger L. 1981. "René Worms, l'organicisme et l'organisation de la sociologie." *Revue Française de Sociologie*, (22): 345–60.

Haberman, Jacob. 1995. "Ludwig Stein: Rabbi, Professor, Publicist, and Philosopher of Evolutionary Optimism." *The Jewish Quarterly Review*, 86 (1/2): 91–125.

Hejl, Peter M. 1995. "The Importance of the Concepts of 'Organism' and 'Evolution' in Emile Durkheim's Division of Social Labor and the Influence of Herbert Spencer." In *Biology as Society, Society as Biology: Metaphors*, edited by Sabine Maasen, Everett Mendelsohn, and Peter Weingart. Dordrecht/London: Kluwer Academic, 155–92.

Institut International de Bibliographie. 1908. *Actes de la Conférence Internationale de Bibliographie et de Documentation. Bruxelles, 10 et 11 Juillet 1908*. Tome premier, Publication no 98. Bruxelles/Paris/Zürich: IIB.

Josephson, Harold (ed.) 1985. *Biographical dictionary of modern peace leaders*. Westport, Conn.: Greenwood Press.

Logue, William. 1983. *From Philosophy to Sociology: The Evolution of French Liberalism, 1870–1914*. De Kalb Northern Illinois University Press.

Maheim, Ernest. 1892. "Emile Vandervelde – Enquête sur les associations professionnelles d'artisans et d'ouvrier en Belgique." *Revue Sociale et Politique*, 2 (2): 182.

Massart, Jean and Vandervelde Emile. 1895a. *Parasitism. Organic and Social*. London: Swan Sonnenschein.

Massart, Jean and Vandervelde Emile. 1895b. "Parasitisme organique et parasitisme social." *Bulletin scientifique de la France et de la Belgique*, 25: 227–94.

Mosbah-Natanson, Sébastien. 2005. Histoire de la sociologie et tradition nationale: le cas français entre 1880 et 1930. Paper read at traditions nationales en sciences sociales, at Amsterdam.

Mucchielli, Laurent. 1998. *La découverte du social: naissance de la sociologie en France (1870–1914)*. Paris: La Découverte.

Novicow, Jacques. 1911. "L'essence de l'association humaine." In *Annales de l'Institut des Sciences Sociales*, edited by René Worms. Paris: V. Giard et E. Brière, 15–43.

Office central des Institution internationales and *le Mouvement sociologique international*. 1907. *Enquête sur les structures sociales. L'Association internationale*. Bruxelles.

Otlet, Paul. 1896. "La Comptabilité et le Comptabilisme." *Annales de l'Institut des Sciences Sociales*, 2 (3): 47–60.

Otlet, Paul. 1898. "Observations à propos de la dernière note de M. Solvay relative au Comptabilisme." *Annales de l'Institut des Sciences Sociales*, 4 (4): 290–94.

Otlet, Paul. 1907. "La loi d'ampliation et l'internationalisme." *Le Mouvement sociologique international*, 8 (4): 133–74.

Otlet, Paul. 1910. *L'Organisation internationale et les Associations Internationales*. Rapport No1. *Congrès des Associations Internationales*. Bruxelles: Office Central des Institutions Internationales.

Otlet, Paul. 1916. *Les Problèmes Internationaux et la Guerre. Tableau des conditions et solutions nouvelles de l'économie, du droit et de la politique*. Publications de l'Union des Associations Internationales. Genève/Paris: Librairie Kundig/ Rousseau & Cie.

Otlet, Paul. 1935. *Monde. Essai d'Universalisme. Connaissance du monde, sentiment du monde, action organisée et plan du monde*. Bruxelles: Mundaneum.

Otlet, Paul and La Fontaine Henri. 1895. *Revue Sociale et Politique*, 5 (1–2): 5–6.

Picard, Edmond. 1897. *L'Institut des Hautes Études à l'Université Nouvelle de Bruxelles*. Paris: Librairie de l'art social.

Rayward, W. Boyd. 2010. *Mundaneum: Archives of Knowledge*. Urbana-Champaign: Graduate School of Library and Information Science University of Illinois.

Rayward, W. Boyd. 2008. Places, Faces, Networks: Paul Otlet in Brussels, 1890–1897 (unpublished conference paper). Paper read at Analogous Spaces: Architecture and the Space of Knowledge, Intellect and Action at Ghent.

Sapiro, Gisèle (ed.) 2009. *L'espace intellectuel en Europe. De la formation des États-nations à la mondialisation XIXe–XXI siècle*. Paris: La Découverte.

Sarton, George. 1913. "L'Histoire de la Science et l'Organisation Internationale." *La Vie Internationale*, 2: 27–40.

Schaeffle, Albert. 1894. "Deutsche Kern- und Zeitfragen. Résumé." *Annales de l'Institut des Sciences Sociales*, (1): 36–64.

Schuerkens, Ulrike. 1996. "Les Congrès de l'Institut International de Sociologie de 1894 à 1930 et l'internationalisation de la sociologie." *International Review of Sociologie*, 6 (1): 7–24.

Solvay, Ernest. 1894. *Comptabilisme et Proportionnalisme Social. Extrait des "Annales de l'Institut des sciences sociales."* Brussel: Institut des sciences sociales.

Sorokin, Pitirim. 1928. *Contemporary Sociological Theories*. New York and London: Harper and brothers.

Speeckaert, Georges Patrick. 1970. "Regards sur soixante années d'activité de l'Union des Associations Internationales." In *L'Union des Associations Internationles. 1910–1970. Passé, Présent, Avenir*. Bruxelles: Union des Associations internationales, 19–52.

Turner, Jonathan H. 2000. "The Origins of Positivism: The Contributions of Auguste Comte and Herbert Spencer." In *Handbook of Social Theory*, edited by George Ritzer and Barry Smart. London: Sage, 30–42.

Van Acker, Wouter. 2011. *Universalism as Utopia. A Historical Study of the Schemes and Schemas of Paul Otlet (1868–1944)*, Department of Architecture and Urban Planning, Ghent University, Ghent.

Van Overbergh, Cyriel. 1907. "La Documentation en matière de sociologie générale." *Le Mouvement sociologique international*, 8 (1): 1–15.

Van Overbergh, Cyriel. 1907. "La Documentation en matière de sociologie générale: Guillaume De Greef." *Le Mouvement sociologique international*, 8 (4): 1–92.

Van Overbergh, Cyrille. 1907. "L'association internationale." *Le Mouvement sociologique international*, 8 (3): 615–927.

Vandervelde, Emile. 1898. "Le Livre III du 'Capital' de Marx et la Théorie de la Rente foncière." *Annales de l'Institut des Sciences Sociales*, 4 (1): 5–19.

Vandervelde, Emile. 1899. "L'Influence des villes sur les campagnes: La Propriété foncière dans les provinces du Luxembourg, de Namur, de la Flandre orientale et de la Flandre occidentale." *Annales de l'Institut des Sciences Sociales*, 5 (5): 752–4.

Viré, Lilian. 1971. "Guillaume De Greef." In *Biographie Nationale*. Bruxelles: Établissements Émile Bruylant, 358–73.

Von Lilienfeld, Paul. 1894–1895. "Pensées sur la science sociale de l'avenir. Résumé." *Annales de l'Institut des Sciences Sociales*, (1): 133–79.

Wils, Kaat. 2001. "De Sociologie." In *Geschiedenis van de wetenschappen in België: 1815–2000*, edited by Robert Halleux. Brussel: Dexia, 305–322.

Wils, Kaat. 2005. *De Omweg van de Wetenschap: het Positivisme en de Belgische en Nederlandse Intellectuele Cultuur, 1845–1914*. Amsterdam: Amsterdam University Press.

Wirt-Cordier, Anne-Marie. 1997. "Solvay, Ernest." In *Nouvelle biographie nationale*. Bruxelles: Académie royale des sciences des lettres et des beaux-arts de Belgique, 304–12.

Worms, René. 1913. *Annales de l'Institut International de Sociologie. Tome XIV Contenant les Travaux du Huitième Congrès, tenu à Rome en Octobre 1912.* Paris: V. Giard & E. Brière.

Worms, René. 1928. *Annales de l'Institut International de Sociologie. Tome XV Contenant les Travaux du Neuvième Congrès, tenu à Paris en Octobre 1927.* Paris: V. Giard & E. Brière.

Chapter 10

Collecting Paper: *Die Brücke*, the Bourgeois Interior, and the Architecture of Knowledge[1]

Nader Vossoughian

It was not until the 20th century that the great research scientist Wilhelm Ostwald, through "Die Brücke" (the Bridge), the international institute founded by him and K.W. Bührer in 1912 in Munich, made a new advance in regulating format sizes.[2] Ostwald likewise proceeded from the Lichtenbergian principle of constant ratio but depended on the measuring in centimeters. The proposed "World-formats" to be adopted worldwide, did not however win many friends: probably because the main format 230x320 mm (letterheads) turned out to be rather unpractical. Nevertheless the work of "Die Brücke," whose work was interrupted by the war, had the very great merit of making people think. The publications of "Die Brücke" are among the most interesting studies in this field. (Tschichold 1928, 96–7)

Collecting is a primal phenomenon of study: the student collects knowledge. (Benjamin *The Arcades Project*, H4, 3)

In a number of recent studies, the rise of standard paper formats in 20th-century Europe has been framed within the context of the history of information (Hapke 2008) and media studies (Krajewski 2006, 2011). In this chapter I want to make the case that it is useful to interpret their development from a spatio-political and architectonic standpoint as well. My argument is that standard paper formats were introduced to stem the crisis of bourgeois subjectivity that emerged with the advent of industrial capitalism. Information reformers such as the creators of Die Brücke (not to be confused with the expressionist school of artists bearing the same name) set about regularizing the dimensions of paper in an effort to alleviate the "crisis of the room" or "Raumnot,"[3] as K.W. Bührer put it, which afflicted the modern collector at the start of the new century. "Raumnot" can be translated in at least two ways either as "spatial crisis" or "crisis of the room." I have selected the latter, despite its awkwardness, for reasons that should become

---

1 I am grateful to Thomas Hapke and W. Boyd Rayward for their incisive feedback and comments.

2 Die Brücke was established in 1911, not in 1912. K.W. Bührer and Adolf Saager were its official founders; Ostwald soon followed.

3 "Raumnot" can be translated in at least two ways, as "spatial crisis" or "crisis of the room." I have selected the latter one, despite its awkwardness, for reasons that should become more apparent over the course of this paper.

apparent. They developed standard paper formats for the purpose of compressing greater amounts of information into ever smaller spaces. They also collaborated with artists such as Emil Pirchan in an effort to visualize this ideal. Die Brücke's example, I argue, allows us to investigate the intimate links that existed between architecture and information theory in early modernism. It also problematizes the theory that the "architecture of knowledge," as Charles van den Heuvel and Boyd Rayward have termed it, only took root after the Second World War.[4] This case study foregrounds the fantasies that accompanied the adoption of paper standards, which were not as democratic or progressive as may at first glance appear. It also casts light on the "long shadow" of the *Belle Époque*, highlighting its liberal and classist aspirations.

Officially formed in 1911 by a merchant and advertising specialist (K.W. Bührer), a chemist and writer (Adolf Saager), and a third unnamed figure, Die Brücke's principal concern was the problem of overspecialization in the natural and humanistic sciences. According to Wilhelm Ostwald, who joined the group soon after its initial formation, Die Brücke "derived its name from its goal of using a specially constructed organ to unify harmoniously and effectively separate intellectual undertakings that emerge on isolated islands"[5] (Ostwald 1912, 244). Its stated goal, as Thomas Hapke puts it, was to become "the central agency in which would be created a comprehensive, collaboratively compiled and illustrated world encyclopedia on sheets of standardized format" (Hapke 2008, 314). Die Brücke saw itself as an "inquiry office for scholars and a message exchange or mediation agency for organizations and individuals" (Hapke 2008, 314). It was dedicated, as Markus Krajewski notes, to "the avoidance of all superfluous energy spent in the entire area of pure and applied mental labors, and in connection with this, liberating the creative mind from the shackles of preliminary mechanical work" (Krajewski 2011, 116). Die Brücke's members lamented the fact that intellectual research was regulated with less oversight than economic or technical work: "While the thoroughgoing organization of all human activity has already taken root in many other fields such as technology and economics," Ostwald notes, "intellectual labor in general, and in the sciences in particular, have been subjected to virtually no level of organization" (Ostwald 1912, 15–16).

They were also eager to coordinate and control the production of information from the bottom up. That is, they wanted to manage how information circulates in and between offices, schools, government agencies, and private citizens, and not just in scientific laboratories. "If one wants to organize, one can only do so if one first intervenes in the unification and coordination of the most everyday, common and thus also least reflective functional routines" (Ostwald 1912, 17). One of Ostwald's aspirations for Die Brücke was that it would produce a "general

---

4   Van den Heuvel and Rayward coined this term for a conference held in Mons at the Mundaneum Institute in 2002. The official title of the meeting was "Paul Otlet: Architecture of Knowledge, Knowledge of Architecture" (Heuvel, Rayward and Uyttenhove 2003).

5   All translations by the author unless otherwise noted.

Color Atlas, which would represent objectively all possible colors and shades of color, brightness and clarity on a scientific basis, which would thereby make possible a corresponding international relationship of all existing shades of color" (Ostwald 1912, 251). He also imagined developing an "international helping language" which he suggested was "the simplest and most elementary processes of intellectual labor that can be organized in the first place and most successfully" (Ostwald 1912, 251; Krajewski, Chapter 6 above).

In practice, Die Brücke's concrete achievements were varied. According to Ostwald, all of them shared in common the fact that they were inspired by the "energetic imperative," which begins with the "organization of the simplest and most common things first and from there building upwards" (Ostwald 1912, 19–20). Between 1911 and 1914, Die Brücke undertook the "systematic collection of press clippings," the building of an archive of musical history, the drafting of bibliographical conventions for indexing and cataloging human knowledge (Bührer and Saager 1911, 9, 13, 14), and the establishment of a center dedicated to the collection of everyday cultural ephemera (e.g., envelopes, tickets, labels, banknotes, stamps identification cards and concert programs) (Bührer 1912b, 9). According to Rolf Sachsse, the group printed "diverse Die Brücke writings, which dealt with the 'cultural mission of the advertisement' and the design of Hotel publicity materials. An archive of musical review clippings was purchased as the first piece of an archive that K.W. Bührer planned … for the purpose of creating a giant, so-called 'small graphics' collection." The group released "around 30 different pamphlets and books with a total circulation of around half a million copies; in conjunction with that, an edited newspaper for six months." In 1913, it also participated in the Bavarian Applied Arts Show, which presented "Bührer's collection of advertisements, discount coupons, and letter stamps as successful examples of the 'organization of intellectual labor'" (Sachsse 2004, 70, 65, 71).

Of all its undertakings, Die Brücke's most significant was its effort to establish internationally sanctioned standard paper formats, as the passage cited at the outset of this paper suggests. Wilhelm Ostwald's writings in this domain contributed significantly to the development of the modern A-Series formats. His so-called World Formats adhere to a few simple rules, the first of which states that doubling or halving the width or length of a given format must yield the dimensions of another. This requirement, Ostwald writes, is dictated by "the nature of the material, namely paper, since under this assumption an efficient distribution of large sheets into small [paper] sizes" can be achieved (Ostwald 1911, 7–8). Second, the width-to-height ratio of all formats should be 1: $\sqrt{2}$ (this is to preserve uniform proportions at all scales). Third, the smallest paper format ought to have a width of exactly one centimeter because "that has already been accepted as the global unit of length" (Ostwald 1911, 9). According to Ostwald, World Format IX (16: 22.6 cm) is a "work format" because it is best suited for "scientific and technical works of all kinds." He thought that "it also appears to the critical eye to be a more rational and comfortable format as the presently normal one" (Ostwald 1911, 9). Format XIII (11.3: 16 cm) is a "comfortable and pretty portable format."

Format X (22.6: 32 cm) is for "letter paper and memos," and Format XI (32: 45.3 cm) is primarily for newspapers (Ostwald 1911, 10).

Industrial efficiency was first among the objectives Die Brücke hoped to accomplish through the adoption of standard paper formats. According to Ostwald, paper is to science as money is to economics because it facilitates the objectification of labor power, i.e., it allows one to disassociate the production of knowledge from its consumption. The "introduction of unified formats," Ostwald observes, "just as happened with the introduction of unified measures, weights, and coins in Germany, will bring about enormous relief in manufacturing and commerce as well as enormous reduction of costs." He continues: "When one considers ... that all intellectual labor must with few exceptions assume written or printed form, and that the organization of intellectual labor must begin with the purely mechanical ordering of written and printed documents, then one recognizes the truth of this statement" (Ostwald 1911, 6).

Second, Die Brücke was committed to the idea that standard paper formats would bring about social harmony, that is, they could unite scholars and researchers across the world. "The unification of labor must enter in such a way ... that the products or individual labor achievements, which are arrived at through solitary labor [*Teilarbeit*], are ordered and brought into a relationship that permits them to be accessible to anyone, according to need, without aggravation or difficulty." Ostwald continues: "In order to bring this general progress into fruition, Die Brücke has been established as an international central organ for intellectual labor" (Ostwald 1912, 249–50).

Finally, the creators of Die Brücke also believed (albeit implicitly) that standard paper formats could remedy some of the contradictions that inhere in modern capitalism. Indeed, the group aimed to rehabilitate the figure of the collector in order to reinvigorate the liberal-bourgeois concept of privacy. As Walter Benjamin notes in *The Arcades Project*, the collector "makes his concern the idealization of objects" (Benjamin 1999, 19). His self-*alienation* is the basis of his self-*realization*. He uses *property* to express *privacy*. Similarly, Die Brücke's collector is a kind of *connector*, one who simultaneously divides and unites the private and public spheres. If the prototypical nineteenth century collector is, following Benjamin, a hoarder; if his home (like his mind) is cluttered with objects; if his identity is split, divided between a private self (which collects and cathects) and a public one (which buys and sells), Die Brücke's collector is by contrast an "organizer"–William H. Whyte's "Organizational Man" in embryo (Whyte 1956). In collecting knowledge about music and science, he connects people from across the world. "The number of collectors is much larger than one usually thinks," Bührer and Saager wrote. "The collector, whether he puts his collection together out of pure joy in collecting or not, will be happy when his pleasurable activity bears practical fruit, when his seemingly idle pleasures simultaneously become useful or when his hobby achieves recognition." Die Brücke, they add, will make collecting pay. It will also render the act of collecting more efficient, offering "every collector the advantage of simplifying his work, saving many fruitless reflections and some bitter disappointments" (Bührer and Saager 11, 8–9, 12).

Of its many publications, Die Brücke's *Raumnot und Weltformat* is the text that captures Die Brücke's identification with the figure of the collector most poignantly. Written by Bührer, its explicit claim is that "book production has grown to a monstrous extent," so that collectors and collecting institutions are ill-equipped to handle this reality (i.e., they lack space), and that standard paper formats could well remedy the problems at hand by economizing on the use of space (Bührer 1912). Implicitly, however, *Raumnot und Weltformat* also argues that the interior is itself in crisis. "Raumnot" refers not just to a crisis of space but also a crisis of the room. This is borne out by the drawings that accompany the publication. Produced by the Austrian illustrator, architect, and stage designer Emil Pirchan, who had studied with Otto Wagner in Vienna, these drawings have been dismissed by Rolf Sachsse because they "caricature ... the aesthetic principles of Wilhelm Ostwald" and mimic "the designs of the Viennese architect and furniture designer Josef Hoffmann" (Sachsse 2004, 86). But it can be argued that they also make Die Brücke's utopian ambitions tangible in a way that their writings alone do not. The drawings depict the collector as the quintessential apologist of the *Gesamtkunstwerk*, the "total art work," which, as Juliet Koss writes, seeks to "sustain and destroy the autonomy of the individual arts" (Koss 2010, xii–xiii). The *Gesamtkunstwerk* contests the boundaries separating art from life, aesthetics from politics. It submits all the arts, from architecture to painting, to a common formal language. Similarly, Die Brücke's collector refuses to acknowledge the distinction between fetish value and use value, architecture and information, structure and ornament, home and work. While he is sober, efficient, and industrious, a picture of the Protestant ethic, he is also a fetishist, someone whose sense of self is intimately bound up with the objects that surround him. He thus also challenges the image of the collector that surfaces in Benjamin's writings in important ways. For while, *pace* Benjamin, his proper "home" is the interior, the private sphere, his goal—contra Benjamin—is also to make privacy pay. Here, the ideal collector can, for instance, provide "all relevant information about the travel relations of all countries without losing any time," as Bührer notes (1912a 16). He can rationalize his private fetishes in ways that service public (or in this case, commercial) aims.

A comparison of a prototypical 19th-century interior with those illustrated in *Raumnot und Weltformat* is useful for understanding these points. Any number of examples of generic interiors could be selected for the purposes of this comparison—Empire Style designs by Percier and Fontaine that Giedion (1948) and Asendorf (1984) have explored, or Victorian homes, as Brown (2008) and Forty (1986) suggest. My example is Sigmund Freud's consulting room and study, which represents one of the most carefully documented bourgeois interiors ever produced in the German-speaking world (Figure 10.1).

According to Diana Fuss, Freud was the paradigmatic collector, for "two months after his father's death in October 1896 [he] began assembling the antiquities that would transform his office into a veritable tomb." He collected

Figure 10.1   Freud's study, 1938 (photograph © Edmund Engelman)

"Egyptian scarabs, Roman death masks, Etruscan funeral vases, bronze coffins, and mummy portraits," and he housed them in his office, which "is located in the back wing of what was originally designed to be part of a domestic residence, in that area of the apartment house typically used as sleeping quarters." The rooms bear out the validity of Benjamin's argument that the 19th-century interior "signifies more than mere shelter; it becomes coextensive with the person of the dweller, a kind of second skin." The rooms are also useful for understanding Pirchan's illustrations in *Raumnot und Weltformat*. According to Fuss, Freud's spaces are "the exteriorized theater of Freud's own emotional history, where every object newly found memorialized a love-object lost" (Fuss, 2004, 33, 34, 3, 391). In like fashion, Pirchan's interiors resemble a neatly organized "world brain." Indeed, the figure of the "world brain" or "*Gehirn der Welt*" is a recurrent motif in Die Brücke's writings, as Buckland notes (2008), and it is especially prevalent here. Like Freud's interiors, these spaces externalize an internal world. They are hermetically sealed. Their function is to organize and store memories (i.e., *informational* memories), and they complement the proportions and even fashion the preferences for the subjects they house.

If, however, Freud's interiors tend toward the Oriental and the exotic, if they delight in the fragmentary and the mysterious, Pirchan's are more akin to logically ordered puzzles in which the relationship *between* parts matters more than the parts themselves. Freud's interiors recall cabinets of *curiosity*, arranged as they are

1. SCHREIBTISCH
MIT SCHACHTELN IM WELTFORMAT

**Figure 10.2   Emil Pirchan, "Writing Table with World Format Boxes"
(reprinted from K.W. Bührer, *Raumnot und Weltformat* 1912)**

to inspire wonder more than they do resonance, to use the terminology of Stephen Greetblatt (1990). By contrast, Pirchan's resemble cabinets of *communication*. They depict the modern interior as a system for arranging information in space. On the walls of Freud's consulting room hang portraits of friends, Pompeiian fragments, and an Egyptian papyrus. On a bookshelf stands a statuette of the Egyptian goddess Neith from Sais; on a table, a hawk-headed Horus, two Jewish cups, some wooden figures and a number of Egyptian funerary objects (Sigmund Freud Museum). In Pirchan's interiors, on the other hand, paper and books, closets and cartons—in a word, vessels of information—monopolize one's view. Here, collecting systems are collected. In his "Writing Table with World Format Boxes" [Schreibtisch mit Schachteln im Weltformat] (Figure 10.2), Pirchan depicts the modern desk as a portable knowledge management system, one that brings "a library into every single home" (1912a 15).

His drawings for a travel office [Verkehrsbüro] present us with facts that double as artifacts. The drawers of information frame the room as a proscenium does a stage. Sheets of paper carry the ceiling the way bricks do load-bearing walls. "Paper work" both defines and negates two intersecting walls (Figure 10.3).

Finally, in the plans for a large library [Gross Bücherei], shelves for standard-format books determine the arrangement of objects in space. Symbolically, they suggest an interior realm in which architecture and information are one, one in which the collector is no longer troubled by the clutter that both defines—and undermines—

2. VERKEHRSBÜRO

Figure 10.3   Emil Pirchan, "Travel Office" (reprinted from K.W. Bührer, *Raumnot und Weltformat* **1912)**

his sense of home. Pirchan's idealized collector miniaturizes and regularizes the objects contained in the domestic realm. As such, he also resolves the "crisis of the room" by improving the room's collecting capacities. Capitalism needs new markets in order to sustain existing ones; similarly, the bourgeois room can afford infinite roominess, at least as Pirchan (and Die Brücke for that matter) imagine it.

Pirchan's illustrations raise an important question, particularly where Die Brücke is concerned: namely, how are we to view the latter's project vis-à-vis other contemporaneous cultural experiments such as those undertaken by the German Werkbund? My view is that they need to be seen in analogous terms. As Massimo Cacciari has observed (1993), the Werkbund aimed to rescue the bourgeoisie from its own excesses. In contrast to the Austrian architect-polemicist Adolf Loos, he argues, it was committed to the Wagnerian ideal of the *Gesamtkunstwerk*, the possibility that the arts and crafts, architecture and interior design, could achieve a harmonious rapprochement under capitalism:

> Where the Werkbund imagines bridges, Loos posits differences. And this holds as true for the general difference between art and handicraft as for the internal differences that make up the structures of the various languages of composition: the

8. GROSS-BÜCHEREI

Figure 10.4   Emil Pirchan, "Large Library" (reprinted from K.W. Bührer, *Raumnot und Weltformat* 1912)

languages that figure in the composition of dwelling and the home, the experience of which constitutes the base on which the Loosian *Baukunst* (architecture) is defined. A fundamental difference exists between the wall, which belongs to the architect, and the furnishing, the overall composition of the interior, which must ensure maximum use and transformation by the inhabitant. This is a difference of languages, which no aura of universal syntax will ever be able to overcome. The wall is form, calculated space-time—it is "abstract." It would be absurd, "Wagnerian," to attempt to reconcile it with the interior, this lived experience, with the space of the multiplicity of languages that make up life. Therefore, the bourgeois, philistine concept of the home—the concept of a totality of dwelling, of a reciprocal transparence between interior and exterior—on which every *Stilarchitektur* has been based up to this point, is intrinsically, logically false. The home is in reality a plurality of languages that cannot be reduced to unities by the deterministic logic of nineteenth-century positivistic utopianism. (Cacciari 106–7)

As Cacciari notes, the most significant flaw of the Werkbund is that it aimed to mask the experience of alienation that defined life under capitalism at the start of the 20th century. It aimed to do so, moreover, by indulging in escapist fantasies, ones that fetishized—and ultimately also aestheticized—bourgeois ideas about domesticity. In an analogous way, Die Brücke used "the crisis of the room," the *Raumnot*, to confront the crisis of capital. It used the figure of the collector to confront the "problem" of collecting. Like the Werkbund, it also privileged the *Gesamtkunstwerk* as a cultural ideal, contrary to what Sachsse has argued (Sachsse 2004, 87). It sought to develop a unified visual identity, one that obfuscated the alienated conditions under which knowledge is produced in capitalist societies. It had its own trademark which "developed between 1911 and 1913 from a conspicuous symbol knitted from the sun, bridge, coat of arms and dual frame into a logo for the invitation to the annual meeting" (Sachsse 2004, 85). Under Pirchan's direction, Die Brücke developed unified conventions for organizing information on standard-format sheets of paper. Given the foregoing, it is probably no coincidence that it drew from the Werkbund's ranks in an effort to strengthen its cause. Both Hermann Muthesius and Peter Behrens were among its circle of supporters; reciprocally, Wilhelm Ostwald participated actively in the Werkbund's 1914 meeting(Campbell 172–3).

Conclusion

Die Brücke's project was not entirely forward-thinking as its history might otherwise suggest. Rather, it was a throwback to an era when the figure of the collector mediated relations between the public and private sphere. Its example reminds us of how our desire to collect information is often connected to our desire to collect things. It also complicates our understanding of the information age at large. Indeed, Die Brücke struggled to present facts as artifacts, informational commodities as formal commodities. The standardization of paper represented one means through which it its members hoped to safeguard bourgeois ideas about privacy. Standard paper formats anticipate the advent of "collecting machines" such as the Internet. They thus also reveal some of the complicated ways in which dreams of the 19th-century have endured well into the twenty first.

References

Asendorf, Christoph. 1994. *Batteries of Life: On the History of Things and their Perception in Modernity.* Berkeley: University of California Press.

Banham, Reyner. 1980. *Theory and Design in the First Machine Age*, 2nd ed. Cambridge: MIT Press.

Benjamin, Walter. 1999. *The Arcades Project*, translated by Howard Eiland and Kevin McLaughlin. Cambridge: Belnap Press of Harvard University Press.

Brown, Julia Prewitt. 2008. *The Bourgeois Interior: How the Middle Class Imagines Itself in Literature and Film.* Charlotte: University of Virgina Press. *Google Play* file.

Buckland, Michael. 2008. "On the Cultural and Intellectual Context of European Documentation in the Early Twentieth Century," *European Modernism and the Information Society: Informing the Present, Understanding the Past*, edited by W. Boyd Rayward. London: Ashgate, Alsdershot, Hants., 45–57.

Bührer, K.W. and Saager, Adolf. 1911. *Allgemeine Gesichtspunkte. Das Keller'sche Musikarchiv*. Ansbach: Fr. Seyboldt's Buchhandlung.

Bührer, K.W. 1912a. *Raumnot und Weltformat*. Munich: Die Brücke.

Bührer, K.W. 1912b. *Weltarchiv der Bücke. Abteilung Kleingraphik*. Munich: Die Brücke.

Cacciari, Massimo. 1993. *Architecture and Nihilism: On the Philosophy of Modern Architecture*, translated by Stephen Sartarelli. New Haven: Yale University Press.

Campbell, Joan. 1978. *The German Werkbund: The Politics of Reform in the Applied Arts*. Princeton, N.J.: Princeton University Press.

Engelman, Edmund. 1976. *Berggasse 19: Sigmund Freud's Home and Offices, Vienna 1938*. Chicago: University of Chicago Press.

Forty, Adrian. 1986. *Objects of Desire: Design and Society 1750*. London: Thames and Hudson.

Fuss, Diana. 2004. *The Sense of an Interior. Four Writers and the Rooms that Shaped Them*. Charlottesville: University of Virgina Press. *Kindle* ebook file.

Greenblatt, Stephen. 1990. "Resonance and Wonder." In *Bulletin of the American Academy of Arts and Sciences*, 43 (4): 11–34.

Hapke, Thomas (1999), "Wilhelm Ostwald, Die Brücke (Bridge), and Connections to Other Bibliographic Activities at the Beginning of the Twentieth Century." In M.E. Bowden, T.B. Hahn, and R.V. Williams (eds) *Proceedings of the 1998 Conference on the History and Heritage of Science Information Systems*. Medford, NJ: Information Today. 139–47.

Hapke, Thomas. 2008. "Roots of Mediating Information: Aspects of the German Information Movement." In *European Modernism and the Information Society: Informing the Present, Understanding the Past*, edited by W. Boyd Rayward. Aldershot, Hants: Ashgate. 307–27.

Heuvel, Charles van den, Rayward, W. Boyd and Uyttenhove, Pieter. 2003. "L'Architechture du savoir: une recherche sur le Mundaneum et les précurseurs européen de l'Internet," *Transnational Associations/Associations Transnationales*, issue 1–2 June, pp. 16–28.

Koss, Juliet. 2010. *Modernism after Wagne.*Minneapolis: University of Minnesota Press.

Krajewski, Markus. 2006. *Restlosigkeit: Weltprojekte um 1900.*Frankfurt a.M.: Fischer Taschenbuch Verlag.

Krajewski, Markus. 2011., *Paper Machines. About Cards & Catalogs, 1548–1929*, translated by Peter Krapp. Cambridge, MA: MIT Press.

Ostwald, Wilhelm. 1911. *Die Weltformate I. Für Drucksachen*. Ansbach: Fr. Seyboldts Buchhandlung.

Ostwald, Wilhelm. 1912. *Der energetische Imperativ*. Leipzig: Akademische Verlagsgesellschaft.

Rice, Charles. 2007. *The Emergence of the Interior; Architecture, Modernity, Domesticity*. London: Routledge. *Kindle* ebook file.

Sachsse, Rolf. 2004. "Das Gehirn der Welt: 1912." In Peter Weibel (ed.) *Wilhelm Ostwald: Farbsysteme. Das Gehirn der Welt*. Ostfildern-Ruit: Hatje Cantz Verlag.

Sigmund Freud Museum. http://www.freud.org.uk/photo-library/category/10046/house-couch-study/ (accessed January 2, 2013).

Tschichold, Jan. 1998. *The New Typography: A Handbook for Modern Designers*, translated by Ruari McLean. Berkeley: University of California Press.

Whyte, William H. 1956. *The Organization Man*. New York: Simon and Schuster.

Chapter 11

Alfred H. Fried and the Challenges for "Scientific Pacifism" in the Belle Époque

Daniel Laqua

"The science of internationalism is of a very recent date. It is based on an understanding of international cooperation, its origins and its nature." With these words, Austrian-German pacifist Alfred Hermann Fried introduced the *Annuaire de la Vie Internationale* (Fried 1909, 23). In over 1,300 pages, this "yearbook" summarized the activities of international associations, the proceedings of international congresses and a multitude of initiatives for cultural and political exchange. The publication was not only international in content, but also in its genesis. While Fried operated from Vienna, the other two driving forces behind the project—Henri La Fontaine and Paul Otlet—hailed from Belgium. At the time, La Fontaine was a socialist senator and prominent pacifist, whereas Otlet was primarily known as a bibliographer. Together, they had long been involved in a range of international ventures and supported the *Annuaire* through their Central Office of International Institutions, which they had established in 1907 (Rayward 1975, 172–3).

The collaboration between Fried and the Central Office sheds light on several interrelated issues. First, it allows us to examine the ideas and assumptions that underpinned pacifism and internationalism in the *Belle Époque*. In contemporary usage, the two terms were closely related. After all, "pacifism"—a neologism coined in 1901—did not necessarily denote a universal and unconditional rejection of war (Holl 1978, 771–3). "Organizational pacifism" or "pacific-ism" may be more appropriate labels for many campaigners who regarded themselves as pacifists at this time. They promoted ideas such as the extension of international law, the creation of new international institutions and the strengthening of arbitration mechanisms (Ceadel 1987, 5). There is wealth of scholarship on the peace movement in its national contexts, covering Germany (Holl 1988 and Riesenberger 1985) and Britain (Laity 2002 and Ceadel 2000) but also Belgium (Lubelski-Bernard 1977). Other studies have adopted an international approach in recognition of pacifists' transnational aims and strategies (Cortright 2008; Brock and Young 1999; Grossi 1994; Cooper 1991). Indeed, pacifism as an idea and movement can also be subsumed under the broader term "internationalism," which describes the recognition and promotion of bonds between groups or individuals from different nations (Geyer and Paulmann 2001; Iriye 1997; Kuehl 1986). The activism of Fried, La Fontaine, Otlet and other internationalists was driven by the conviction that the world was moving towards ever-closer interdependence. Their perception of this process explains the confidence with which they pronounced their views in an era of arms races and heightened diplomatic tensions.

The second key issue was the growing number of international associations in the *Belle Époque*. Peace activists consolidated their structures for cultural and political exchange in this period. From 1889, Universal Peace Congresses took place on a near-annual basis, and from 1891–1892, a new body—the International Peace Bureau (IPB) in Bern—coordinated such efforts. La Fontaine was involved in the IPB from the start and served as its president from 1907 until his death in 1943. Meanwhile, Fried attended IPB meetings and reported on the discussions at the Universal Peace Congresses. Historians and political scientists alike have noted the plethora of new international associations that sprang into existence in the decades before the First World War, with the IPB being a case in point (Herren 2009; Iriye 2002; Rasmussen 1995; Lyons 1963). In noting the growing number of these organizations, some scholars draw on information compiled by Fried, La Fontaine and Otlet back in the early twentieth century (e.g. Boli and Thomas 1999). Internationalists sought to validate their efforts by demonstrating the extent of "international life"—and, in so doing, generated data that has informed the work of subsequent generations.

Transnational cooperation was the third major aspect of the Vienna-Brussels axis. The joint efforts of Fried and the Central Office illustrate the potentials and pitfalls of working across national boundaries. The protagonists of this chapter were located at the heart of several international undertakings. In 1892, Fried co-founded the Deutsche Friedensgesellschaft which became the most prominent German peace association. Having returned to his hometown of Vienna in 1903, Fried remained prominently involved in pacifism, especially as editor of the periodical *Die Friedens-Warte*, which he had founded in 1899. In recognition of his efforts, he was awarded the Nobel Peace Prize of 1911. The Nobel Prize committee praised *Die Friedens-Warte* as "the best journal in the peace movement, with excellent leading articles and news of topical international problems" (Haberman 1999, 238). Two years later, La Fontaine received the same accolade, with the secretary of the Nobel Committee asserting that "no-one … has contributed more to the organization of peaceful internationalism" (Ragnvald Moe in Haberman 1999, 270). Meanwhile, La Fontaine's long-time associate Otlet dedicated his life to bibliography and international organization. In recent decades, he has been reappraised as a pioneer in information science or even been portrayed as a prophet of the worldwide web. The pioneering study of Otlet's life (Rayward 1975) offers a multifaceted picture of his internationalism. Subsequent publications have shed further light on his life and his transnational endeavors (Van Acker 2012; Levie 2006; Rayward 1994 and 2003). In contrast to the considerable body of literature on Otlet, there has been no complete biographical study of La Fontaine, although two edited volumes (Mundaneum 2002; Gillen 2012) have discussed various aspects of his activism. While the biographical studies of Fried and Otlet do mention the *Annuaire*, they remain silent on the exact nature of the collaboration between Brussels and Vienna (Schönemann-Behrens 2011, 168; Göhring 2006, 119–20; Levie 2006, 120; Rayward 1975, 175).

The efforts surrounding the *Annuaire de la Vie Internationale* illustrate the ambitions and optimism that animated internationalism before the Great War. The first two sections of this chapter focus on shared concepts and consider the Austro-Belgian venture as an instance of successful transnational cooperation. Yet this case study also reveals the limits of such endeavors. As I will subsequently show, cooperation between Fried and the two Belgian founders of the Central Office of International Institutions was hampered by disagreements and tensions at several levels. Furthermore, as the response to the outbreak of war in 1914 illustrates, pacifists—despite their seemingly idealistic stance—were subject to national constraints that undermined their efforts. The aim of "world peace" was inherently transnational, but this did not mean that pacifists found collaboration unproblematic. With regard to the "patriotic pacifism" of the period before the First World War, Sandi Cooper (1991, 5) has noted the "profound political and national differences ... that reflected the political and cultural values of the activists."

Concepts: Pacifism as Movement and Science

Fried outlined his ideas about war and peace in a number of books, pamphlets and articles. While the Austrian activist himself described his stance as "revolutionary pacifism," the historian Roger Chickering (1975, 102) has viewed it as "almost entirely unoriginal." This assessment may appear harsh, yet Fried's writings certainly went with the grain of pacifist discourse. There were two principal dimensions to his argument. First, he stressed that war was unprofitable, crippled the economy even during peacetime and would become increasingly unlikely because of the enormous devastation that it would cause. For this analysis, he drew upon the work of the Polish industrialist, Jan Bloch, whose study on the "war of the future" was published in 1898. Its abridged English version resonated widely in pacifist circles. The Austrian baroness Bertha von Suttner—herself a celebrated figure in the European peace movement—praised the factual foundations of Bloch's work (2010, 78):

> He [Bloch] did not present himself as an enthusiast and prophet, but as an assiduous scholar—he did not demand faith, but examination ... He said to the experts ... : war has become impossible—this does not mean impossible to wage, as human unreason knows no boundaries, but impossible to bring to a decisive end.

Suttner's comments exemplified the view that Bloch's study "should be treated as a scientific theory" (Welch 2000, 274). In Britain, Norman Angell subsequently developed ideas about the economic futility of war in his famous book on *The Great Illusion* (Angell 1911). Angell and Fried shared another influence: both translated the pacifist writings of Yakov Novikov who—like Bloch—came from the Russian Empire. Novikov had stressed the detrimental effects of military conflict and offered European "federalism" as the solution (Novikov 1901 and 1912).

The second dimension of Fried's argument centered on the view that the world had become increasingly integrated and required appropriate institutions. "Mutual dependence between states," he concluded, had become "so extensive ... that the smallest disturbance of political business would have enormous consequences (Fried 1910a, 201)." According to his critique, links in the economic and cultural sphere were not yet adequately matched in the sphere of international relations (Fried 1905, 47). As a solution, he called for the introduction of robust international norms and mechanisms, exemplified by compulsory arbitration and an international tribunal (Fried 1905, 17–18). The concern to establish a new international order by such means was summarized by the motto of *Die Friedens-Warte*: "organize the world" (*"Organisiert die Welt"*).

Fried claimed that his activism was not driven by ideological or moral concerns, but based on scientific reasoning. He argued that pacifists had understood the direction of historical change and therefore held evidence-based answers to this process. To Fried, growing global interdependence was a provable fact, leading him to adopt the term "scientific pacifism." As a case in point, he published his *Handbuch der Friedensbewegung* in 1905, listing peace associations, cases of international arbitration and general information on international movements. Further editions followed in 1911 and 1913, as did his broader survey *Das internationale Leben der Gegenwart* in 1908 and the first issues of the *Annuaire*. In studying "contemporary international life," Fried listed the increasing number of international congresses and associations, thus seeking to prove the interconnectedness on which his argument was based. In this quest, he drew on financial support from the International Institute of Peace. This organization had been founded in 1902 by Albert I of Monaco. It engaged in educational work but also "offered bibliographical services to the movement" (Cooper 1991, 20). The prince had encountered Bertha von Suttner at the Universal Peace Congress of 1900. Suttner subsequently developed her links with Monaco and ensured support for her friend and fellow campaigner, Fried (Hamann 1986, 311–17).

The endeavors of La Fontaine and Otlet were nourished by concepts that resembled Fried's. The Belgians described the world as being characterized by a movement for closer global integration: they referred to a "law of amplification" that manifested itself in a growing number of international associations and congresses (La Fontaine and Otlet 1907). Their analysis was backed up by their activism. La Fontaine was a key figure in the foundation of the Société belge de l'Arbitrage et de la Paix in 1889. Underlining the significance of transnational links, the Belgian society was set up as an affiliate of the International Arbitration and Peace Association (Lubelski-Bernard 1977, 364–5). Both La Fontaine and Otlet shared Fried's practical concern with documentation. In 1895, they had set up the International Institute of Bibliography. This institution promoted the Universal Decimal Classification as an extended version of the Dewey Classification. It also pursued the aim of a Universal Bibliographical Repertory—a card catalogue that was to assemble "all of the world's knowledge" (La Fontaine and Otlet 1895, 6). Such activities revealed an encyclopedic streak that extended to the realm of peace activism. Indeed, La Fontaine had published a small bibliography on peace-related publications before co-founding the institution with Otlet, and he later documented instances of international arbitration (La Fontaine 1891 and 1902). He also consulted Fried on such efforts, as evidenced by a letter exchange from 1905.[1]

Another phenomenon brought Fried, La Fontaine and Otlet together: the considerable number of international associations and congresses in La Fontaine and Otlet's own country. Belgium's prominent role as a site of internationalism was partly the result of state support for such efforts. As Madeleine Herren has highlighted with regard to the pre-1914 period, "states on the periphery of power" used "strategies of internationalization" to develop their profile in the international arena (Herren 2000, 5). In 1905, Belgian policy along these lines culminated in the World Congress of Economic Expansion. Taking place in the city of Mons, the event shared features with many international events of this period, yet occurred in delicate circumstances, as the rule in King Leopold's Congo Free State had become subject to particular international scrutiny. According to Herren (2000, 14 and 174), the congress was a "skilful mixture between an economic summit and a justification of imperial rule." Taking up one of the event's resolutions, La Fontaine, Otlet and Catholic civil servant Cyrille Van Overbergh planned the creation of the Central Office of International Associations, which began operations in 1907 with a "primary focus ... [that] was seen as documentary in character" (Rayward 1975, 175).

Collaboration: The *Annuaire* as an Example for Transnational Exchange

In light of these shared concerns, collaboration between Fried and the two Belgian activists seemed a logical step. Both sides had something to gain: whereas Fried did not have a firm institutional base in Vienna, La Fontaine and Otlet could rely on their International Institute of Bibliography with its longstanding involvement in documentation efforts. In this respect, the Belgians were able to provide a setting and infrastructure through which Fried's documentary work could be extended. Meanwhile, by teaming up with Fried, the activists from Brussels avoided direct competition with an existing venture and at the same time established a flagship project for their Central Office of International Institutions which was still staking out its agenda in 1907. Fried's Monaco funds are likely to have been a supporting argument for this partnership.

---

1 La Fontaine to Fried, January 14, 1905 and February 8, 1905, Fried-Suttner Papers: Alfred Hermann Fried (IPB/FSP/AHF), box 66 (Correspondence La Fontaine), League of Nations Archives, United Nations Library, Geneva [henceforth *MS AHF*].

Fried initially expressed the "opinion that, given our longstanding acquaintance and the higher aims on which our work is based, a contract is not necessary."[2] Indeed, La Fontaine and Fried knew each other through their involvement in pacifist circles—and it was La Fontaine, rather than Otlet, who subsequently served as Fried's main Belgian contact. Nonetheless, to clarify various practical issues, Fried, La Fontaine and Otlet signed a formal accord in 1907.[3] They agreed to prepare the *Annuaire de la Vie Internationale* as an expanded version of Fried's earlier efforts. The three were to serve as joint editors, with the Central Office providing the central infrastructure. The document stipulated that, if the latter institution was not yet fully operative, the International Institute of Bibliography was to take charge of the practical work. Funding would still come from the Monaco Institute (to the sum of 2,000 francs), with Fried receiving a remuneration for his contribution.[4]

In many respects, the *Annuaire de la Vie Internationale* appears as a successful transnational venture. When it finally appeared in 1909, it was five times the size of Fried's preceding volumes (Rayward 1975, 175). Following introductory essays by Fried, Otlet and La Fontaine, the book covered several diplomatic conferences in significant depth, notably the Second Peace Conference, held in 1907 at The Hague. Further sections listed the activities of intergovernmental bureaus and instances of "international legislation," by which the authors meant different developments in international law. The book's final part was dedicated to "international private life." It surveyed an array of associations, ranging from the natural and social sciences to humanitarian efforts. In a letter to La Fontaine, Fried confirmed that he was "pleased that you were able to transform my idea into something which I could not have done all by myself. The book will undoubtedly perform a valuable service. It will show to the world how far it is already organized."[5] The book's physical appearance—bulky as it was—would underscore the key point: the sheer extent of the "interconnection of all human activity." Fried praised Otlet's essay in the *Annuaire*, describing it as "the first comprehensive survey of the new science which I would like to call 'Internationology'". His comment referred to Otlet's essay of over 130 pages, in which the Belgian bibliographer summarized the history of international organizations and developed a typology for such bodies (Otlet 1909). As a whole, the *Annuaire* was a tangible expression of Fried's "scientific pacifism." It was premised on the notion that international life could be scientifically documented. In the book, La Fontaine spelt out this view in an essay on "Documentation and Internationalism" (La Fontaine 1909).

Yet Fried and his Belgian associates were not the only activists engaged in such efforts. Not long after the first Belgo-Austrian *Annuaire*, the Dutch internationalist Pieter Hendrik Eijkman published his own surveys—*L'internationalisme medical*

---

2   Fried to La Fontaine, November 16, 1907, *MS AHF*.

3   "Convention entre M. A.H. Fried et MM. H. La Fontaine et P. Otlet," La Fontaine Papers, Archives of the Union of International Associations, Brussels [henceforth *MS UIA*].

4   "Projet de budget," *MS UIA*.

5   Fried to La Fontaine, September 18, 1909, *MS UIA*.

and *L'internationalisme scientifique*—conceiving them as the launch pad for his "Foundation for Internationalism" (Eijkman 1910 and 1911). Initially, Fried was impressed. In a postcard to La Fontaine, he stated that "The Hague has overtaken us. We must now pay attention that we are not beaten altogether."[6] La Fontaine was less enthusiastic, regarding Eijkman's efforts not merely as competition, but as derivative: "Can his work really be compared to ours? The information that he provides is, in its great majority, taken from our *Annuaire*; he even reproduces certain fatal errors from a work as considerable as ours."[7] La Fontaine also pointed out in the same note that Fried's "excessive praise" for Eijkman had proven "painful" to both himself and Otlet. In response, the Viennese pacifist explained that he had been unaware of Eijkman's "unscrupulous" use of the Belgians' work. He conceded that his judgment may have been blurred by his "great interest in the development of internationalism," and by the initial joy of holding Eijkman's *L'Internationalisme scientifique* in his hands.[8]

La Fontaine's hostility to Eijkman's project was more than a matter of intellectual pride. Internationalists competed for a limited amount of financial resources. In his quest to make The Hague a "world capital," Eijkman sought subsidies from the Carnegie Endowment for International Peace.[9] However, the Carnegie Endowment was also a key source of revenue for the Belgian activists. From 1912, it provided an annual subsidy of 15,000 US Dollars (or 75,000 Belgian Francs) to support the Central Office of International Institutions and the publication of its periodical *La Vie Internationale*. This was but one of several pacifist ventures supported by the Carnegie Endowment. In the same year, the foundation granted 24,000 US dollars to the International Peace Bureau, over which La Fontaine presided at the time. The main bulk was intended for the bureau itself, with 4,000 US dollars going to national peace societies.[10] Both Fried and La Fontaine were associated with the Carnegie Endowment's European Centre in Paris. They were also among the secretaries of Conciliation Internationale, a Carnegie-backed peace organization led by a prominent French pacifist, Baron d'Estournelles de Constant (Rietzler 2005, 22). Carnegie had started to fund Fried's efforts in 1908, following an encounter between the American philanthropist and Bertha von Suttner (Hamann 1986, 374).

---

6   Fried to La Fontaine, January 4, 1912, *MS UIA*.

7   La Fontaine to Fried, January 31, 1912, *MS UIA*.

8   Letter Fried to La Fontaine, February 5, 1912, La Fontaine Papers, box HLF 064 (correspondance F), Mundaneum, Mons [*MS HLF-F*].

9   Cf. "Correspondence and Documents of Messrs. Eijkman and Horrix, representing the Preliminary Office of the Foundation for the Promotion of Internationalism," Carnegie Endowment for International Peace Records, part III: Division of Intercourse and Education (henceforth *MS CEIP*), vol. 79: "Academy of International Law in The Hague," Rare Books and Manuscript Library, Columbia University.

10   UAI, "Faits rélatifs au développement de l'Union des Associations Internationales," May 5, 1912, La Fontaine Papers, box 218, Mundaneum, Mons; La Fontaine to Fried, July 7, 1912, *MS AHF*. These sums also feature in the records of the Carnegie Endowment (Rietzler 2005, 26).

Money was a major concern for many internationalists and the Belgian activists felt underfunded even after the granting of Carnegie support. They had initially hoped for 375,000 francs and, in expectation of this subsidy, had considered launching a German-language publication under Fried's editorship.[11] The much lower award of 75,000 francs put paid to such plans. As La Fontaine argued, the "insufficient sum" from the Carnegie Endowment meant that the Brussels activists had to downscale their ambitions and spend more of their time without financial compensation.[12] Nonetheless, it was this financial support which ultimately allowed Otlet, La Fontaine and their compatriot Van Overbergh to launch the journal *La Vie Internationale* in 1912—a publication that was driven by similar ideas to those underlying the *Annuaire*. Such examples testify to the increasingly prominent role of American philanthropy in European politics and culture (Rietzler 2013; Krige and Rausch 2012; Parmar 2011). However, they also indicate how any shifts in American grant-giving would have direct implications for European internationalists.

Conflict: The Obstacles to Transnational Cooperation

Funding-related disappointments suggest that not everything was rosy in the international world of European pacifism. Indeed, it is possible to view the Vienna–Brussels axis from a very different angle: namely as a case of shared convictions and ambitions being undermined by practical obstacles and personal frustrations. The correspondence between Fried and La Fontaine exemplifies the manifold challenges for transnational collaboration. One key conflict was over inadequate communication. Fried frequently referred to the lack of correspondence from Brussels. In May 1908, a letter started with the words "Finally a sign of life!"[13] In December 1908, a postcard reminded the Belgians of Fried's questions regarding the status of the *Annuaire*.[14] Around four weeks later, an exasperated Fried claimed that he began to view it as "a personal insult" that his repeated communications had not met with a response.[15] Compelled to react, La Fontaine excused the silences with the health of his in-laws, an influenza from which he and his wife had suffered, as well as problems with the printers of the *Annuaire*.[16]

Further evidence suggests, however, that La Fontaine and Otlet, with their multiple commitments in the fields of politics, bibliography and international activism, had overburdened themselves. Indeed, communication problems persisted throughout 1909. In April, another letter from Fried complained that

---

11   Fried to La Fontaine, January 18, 1912, in both *MS UIA* and *MS AHF*; La Fontaine to Fried, January 31, 1912, *MS UIA*.

12   La Fontaine to Fried, January 10, 1912, *MS UIA*.

13   Fried to La Fontaine, May 26, 1909, *MS UIA*.

14   Fried to La Fontaine, December 19, 1908, *MS HLF-F*.

15   Fried to La Fontaine, 28 January 1909, *MS HLF-F*.

16   La Fontaine to Fried, January 30, 1909, *MS AHF*.

effective correspondence was "impossible if one has to wait two months for an answer."[17] In June, Fried wondered whether "it might still be possible to get an answer from you."[18] One month later, he described it as "not in order" that his repeated queries had been ignored.[19] In November, he explained that he did not expect a personal letter, but at least an answer via La Fontaine and Otlet's support staff.[20] Even though the three activists served as joint editors of the *Annuaire*, it is clear that the circumstances posed significant obstacles for Fried's involvement.

The inherent problems of the Brussels office did not only affect the correspondence with Fried, but also delayed the *Annuaire*'s publication. The first Vienna–Brussels edition was supposed to cover the year 1907—yet by March 1909, it still had not appeared. Fried concluded that two editions would have appeared if he had continued to work by himself. Given the slow pace at which the project progressed, he expressed his regret at having transferred the Monaco funds.[21] Fried reiterated these points in June 1909:

> If the *Annuaire* does not get published in July, I will give up hope altogether. It seems that you have miscalculated the whole format. You are not publishing a yearbook, but a handbook of international life. Had I stayed in Monaco, I would have published two volumes by now. I will have to speak to you in Stockholm [host city of the Universal Peace Congress of 1910] whether I shouldn't go back to publishing it by myself.[22]

La Fontaine admitted his own frustration at these delays and the amount of work that the *Annuaire* had caused him. Yet he promised that the outcome would obliterate these disappointments. The publication would "cause a sensation: that is what is essential."[23]

La Fontaine's optimism about the public response was barely justified. The *Annuaire* was finally printed in autumn 1909, covering the years 1908 and 1909, yet Fried soon noted its poor distribution. Several people had told him that the Brussels office had not answered their queries and a French bookseller had informed him of the difficulty of obtaining copies.[24] Fried concluded that the *Annuaire* was published at the exclusion of the general public: "Nobody knows about the book." La Fontaine apologized by referring to "insurmountable difficulties" in Brussels.[25] By 1910, the promotion of the *Annuaire* was complicated by the fact that the

---

17   Fried to La Fontaine, April 7, 1909, *MS AHF*.
18   Fried to La Fontaine, June 30, 1909, *MS UIA*.
19   Fried to La Fontaine, July 6, 1909, *MS HLF–F*.
20   Fried to La Fontaine, November 17, 1909, *MS UIA*.
21   Fried to La Fontaine, March 23, 1909, *MS UIA*.
22   Fried to La Fontaine, June 30, 1909, *MS UIA*.
23   La Fontaine to Fried, July 13, 1909, *MS AHF*.
24   Fried to La Fontaine, November 27, 1909 and February 19, 1910, both in *MS UIA*.
25   La Fontaine to Fried, April 20, 1910, *MS AHF*.

Central Office was involved in another large-scale project: organizing the first World Congress of International Associations in Brussels. Held in May that year and coinciding with the Brussels World's Fair, it brought together representatives of 137 international organizations, involving politicians, activists and academics such as the Nobel laureate, Wilhelm Ostwald. The event resulted in the foundation of the Union of International Associations, which sought to be focal point and lobby for various groups, societies and institutions and for which the Central Office served as a secretariat (Herren 2000, 186; Rayward 1975, 179–83). Having thus extended their remit, it is clear why La Fontaine and Otlet were slow to prepare a follow-up to the *Annuaire*. Fried, evidently, was disappointed. Evoking their longstanding acquaintance and cooperation, he urged them to establish a *modus vivendi*.[26] La Fontaine responded by proposing a division of labor, with Fried exclusively focusing on international events. At the same time, La Fontaine stressed that practical reasons required them to publish the "yearbook" on a biennial rather than annual basis.[27]

There was a further dimension to the tensions involving Fried and the Central Office. As a full-time writer and activist, Fried frequently struggled financially (Schönemann-Behrens 2011, 145–6 and 186). Expressing concerns about his personal situation, he asked for an advance in July 1908 and, five months later, stressed that the *Annuaire*'s timely publication had important material implications for him.[28] In April 1909, Fried drew attention to the "severe embarrassment" caused by the delays in paying him.[29] Yet the Belgians often suffered from limited funds themselves, as La Fontaine admitted.[30] These concerns explain why external funding through foundations or other organizations was such a major objective. In coordination with Fried, La Fontaine contacted Abbé Pichot, a French clergyman and representative of the Institute of International Peace. Pichot was a Catholic Dreyfusard; he lost his job as professor of mathematics in France when supporting Dreyfus during the *affaire* (Delmaire 2000, 28). La Fontaine asked him to intervene with the Prince of Monaco to see if his subsidy could be expanded seeing that the *Annuaire* now covered two years and that Fried's "modest undertaking" had been developed by the two Belgians.[31] Restating this request, La Fontaine stressed that the development was "not without great expense for our poor office."[32] Pichot, however, regarded such a request as "inopportune."[33]

Such concerns account for the hopes vested in attracting Carnegie funding. Although the Carnegie grant exceeded the Monaco funding by far, Fried reacted

---

26 Fried to La Fontaine, January 21, 1911, *MS UIA*.
27 La Fontaine to Fried, February 2, 1911, *MS AHF*.
28 Fried to La Fontaine, June 11, 1908 and November 16, 1908, *MS HLF-F*.
29 Fried to La Fontaine, April 17, 1909, *MS HLF-F*.
30 See e.g. a brief reference in La Fontaine to Fried, June 4, 1909, *MS AHF*.
31 La Fontaine to Abbé Pichot, October 30, 1909, *MS UIA*.
32 La Fontaine to Pichot, November 24, 1909, *MS UIA*.
33 Pichot to La Fontaine, November 1909 (undated letter), *MS UIA*.

with disappointment about the news about the American subsidy, describing it as a "cold shower."[34] Meanwhile, the personal toll on La Fontaine became evident in a letter to the Carnegie Endowment in November 1911. As he pointed out, the Central Office in Brussels had "published 4,148 pages, all being proofread by me. Every day I proofread." As a result of this situation, he even suggested a possible end to his activism to document international life: "My friend Otlet and I work 12–14 hours a day. No theatre, no book, no social life. We cannot go on. I have decided to give up ... Only Carnegie can help us."[35]

Continuation: Joint Efforts in the Face of Tensions

Clearly, shared aims and ideas did not prevent strains at different levels. However, despite their frustrations, Fried persevered. Early in 1911 he affirmed his willingness to continue working with La Fontaine and Otlet on the *Annuaire*.[36] By early 1912, however, he feared that the two Belgians might "push him aside" and insisted on adherence to their written accord.[37] He acknowledged the logistical problems involved in editing a publication from two different countries, but stressed the insignificant nature of such obstacles in comparison to "other, far greater challenges" for pacifists. In return, La Fontaine expressed himself to be "profoundly pained" by Fried's letter as he had never intended to marginalize him.[38] He suggested that Fried could write the introduction of the next edition, although the main work would continue to take place in Brussels. Despite his diminishing role in the transnational venture, Fried praised the next volume of the *Annuaire* as "an extraordinary achievement: you have really created something wonderful." He also reiterated his enthusiasm for a potential German edition.[39] Fried's apparent satisfaction coincided with a period of good fortune for the Brussels efforts: Rayward (1975, 195) has described the period between 1910 and 1914 as "the years of greatest success for Otlet and La Fontaine" who were "secure, happy and relatively prosperous."

Even at times when their joint efforts for the *Annuaire* were subject to disagreements and disappointments, La Fontaine and Fried maintained a conciliatory dialogue in other regards. The International Peace Bureau was one framework in which such exchanges occurred. In October 1909, for instance, Fried attended an IPB session in the Belgian capital, met up with La Fontaine and

---

34   Fried to La Fontaine, May 29, 1911, *MS UIA*.

35   La Fontaine to Nicolas Murray Butler, November 21, 1912, in *MS CEIP*, vol. 178: "L'Office Central, Brussels, 1911–1915." I am indebted to Pierre-Yves Saunier for sharing his notes on vols. 178 and 179 of the CEIP files.

36   Fried, February 10, 1911, *MS UIA*.

37   Fried, January 18, 1912, *MS UIA*.

38   La Fontaine, January 31, 1912, in both *MS UIA* and *MS AHF*.

39   Fried to La Fontaine, January 14, 1913, *MS HLF-F*.

Otlet on this occasion, and returned to Vienna with "fond memories."[40] The three men also solicited contributions from each other for their respective periodicals *Die Friedens-Warte* and *La Vie Internationale*.[41] At a personal level, La Fontaine was touched by the "kind words" that *Die Friedens-Warte* had dedicated to him on the occasion of his 60th birthday.[42] Such examples suggest that occasional tensions did not put an end to mutual esteem.

Furthermore, cooperation continued to be desirable in the quest for external recognition. In 1910 La Fontaine asked Fried to nominate the International Peace Bureau—over which the Belgian presided at the time—for the Nobel Peace Prize. He argued that its work had been set back by the lack of sufficient funds, which could be remedied through Nobel money.[43] The IPB was indeed awarded the prize that year, although the Nobel Nominations Database suggests that Fried had not followed up La Fontaine's request.[44] A year later, Fried returned to the issue of nominations, pointing out that he would back La Fontaine: "You know that you have been my candidate for two years and that I will also nominate you this year."[45] When it came to joint projects and exchanges, we cannot always view them through the lens of "conflict versus cooperation": attitudes vacillated according to external necessities or circumstances.

Collapse: The Impact of the First World War

"Scientific pacifism" postulated that ever-greater global interdependence would make wars increasingly unlikely. The outbreak of the First World War manifestly contradicted such optimism. However, as far as their theories were concerned, neither Fried nor the Belgian activists admitted defeat. In 1915, Fried argued that the war had not discredited his arguments. In contrast, it had proved that the states had made inadequate attempts to organize international life and thus vindicated pacifists' demands for international rules and institutions. The book that presented this argument—*Europäische Wiederherstellung*—sought to suggest a remedy for this state of affairs by outlining a new international order. Meanwhile both La Fontaine and Otlet devised their own schemes for post-war Europe. In his book *The Great Solution: Magnissima Carta*, La Fontaine expressed ideas

---

40   Fried to Central Office of International Institutions, October 13, 1909, *MS UIA*. See also Fried's letter to La Fontaine prior to the meeting, September 18, 1909, *MS UIA*.

41   See e.g. La Fontaine inviting a contribution from Fried on May 8, 1912, *MS AHF*.

42   Undated, La Fontaine to Fried, c. April/May 1914, *MS AHF*.

43   La Fontaine to Fried, January 15, 1910, *MS AHF*.

44   "The Nomination Database for the Nobel Peace Prize, 1901–1956," http://www. nobelprize.org/nobel_prizes/peace/nomination/database.html [accessed 20 June 2012].

45   Fried to La Fontaine, January 14, 1913, in both *MS AHF* and *MS HLF-F*. The Nominations Database lists Fried as nominating La Fontaine in 1912 and 1913, but – in apparent contradiction to his claims – not 1911.

that resembled Fried's, and reiterated the notion that international life had already become "strongly intensified" before the war (La Fontaine 1916, 75). According to him, it was necessary to increase the awareness of this process and to create appropriate institutions "which can easily adapt themselves to circumstances and contribute to their own improvement" (La Fontaine 1916, 6). Otlet also called for the creation of world organization. He presented a draft "peace treaty" and charter of rights on which a new international order could be based, specifying his vision in several publications during the war years (Otlet 1914, 1915 and 1917).

Such examples suggest that in theoretical terms, the war itself did not trigger the evaporation of pacifist narratives. At a practical level, however, the implications of international conflict became evident as early as the summer of 1914. Bertha von Suttner had died in June 1914 and Fried took charge of preparations for the next Universal Peace Congress, which was scheduled to take place in Vienna. However, on June 28, the Austrian Archduke Franz Ferdinand was assassinated in Sarajevo. His death triggered the July Crisis, a month characterized by the escalation of international tensions. After Austria's declaration of war on Serbia on July 28, the pacifists decided to cancel their congress in Vienna. In light of the international situation, La Fontaine convened a meeting in Brussels, seeking to bring together the International Peace Bureau's leadership to coordinate the peace movement's reaction. Fried, however, was unable to attend: "Through the strains of the congress preparations and the horrendous political situation, the condition of my health is so shaken that I must absolutely give myself a few weeks of rest."[46] The IPB met on July 31, 1914 with the Austrian Peace Society being the only one of the "major national societies" to be unrepresented (Cooper 1991, 187). On August 4, Belgium fell victim to German aggression. Two days later La Fontaine expressed his anger in a letter to Fried:

> What a deplorable situation has your government caused for Europe and for my poor and vigilant country!! We are shocked but at the same time proud to have responded honestly to the perfidious proposition of the *Friedenskaiser* (Peace Emperor). May our good right give victory to our army and our allies! ... This time, Belgium has really merited, through its vigilance and its willingness for sacrifice, to become the world centre.[47]

The reference to the *Friedenskaiser* was an implicit criticism of Fried who previously had expressed his hope that Wilhelm II might be a force for peace (Fried 1910). Despite the reproachful tone of his letter, La Fontaine's post-scriptum indicated a willingness to contemplate further cooperation: "We should already reflect on what one could do after the war. If you have any ideas, communicate them to me." In November, however, La Fontaine addressed direct criticism at Fried, whom he reproached for not having protested against the violation of

---

46 Fried to La Fontaine, July 29, 1914, *MS AHF*. Cf. Grossi 1994, 387.
47 Letter La Fontaine to Fried, August 6, 1914, *MS HLF-F*.

Belgium's neutrality. Fried, in his turn, believed that La Fontaine "had still not regained his objectivity."[48]

Given the Belgian experience of invasion and occupation, La Fontaine initially rejected calls for an International Peace Bureau meeting with activists from the warring nations (Cooper 1991, 191). When such a meeting was eventually held in January 1915, Fried felt that La Fontaine's attitude to the German and Austrian delegates was marked by "hatred bordering on brutality." The Austrian was hurt about the attitude of his former associate who, it seemed to him, "wanted to make us pacifists to some extent co-responsible for Belgium's fate."[49] The historian Sandi Cooper has drawn attention to the acrimonious nature of this meeting, at which La Fontaine stated that "As much as I am … an internationalist, I am also … Belgian" and he asked the German and Austrian delegates to "acknowledge the wrong that [their nations] have done to us" (Cooper 1991, 196).

During the war, La Fontaine continued to seek funds for the Brussels-based periodical *La Vie Internationale* and the next issue of the *Annuaire*. In a letter to the Carnegie Endowment for International Peace, he argued that an end to American funding would be a "disaster."[50] The foundation had, in fact, decided to suspend its support for the Brussels office and the International Peace Bureau, describing this step as a result of the international situation. Yet the war allowed the foundation to obscure an earlier policy shift. As Katharina Rietzler has shown, the Carnegie Endowment had already decided in late 1912 to phase out direct funding for pacifist organizations (Rietzler 2005, 32). Already by this point, some American funders viewed support for the likes of La Fontaine and Fried as unproductive as they viewed them as "sentimentalist propagandists that did not appeal to the political elite" (Rietzler 2005, 47).

The American decision to stop support for La Fontaine's project caused him the "greatest depression." He complained about his treatment: "I protest and reserve my right to explain publicly how pacifists were unable to get aid and help when the time came"[51] In a subsequent letter, he pointed out that the *Friedens-Warte* still drew on Carnegie support, but claimed that, in contrast to the continuing internationalist spirit of *La Vie Internationale*, it had adopted a "nationalistic tone."[52] This assessment reveals the continuing critical attitude that La Fontaine had adopted towards Fried. The Austrian pacifist argued that the domestic situation had limited his scope of action – in fact he subsequently left Vienna and moved to Switzerland, where he adopted a more outspoken stance. Nonetheless,

---

48 Diary entry of November 11, 1914 (Fried 2005, 52).

49 Diary entry January 11, 1915 in Fried, *Mein Kriegstagebuch*, p. 64; cf. Grossi, *Le pacifisme européen*, p. 395.

50 La Fontaine to Murray Butler, December 3, 1914, in *MS CEIP*, vol. 178: 'L'Office Central, Brussels, 1911–1915.'

51 La Fontaine to Murray Butler, December 3, 1914 (see above note 50).

52 La Fontaine to Murray Butler, March/April 1916, in *MS CEIP*, vol. 179: 'L'Office Central, Brussels, 1915–1922.'

the partnership was not rekindled. In 1919, Fried received La Fontaine's letter from August 1914 with a delay of over five years. In response, he stressed that he viewed criticism of his war-time stance as misguided: "I cannot, for God's sake, assume that people of your educational standard make me responsible for the war, for the attack on Belgium, for all the terrors of Prussian militarism!"[53] As he pointed out, German nationalists had frequently attacked him during the war, and his published war diaries revealed his critical attitude to the German and Austrian policies. Indeed, as his biographer Peter Schönemann-Behrens has stressed (2011, 251–5), the nationalist press treated Fried in a very hostile manner and the German authorities prohibited the continued publication of *Friedens-Warte* in April 1916. Fried's post-war letter to La Fontaine ended on a nostalgic note, appealing to the "memory of the old times, the days of the Antwerp congress 25 years ago." The personal rifts, such as that between La Fontaine and Fried, underline the contradiction that existed between the narrative about ever-greater international life and the challenge of putting this into practice at a more personal level.

Conclusion

This Austro-Belgian case study has highlighted the practicalities and impracticalities of working for a shared cause across national borders. Language is one aspect that was relevant in this context. As Brigitte Hamann has observed, Fried's French was poor, as evidenced by his difficulties in translating Novicov's work on a European federation from French into German (Hamann 1986, 326). The two Belgians, however, understood both German and English. A working pattern emerged in which Fried wrote in German and La Fontaine responded in French, though making occasional forays into German. Such communication was easier in writing than face-to-face. Noting the prevalence of French at the International Peace Bureau's meetings, Fried pointed out that "us non-French" rarely had opportunities "to express our opinions to the full extent."[54]

The history of the *Annuaire de la Vie Internationale* also sheds light on the nature and workings of internationalism. For activists such as Fried, La Fontaine and Otlet, there were no clear dividing lines between political, cultural and scientific exchange. Today, the political dimension of a publication such as the *Annuaire* may be self-evident as it makes the case for closer international cooperation. To its editors, however, the "yearbook" was a scholarly undertaking. For instance, Otlet's long essay in the 1909 volume drew upon a survey of international organizations. By basing his arguments on the gathering of "facts," he affirmed the scientific basis of his observations – just as Fried claimed that pacifism was rooted in scientific principles. At the same time, the *Annuaire* showed how internationalism encompassed political, social, cultural and economic notions.

---

53   Fried to La Fontaine, November 17, 1919, *MS HLF-F.*
54   Fried to La Fontaine, November 30, 1911, *MS AHF.*

While affirming unity in terms of their promotion of internationalist conceptions, activists experienced manifold rifts in their collaboration.

When examining transnational activism, historians often stress ideological divisions and disagreements. This case study, however, draws attention to a different kind of challenge: the maintenance of regular correspondence. This proved particular difficult with activists such as La Fontaine and Otlet, who were prone to involvement in several ventures at the same time. In light of the obstacles to communication encountered by Fried, it seems surprising that the Nobel Committee praised La Fontaine's "outstanding talent for administration" (Haberman 1999, 270). Despite the inefficiency of the Central Office and Fried's apparent irritability, the Vienna-Brussels axis survived for several years. Already before the war, it had stalled to some extent, yet the war itself caused a triple rupture in the form of personal rifts, practical challenges and funding inadequacies.

References

Angell, Norman. 1911. *The Great Illusion: A Study of the Relation of Military Power in Nations to their Economic and Social Advantage*. London: Heinemann.

Bloch, Jan. 1899. *Is War Now Impossible? Being an Abridgment of 'The War of the Future in its Technical, Economic and Political Relations'. With a Prefatory Conversation with the Author by W. T. Stead*. London: Grant Richards.

Boli, John and George Thomas. 1999. "INGOs and the Organization of World Culture." In *Constructing World Culture: International Non-Governmental Organizations Since 1875*, edited by idem. Stanford: Stanford University Press, 13–49.

Brock, Peter and Nigel Young. 1999. *Pacifism in the Twentieth Century*. New York: Syracuse University Press.

Ceadel, Martin. 1987. *Thinking about Peace and War*. Oxford and New York: Oxford University Press.

Ceadel, Martin. 2000. *Semi-Detached Idealists: The British Peace Movement and International Relations, 1854–1945*. Oxford: Oxford University Press.

Chickering, Roger. 1975. *Imperial Germany and a World Without War. The Peace Movement and German Society, 1892–1914*. Princeton: Princeton University Press.

Cooper, Sandi. 1991. *Patriotic Pacifism: Waging War on War in Europe, 1815–1914*. New York and Oxford: Oxford University Press.

Cortright, David. 2008. *Peace: A History of Movements and Ideas*. New York: Cambridge University Press.

Delmaire, Danielle. 2000. "Antisémitisme des catholiques au vingtième siècle: de la revendication au refus." In *Catholicism, Politics and Society in Twentieth-Century France*, edited by Kay Chadwick. Liverpool: Liverpool University Press, 26–46.

Eijkman, Pieter. 1910. *L'internationalisme médical*. The Hague: Bureau préliminaire de la Fondation pour l'Internationalisme.

Eijkman, Pieter. 1911. *L'internationalisme scientifique (Sciences pures et lettres)*. The Hague: Bureau préliminaire de la Fondation pour l'Internationalisme.

Fried, Alfred Hermann. 1905. *Handbuch der Friedensbewegung*. Berlin and Leipzig: Verlag der Friedens-Warte.

Fried, Alfred Hermann. 1908. *Das internationale Leben der Gegenwart*. Berlin and Leipzig: Verlag der Friedens-Warte.

Fried, Alfred Hermann. 1909. "La Science de l'Internationalisme." In *Annuaire de la Vie Internationale*, edited by Alfred Hermann Fried, Henri La Fontaine and Paul Otlet. Office Central des Institutions Internationales: Brussels, 23–8.

Fried, Alfred Hermann. 1910a. "Von den Delegationen." *Die Friedens-Warte* 12: 201–5.

Fried, Alfred Hermann. 1910b. *Der Kaiser und der Weltfrieden*. Berlin: Maritima.

Fried, Alfred Hermann. 1916. *Europäische Wiederherstellung*. Zürich: Füssli.

Fried, Alfred Hermann. 2005. *Mein Kriegstagebuch. 7. August 1914 bis 30. Juni 1919*. Bremen: Donat.

Geyer, Martin and Paulmann Johannes. (eds) 2001. *The Mechanics of Internationalism: Culture, Society and Politics from the 1840s to the First World War*. Oxford: Oxford University Press.

Gillen, Jacques (ed.) 2012. *Henri La Fontaine, Prix Nobel de la Paix en 1913. Un belge épris de justice*. Brussels: Editions Racine.

Göhrig, Walter. 2006. *Verdrängt und vergessen: Friedensnobelpreisträger Alfred Hermann Fried*. Vienna: Kremayr and Scheriau.

Grossi, Verdiana. 1994. *Le pacifisme européen: 1889–1914*. Brussels: Bruylant.

Haberman, Frederick. (ed.) 1999. *Nobel Lectures in Peace (1901–1925)*. new edn. Singapore: World Scientific Publishing.

Hamann, Brigitte. 1986. *Bertha von Suttner: Ein Leben für den Frieden*. Munich: Piper.

Herren, Madeleine. 2000. *Hintertüren zur Macht: Internationalismus und mordenisierungsorientierte Außenpolitik in Belgien, der Schweiz und den USA, 1865–1914*. Munich: Oldenbourg.

Herren, Madeleine. 2009. *Internationale Organisationen seit 1865: Eine Globalgeschichte der internationalen Ordnung*. Darmstadt: Wissenschaftliche Buchgesellschaft.

Holl, Karl. 1978. "Pazifismus." Vol. 4 of *Geschichtliche Grundbegriffe: historisches Lexikon zur politisch-sozialen Sprache in Deutschland*, edited by Otto Brunner. Stuttgart: Klett, 771–3.

Holl, Karl. 1988. *Pazifismus in Deutschland*. Frankfurt: Suhrkamp.

Iriye, Akira. 1997. *Cultural Internationalism and World Order*. Baltimore and London: Johns Hopkins University Press.

Iriye, Akira. 2002. *Global Community: The Role of International Organizations in the Contemporary World*. Berkeley and Los Angeles: University of California Press.

Krige, John and Rausch Helke. (eds) 2012. *American Foundations and the Coproduction of World Order in the Twentieth Century*. Göttingen: Vandenhoeck & Ruprecht.

Kuehl, Warren. 1986. "Concepts of Internationalism in History." *Peace and Change*, 11: 1–10.

La Fontaine, Henri. 1891. *Essai de bibliographie de la Paix*. Brussels: Lombaerts.

La Fontaine, Henri. 1902. *Pasicrisie internationale: histoire documentaire des arbitrages internationaux, 1794–1900*. Bern: Stämpfli.

La Fontaine, Henri. 1909. "Documentation et l'Internationalisme." In *Annuaire de la Vie Internationale*, edited by Alfred Hermann Fried, Henri La Fontaine and Paul Otlet. Brussels: Office Central des Institutions Internationales, 166–81.

La Fontaine, Henri. 1916. *The Great Solution: Magnissima Charta*. Boston: World Peace Foundation, 1916.

La Fontaine, Henri and Otlet Paul. 1895. *Sur la création d'un Répertoire Bibliographique Universel. Conférence bibliographique internationale*. Brussels: Larcier.

La Fontaine, Henri and Otlet Paul. 1907. "La loi d'ampliation et l'internationalisme." *Le Mouvement sociologique international*, 1: 133–62.

Laity, Paul. 2002. *The British Peace Movement, 1870–1914*. Oxford: Clarendon.

Levie, Françoise. 2006. *L'homme qui voulait classer le monde: Paul Otlet et le Mundaneum*. Brussels: Les Impressions Nouvelles.

Lyons, Francis. 1963. *Internationalism in Europe 1815–1914*. Leiden: Sythoff.

Lubelski-Bernard, Nadine. 1977. "Les mouvements et les idéologies pacifistes en Belgique 1830–1914." PhD diss., Université Libre de Bruxelles.

Mundaneum. 2002. *Henri La Fontaine: Tracé[s] d'une vie. Un Prix Nobel de la Paix, 1854–1943*. Mons: Mundaneum.

Nehring, Holger. 2009. "Pacifism." In *The Palgrave Dictionary of Transnational History*, edited by Akiria Iriye and Pierre-Yves Saunier. Basingstoke: Palgrave, 803–6.

Novikov, Yakov. 1912. *War and its Alleged Benefits: With an Introduction by Norman Angell*. London: William Heinemann.

Nowicow, Jacques [Novikov, Yakov]. 1901. *Die Föderation Europas: Autorisierte Übersetzung von A.H. Fried*. Berlin and Bern: Edelheim.

Otlet, Paul. 1909. "L'organisation internationale et les associations internationales." In *Annuaire de la Vie Internationale*, edited by Alfred Hermann Fried, Henri La Fontaine and Paul Otlet. Brussels: Office Central des Institutions Internationales, 29–166.

Otlet, Paul. 1914. *La Fin de la Guerre. Traité de paix générale basé sur une Charte Mondiale déclarant les droits de l'humanité et organisant la confédération des états*. Brussels: Oscar Lamberty.

Otlet, Paul. 1915. *Etude des conditions à réaliser par le prochain Traité pour assurer une Paix solide et durable. Programme minimum*. The Hague: s.n.

Otlet, Paul, 1917. *Constitution mondiale de la Société des Nations. Le nouveau droit des gens*. Geneva and Paris: Atar and Cie.

Parmar, Inderjeet. 2011. *Foundations of the American Century: The Ford, Carnegie and Rockefeller Foundations in the Rise of American Power*. New York: Columbia University Press.

Rasmussen, Anne. 1995. "L'Internationale scientifique, 1870–1918." PhD diss., EHESS Paris.

Rayward, W. Boyd. 1975. *The Universe of Information: The Work of Paul Otlet for Documentation and International Organisation*. Moscow: FID.

Rayward, W. Boyd. 1994. "Visions of Xanadu: Paul Otlet (1868–1944) and Hypertext." *Journal of the American Society of Information Science*, 45: 235–50.

Rayward, W. Boyd. 2003. "Knowledge Organisation and a New World Polity: The Rise and Fall and Rise of the Ideas of Paul Otlet." *Transnational Associations/ Associations Transnationales*, 1–2: 4–15.

Riesenberger, Dieter. 1985. *Geschichte der Friedensbewegung in Deutschland*. Göttingen: Vandenhoeck & Ruprecht.

Rietzler, Katharina. "American Internationalists in Europe: The Carnegie Endowment for International Peace in the Early Twentieth Century" (MA thesis, University College London, 2005).

Rietzler, Katharina. [in press] 2013. "From Peace Advocacy to International Relations Research: The Transformation of Transatlantic Philanthropic Networks, 1900–1930." *Shaping the Transnational Sphere: Experts, Networks, Issues, 1850–1930*, edited by Bernhard Struck, Davide Rodogno and Jakob Vogel. New York: Berghahn.

Schönemann-Behrens, Petra. 2011. *Alfred H. Fried: Friedensaktivist – Nobelpreisträger*. Zürich: Römerhof.

Suttner, Bertha von. 2010. "Der Krieg und seine Bekämpfung." In *Frieden und Friedensbewegungen in Deutschland 1892–1992: Ein Lesebuch*, edited by Karlheinz Lipp, Reinhold Lütgemeier-Davin and Holger Nehring. Essen: Klartext. Extract from Bertha von Suttner, Der Krieg und seine Bekämpfung (Berlin: Continent, 1904), 77–9.

Van Acker, Wouter. "Universalism as Utopia: A Historical Study of the Schemes and Schemas of Paul Otlet (1868–1944)." PhD diss., Ghent University.

Van Acker, Wouter and Geert Somsen. 2012. "A Tale of Two World Capitals: The Internationalisms of Pieter Eijkman and Paul Otlet." *Revue belge de philologie et d'histoire / Belgisch tijdschrift voor filologie en geschiedenis*, 90, 4: 1389–1409.

Welch, Michael. 2000. "The Centenary of the British Publication of Jean de Bloch's *Is War Now Impossible?* (1899–1999)." *War in History*, 7: 237–94.

Wright, Alex. 2008. 'The Web That Time Forgot', *New York Times*, June 17.

Chapter 12
Global Government through Science: Pieter Eijkman's Plans for a World Capital[1]

Geert J. Somsen

Introduction

In the Spring of 1905, the Dutch physician Pieter Eijkman and his assistant Paul Horrix published a plan to establish a "World Capital" on the outskirts of The Hague. This "Intellectual World-Centre" or "Athens of the Future" was meant to stimulate world peace and advance scientific research at the same time. Grouped around a "Monument for International Brotherhood," the city included a "Peace Palace" housing an international court of justice, an international congress hall, and three "International Academies" for various branches of science. Every academy also possessed a separate "Institute" with state-of-the-art research facilities where "the most eminent scientists of all civilized nations" would come to work for "say, one week a year" (anonymous 1906, 108).[2] Eijkman expected the World Capital to expand further over time to include more and more disciplines as well as the headquarters of major international scientific organizations—most importantly the all-encompassing International Association of Academies which had been established six years before. The rising young architect Karel de Bazel had turned the grand vision into a detailed design (Figure 12.1) and after 1905 he began to publish articles about it in architectural journals while Eijkman, at the same time, was publishing essays and interviews about it in other vehicles (Eijkman 1905a, 1908a; Eijkman and Horrix 1907; anonymous 1905; Netscher 1906; De Bazel 1906; See also Reinink 1965, 118–23; Trapman 1999, 122–43). In Figure 12.1 below, the North Sea coast is on top. The Hague is on the left and at the bottom, and the World Capital stretches to the right from point R. The Monument for International Brotherhood is at X; the Peace Palace at A; and the three International Academies are at Z, W, and to the right of B. The research institutes were all located in the Worker's Garden City to the right of W.

In several ways, Eijkman's World Capital is reminiscent of the Cité Mondiale, envisioned by Paul Otlet and designed by him and others in several forms between 1910 and 1935 (Somsen and Van Acker, 2012). Both Eijkman's and Otlet's

---

1   A version of this chapter has been published as Somsen (2009) and reuse of the material is with the permission of Peeters Publishers.

2   This and all subsequent translations of non-English quotes are mine.

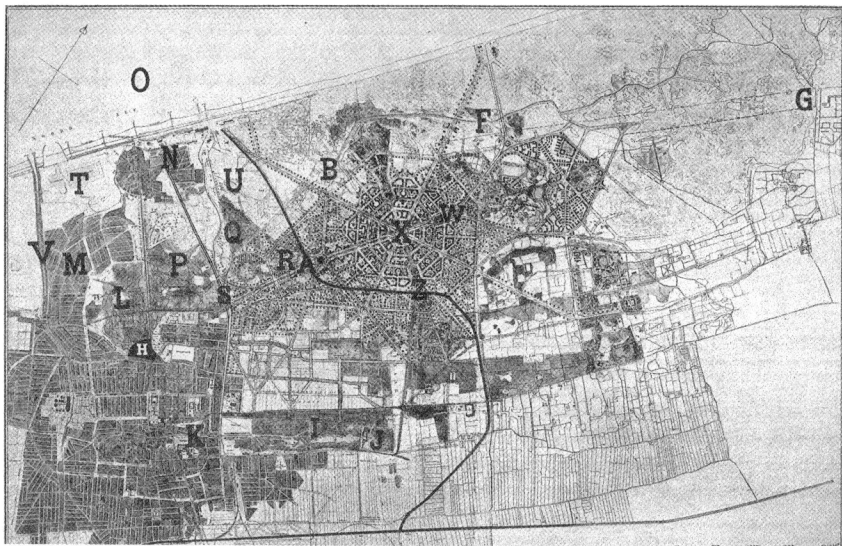

Figure 12.1   Eijkman's world capital designed by Karel de Bazel
Source: Eijkman en Horrix, *Internationalisme en de Wereld-Hoofdstad.*

cities were meant as international hubs; both were centered around science; and both embodied the idea of the advancement of mankind through the spread of knowledge. But there are also some significant differences. First, Eijkman's design was more directed at the practice of science than at its results. It was meant as a home for scientific organizations and actual research while Otlet's scheme was centered on a warehouse of knowledge (or a directory to that knowledge), the Mundaneum. Second, the institution that took center stage in Eijkman's plans was absent in Otlet's: the Peace Palace for the International Court of Arbitration. While Otlet was happy to locate this court elsewhere, it was an essential part of Eijkman's conceptions. Finally, Eijkman's plans predated those of Otlet's and hence they have to be understood from their own local contexts. In the following I will try to explain in terms of these contexts the World Capital's most salient features: its combination of science and arbitration, its grandiose lay-out, and its location in the suburbs of The Hague, of all places. I will associate these characteristics with the state of the Dutch peace movement, with a more general *Kulturkampf* in the Netherlands, and with the professional outlook of the World Capital's originator.

Doctor Eijkman

Pieter Hendrik Eijkman was born in 1862 in Zaandam as the eighth child of a school master (Netscher 1906, 180–81; Van der Vies 1914; Trapman 1999, 123–5).

Three of his siblings also pursued scientific careers, most famously his brother, Christiaan, who won the Nobel prize for medicine in 1929 "for his discovery of the antineuritic vitamin" (the prize was shared with the English physiologist, F.G. Hopkins). Pieter Eijkman studied medicine too and set up practice in Koog-Zaandijk, near his old home town. But his range of activities soon went beyond regular general practice to include initiatives in public health. Eijkman organized the establishment of public baths and school baths in the city. He initiated the creation of holiday homes for workers, and he secured the introduction of the provision of public nursing for the sick and the elderly. After several years he left his practice to study natural medicine with Father Kneipp in Wörishoven, Bavaria, and later in Munich and Dresden. Sebastian Kneipp was a Bavarian priest who had pioneered the introduction of several spa-based forms of therapy (water, open air, sunbathing, herbs, movement, diet, and conscious living). On Eijkman's return to the Netherlands in 1894, he founded "Natura Sanat" in a suburb of The Hague near Scheveningen. This was the first "physiatric institute" in the Netherlands. Here he provided hydrotherapy, lightbathing and other forms of natural medicine. He experimented with x-ray healing of cancer patients, and he published papers on various other investigations he had made in medicine and physical anthropology (Eijkman 1896; 1902a; 1902b; 1902c; 1902d; 1903a; 1903b; 1905b; 1909).

Eijkman's Physiatrische Inrichting „Natura Sanat"

Figure 12.2   Eijkman's clinic Natura Sanat (left) and house (right) in the opulent Van Stolkpark

Source: Eijkman, *Koud Water voor Gezonden en Zieken.*

Eijkman's assistant in the clinic who also worked with him on his plan for the World Capital was Paul Horrix,· a descendant of a wealthy family of furniture manufacturers. Horrix was fascinated less by business than by natural healing, spiritualism, vegetarianism, freethinking and pacifism. It was a set of interests characteristic of Dutch turn-of-the-century progressive elites. Typically Horrix strongly admired Frederik van Eeden, a leading poet, psychiatrist, and opinion maker who had grand ideas and a strong aversion to petty bourgeois thinking (Horrix 1956, 108–9). Van Eeden had ventured into aestheticism, mysticism, anarchism, and socialism, and even set up a commune called Walden after Thoreau. As we will see later, Eijkman and Horrix's internationalist endeavors also fit into this general picture of the progressivism of the times.

According to one source, Eijkman had conceived his World Capital shortly after he came to The Hague, but the plan only became known after 1904 when he and Horrix started to discuss it with influential figures and sent papers about it to periodicals for publicity purposes (Björnson 1907, 6; see also Reinink 1965, 117). These publications attracted much attention. The Dutch daily newspapers' response was predominantly negative, as was that of most politicians (Brooshooft 1906). An important exception was Prime Minister Abraham Kuyper. The newspaper, *De Maasbode*, reported that Kuyper was "heart and soul" into the idea (Horrix 1956, 109), but his foreign policy ambitions were not widely shared and he had already left office in August 1905 – too early to be of much help (Stead 1907). More enduring support came from well-known architects as well as from leading pacifists like the British journalist William Stead and the Austrians, Alfred Fried and Bertha von Suttner (Eijkman 1910 and 1911). The Royal Institute of British Architects even sent a delegation to The Hague that reported favorably on the project. Eijkman and Horrix also attracted the attention of Andrew Carnegie, the Scottish-American steel tycoon and philanthropist, who had just begun to spend his enormous wealth on pacifist projects. They visited Carnegie in March 1905, and managed to get an interview with him after American newspapers began to report favorably on their plan. He seemed to be genuinely interested and the three of them ended up taking the same ship back to Europe, frequently discussing "world issues" (Brooshooft 1906, 183–5; Netscher 1906, 253–6). However, Carnegie's interest did not translate into financial help towards the 20 million dollars Eijkmann and Horrix said they needed.

As part of their continuing campaign Eijkman and Horrix started a "Foundation for Internationalism," of which they themselves formed the "preliminary bureau." They edited a *Review of Internationalism*, which also appeared in French, German and Dutch. They travelled around giving lectures. They organized picnics on the prospective site they had identified for the city. They acquired support signatures from hundreds of prominent European intellectuals (including Richard Strauss, Auguste Rodin, and 18 Leiden professors), and they collected stacks of data on any serious kind of international organization. Eijkman published these data in the books *l'Internationalisme Médical* (Eijkman 1910) and *l'Internationalisme Scientifique*. In the latter, Eijkman continued to call for a concentration of international scientific and legal organizations in The Hague (Eijkman 1911, 105–8). However, by 1911 there was little hope left for realizing the World Capital plan. It had never received the required support from the Dutch government, whose policies actually went against it on some points. The particularly contentious issue was the location of what became the Peace Palace which the government chose to locate on a spot that was incompatible with the World Capital (Brooshooft 1906; Netscher 1906, 249–64). In 1914 Eijkman fell ill and died in a German spa in May of that year. Horrix lost most of his money and died poor in 1929.

World Peace ca. 1900

One of the most important contexts of the World Capital initiative, which will also explain why The Hague was chosen for its location, was the development of the peace movement around the turn of the century. Pacifist organizations had existed in most European countries since the early nineteenth century, but by 1900 something of an international movement had grown out of them. Peace advocates like Von Suttner and Stead were well known across European borders. In 1889 the Interparliamentary Union was founded, an association of parliamentarians from around the globe. The Union held annual conferences, mainly in the interests of promoting international arbitration for the pacific resolution of disputes between the states rather than by recourse to war. As a whole, the pacifist movement was predominantly liberal and elitist – and fairly unsuccessful in reaching those with real power in the world (Eyffinger 1999, 42–69).

It was much to everybody's surprise, therefore, that in August 1898, the Russian czar sent a proposal to his fellow rulers to hold a Peace Conference in order to discuss arms reduction and arbitration. Apparently, the pacifist movement had struck a chord with Nicolas II, and leaders like Von Suttner hailed the initiative as a victory of their ideals. Most heads of state responded skeptically however—at least in private, because it was difficult to publicly refuse cooperation with such a benign and honorable cause. In England the Prince of Wales called the proposal "the greatest nonsense and rubbish I have ever heard of," and Kaiser Wilhelm II spoke of a "Conferenzkomödie" (Eyffinger 1999, 25–39). The pacifists therefore increased public media pressure (for example Stead started a European "Peace Crusade" which was widely publicized), and when the Russians issued a second invitation, all governments eventually complied.

From the outset it was clear that the location of the conference could not be in any of the capitals of the great powers, such as St. Petersburg, since this would be unacceptable to the others. The site had to be a city in a smaller country. As a home to many international events and organizations, Brussels was the primary candidate, but eventually the Russians chose The Hague. Historians are not entirely clear why they did so. One factor mentioned is that the czar's councillor, De Martens, had had some good experiences with previous international meetings there. The Netherlands was also associated with the seventeenth-century international law pioneer, Hugo Grotius. And perhaps Nicolas was influenced by his family ties to the young Dutch queen, Wilhelmina, his third cousin (Eyffinger 1988, 17–18). There is also a rumor that king Leopold II was struggling with a foot problem when the Belgian candidacy as the conference locale was brought to his attention. Eijkman claimed to have heard this from Stead (Eijkman 1908a, 28). Whatever really tipped the balance, The Hague it would be.

Wilhelmina herself was not at all pleased with this choice, for she felt it marked her country with "the stamp of insignificance" (quoted in Fasseur 1998, 393). It was true that the Russians had chosen the Netherlands for its weakness on the world stage, but the queen preferred to see herself as the ruler of a great colonial empire, whose integrity, moreover, was better defended by military reinforcement than by peace talks. But the Dutch government decided to accept the invitation, and Wilhelmina was forced to lend her own palace for the conference meetings. In protest, she refused to open or even attend any of the sessions, and the organizers had to look for another chair.

Once the Peace Conference had started, on May 18, 1899, the great powers proved to be no less skeptical than Wilhelmina and on the issue of arms control. They turned down every Russian proposal. The subject of arbitration met with no great enthusiasm either, but in order to avoid a complete failure, or perhaps to save face for Nicolas, the conference did in the end consent to establish a Permanent Court of Arbitration to be located in The Hague. The court started as a paper tiger, but in subsequent years some powerful Americans began to develop a genuine interest in its functioning. The administration of Theodore Roosevelt brought the first case before it in order to start operation in 1902 (a minor dispute with Mexico). Congressman Richard Bartholdt became an ardent promoter of arbitration in general—as well as the president of the Interparliamentary Union. And he and Roosevelt called for a second Peace Conference in The Hague, which would indeed be held in 1907. Meanwhile the head of the American delegation at the 1899 conference, Andrew Dickson White, persuaded Carnegie to support this kind of pacifism, and in 1903 Carnegie donated $1.5 million to build a "Peace Palace" in order to accommodate the Permanent Court of Arbitration (Eyffinger 1988, 28–35, 49–52).

It was at this moment that Eijkman stepped in. The Dutch government was still largely reluctant to accept the country's growing status as a center of pacifism. The foreign affairs minister, Robert Melvil van Lynden, had at first even declined Carnegie's offer to build a Peace Palace (Brooshooft 1906, 173). But Eijkman argued that the Netherlands should seize the opportunity and actively expand its new role by surrounding the Peace Palace with an entire set of international institutions—the World Capital. In order to advance these ideas, he and Horrix approached the committee overseeing Carnegie's bequest. As already mentioned they visited the philanthropist in the United States, and they launched their campaign of interviews, publications, signatures, picnics, and so on.

On the occasion of the Second Hague Conference in 1907, Eijkman and Horrix even opened a club for the delegates, the *Cercle International*, where tea was served every day at five and conference officials could talk informally with each other and with journalists. An article, "Le Cercle International," in Stead's daily *Courrier de la Conférence de la Paix*, noted that journalists and others were initially banned from the formal meetings of the conference, though subsequent issues of the newsletter suggested that some visitors were later admitted. [3] The club became an important meeting place for arbitration enthusiasts such as Bartholdt, and other leading pacifists like Von Suttner, Fried, and George Francis Hagerup, the leader of the Norwegian delegation and president of the Nobel Peace Prize committee. There were lectures, debates and photo shoots at the *Cercle*, and Stead used it as the basis for publishing his *Courrier de la Conférence de la Paix* – including articles on his hosts' World Capital plan.

Battles Home and Abroad

Through initiatives like the *Cercle*, Eijkman tried to hook up with the peace movement and the sudden rise of The Hague as a seat for international institutions. But other locations competed to host the headquarters of these institutions. There were plans for world capitals in Washington and Paris. The Belgian senator, Henri de la Fontaine, even proposed to build a "District of Columbia of the United States of the World" on the Waterloo battlefields outside of Brussels (Brooshooft 1906, 186; Stead 1907, 7). Most of these places were far more attractive than The Hague. In an article discussing what the various rival locations offered, Stead summed up many of The Hague's advantages, but added that it could also be depicted negatively. From the viewpoint of other European cities,

> The Hague is only a sleepy country village, infested with malaria and cut through by canals with fairly unpleasant odors, far removed from Paris—the centre of the world's desires—deprived of the internationalist spirit, like a kind of Siberia on a rough seashore, where it would be cruel to exile the elite of humankind (Stead 1907, 7).

If Eijkman wanted to advance his plans he had to develop arguments that could overcome the force of such negative depictions and also to find ways to discredit the competing locations.

This was easiest to do in the case of the capitals of the great powers since the choice of one these would never be fully acceptable to the others. Fiercer competition came from other small countries, like Belgium and Switzerland, which already hosted far more international organizations than the Netherlands, and recently also from Norway and Sweden. As small states, all of them shared the advantages of being neutral and insignificant on the world stage, so Eijkman used other points against them in his speeches and publications. Switzerland, he argued, was too unstable, because it was a republic and as such was subject to frequent

---

3   "Le Cercle International" *Courrier de la Conférence de la Paix*, June 15 1907.

regime change. Moreover, it had recently become a safe harbor for terrorists, like the Italian anarchist who had murdered the Austrian empress, Elisabeth, in Geneva in 1898 (Eijkman 1908a, 27–8). According to Andrew Dickson White, this was also the reason why the Russians had not considered Swiss cities for the first Peace Conference in 1899 (White 1905, Vol. II, 250).

Belgium was much more difficult to deal with. It had a strong tradition in hosting international events and organizations which were fully backed administratively and financially by the King, Leopold II, and by private benefactors like Ernest Solvay. As a consequence, Brussels outnumbered any other city as a seat for international institutions. Still, Eijkman argued, the country's eagerness and pride as a world centre could put off other nationals. Also the Belgian insistence on speaking French, a world language, did not always serve Belgium well outside of the francophone sphere. The fact that the Dutch did not expect anybody to speak *their* barely known language actually enhanced their accessibility, for "the Germans, English and Americans feel much more at home in the polyglot Netherlands than in Belgium, which for them speaks a foreign language" (Eijkman 1908a, 31–2). The commitment of Belgian intellectuals and officials to French as an official national language despite the fact that a large section of the Belgian population spoke Flemish, a version of Dutch, implied for Eijkman an inevitable orientation toward France that suggested that Belgian internationalism could not always be considered to be entirely neutral. With arguments like these, Eijkman turned the colorlessness of the Netherlands into a rhetorical advantage. It was the *in*significance of the Dutch language, the country's *lack* of eagerness to play a role on the world stage, and The Hague's *distance* from the excitement and political turmoil of other European cities, that made it such an eligible international meeting ground. Stead had actually made the same point in the article that highlighted The Hague's dullness. It was the Dutchmen's typical "solid, calm, serious way of behaving" that made them so attractive as neutral mediators (Eijkman 1908a, 28; Stead 1907, 6–7).

But Eijkman was not only competing with foreign rivals, he was also fighting a battle at home. Perhaps the Dutch lack of eagerness to become an international centre could be used as a selling point abroad, but within the Netherlands Eijkman needed support. One option he had was to stress the economic benefits to be gained from international activity. Hotels and restaurants had enormously profited from the Peace Conferences (their prices had risen so high that the *Daily Telegraph* had compared them to the horrors of war). The World Capital would clearly bring in more of such money. But while Eijkman included these expectations in the financial scheme for his plan, he tended not to stress economic arguments in trying to win people over to it. Instead he sometimes argued that being the seat of important international institutions would guarantee the security of the Netherlands, guarantee it more effectively in fact than attempts at providing an adequate military defense, an area in which the Netherlands would never really be able to compete with great powers (Eijkman 1908a, 8). But Eijkman's supporters tended to emphasize the national defense argument more than he did. In an article,

"Plannen van Internationalisme," the Architecture journal, *De Opmerker*, for example, claimed that it was possible that a World Capital in the Netherlands might guarantee in perpetuity "our nation's autonomous existence."[4] Journalist, P. Brooshooft, mentioned the same point as the first advantage of Eijkman's plans (Brooshooft 1906, 186). And law professor Cornelis van Vollenhoven would use similar arguments in his calls for a Dutch-based international peacekeeping force (Van Vollenhoven 1910, 185–204).

For Eijkman, however, it was not economics or security that most recommended his scheme but first and foremost the national *honor* that was to be derived from acting in the role of an international host. Receiving the world's nations and leading the way toward peace would not belittle the Dutch, as Wilhelmina had thought, but on the contrary would command respect and esteem from other countries for the Netherlands. Eijkman repeatedly stressed that internationalism was not at all at odds with nationalism, but actually enhanced a nation's pride and standing (Eijkman 1908a, 7). Arbitration was the highest fruit of human civilization, and it would be greatly to the honor of the Netherlands if it could become the site at which international arbitration took place. In "The Capital of the World," Eijkman's publisher and supporter, Nico van Suchtelen, reminded Dutch leaders of the enormous privilege that had come their way.

> All I want to say is that the Idea of Internationalism is alive; it lives and works in the spirits of all nations. The world capital, symbol of its magnificent beauty, instrument toward its harmonious perfection, is no longer the fantastic dream of individuals, but an opportunity, acknowledged and desired by the most profoundly sensitive and the farthest-seeing of all people. You rich and great minds of the Netherlands, when your brothers of all nations will soon come to you in order to realize the grand Dream of Unity with *your* help and in *your* country, how will you receive them? Dutch Government, if Humanity will shortly offer to you its finest creation, the highest symbol of its culture, how will you accept this gift? (Van Suchtelen 1907, 19)

These were the pertinent questions. To perceive the great task and to appreciate the honorable opportunity that had befallen the Netherlands required a vision and a greatness of mind that Eijkman found scarce among his compatriots. Many of his and his allies' writings incessantly criticized the small-mindedness of the Dutch people—their lack of ambition, their limited perspective, their aversion to undertake anything grandiose—"that tininess!," as Eijkman repeatedly exclaimed in interviews. "God, everything is always done so tinily here, so ti-ni-ly!" And with respect to the Dutch government: "Are such Dutchmen actually worthy of a Peace Palace? They have no guts, no feeling for greatness left ... the village people of Europe is what they are! ... that smallish behavior, that tiny acting, that clumsy hustling before the face of the entire world!" (Netscher 1906, 181,

---

4   "Plannen van Internationalisme," De Opmerker 40, 1905, 308.

253). The Architecture journal, *De Opmerker*, agreed and spoke of "our national cautiousness."[5] It was time for a new type of Dutchman, with a greater perspective and bolder plans—a *"Groot-Hollander,"* "who looks over the dikes and the little village towers, for whom the Kalverstraat is not the busiest street and the Dam is not the largest square in the world, [and] who dares to think further than the Frisian islands and the Scheldt" (Netscher 1906, 179). This is how Eijkman saw himself and how others portrayed him. Theirs was the task to wake up the nation to start thinking and acting big.

Yet for all its rhetoric of a fresh beginning, the attitude that Eijkman and his colleagues displayed had been around for a number of years and was clearly evident by the turn of the century. It was the same mentality expressed by the literary "movement of eighty" (named after its emergence in the 1880s) when it condemned pre-existing Dutch poetry as the art of small-time country parsons. It was also the outlook an historian like Johan Huizinga demonstrated when he characterized earlier Dutch culture as that of narrow-minded shopkeepers whose only value was usefulness. And it was the type of discourse that early twentieth century Dutch scientists employed when they called their own times the "Second Golden Age" of Dutch science (the first Golden Age being the seventeenth century). They described the time before their arrival on the scene as a period of stagnation and lack of excellence. Recently, Willem Mijnhardt has warned against taking these characterizations—with their story of nineteenth century stagnation and turn-of-the-century revival—at face value. Instead, he argues, these utterances should be seen as a form of the self-fashioning of a new cultural elite that was more professional and more exclusive than that of the previous generation. Throughout most of the nineteenth century, Dutch intellectuals had considered the dissemination of useful knowledge to the lower classes as their principal task. But around 1900, this mission of social uplift came to be seen as lowly, uncreative, small-minded, and un-grand. Great works and excellence became the new objectives for purveyors of culture. The yardstick was no longer practical utility but international acclaim (Mijnhardt 2004, 36–7).

It is clear that Eijkman's plans fit this picture. They were ambitious, they were international, and they were geared against the small-mindedness that was presumed to prevail among the Dutch. Nor is it a coincidence that his ideas corresponded to the new cultural ideals of the time because Eijkman and Horrix were close to the intellectual circles in which these ideals were emerging. Their publisher Van Suchtelen was a friend and former collaborator of Van Eeden, whom Horrix admired so much and who was perhaps the leading member of the avant-garde "movement of eighty." Albert Verwey, another prominent member, supported the World Capital plans through his journal *De Beweging* ("the Movement"). In addition, Eijkman and Horrix were in contact with self-conscious representatives of the "Second Golden Age" of Dutch science. Hendrik Antoon Lorentz, for example, Nobel Laureate for physics in 1902 and a major international scientific

5   See note 4 above.

figure, adopted Eijkman's ideas in publications such as "International Science promotes Peace" (Lorentz 1913). And Heike Kamerlingh Onnes, who would win the Nobel prize for physics in 1913, was one of the Leiden professors who lent their signatures to the World Capital plan. Through these contacts, Eijkman and Horrix were very much part of the new intellectual and cultural milieu of the time. Thus the World Capital plan was at the same time part of the international movement for the development of pacifism and a reflection of profoundly local concerns: the self-fashioning of a new intellectual elite and its cultural battles within the Netherlands.

Scientific Internationalism

While local culture wars may explain the grandiose nature of Eijkman's plan and the developments in pacifism may clarify its commitment to the Netherlands and its inclusion of the idea of the Peace Palace, neither of these contexts makes clear why *scientific* institutions had such a prominent place in the World Capital, why a centre of international brotherhood was also a center for the development of science. In order to understand this, we need to look at another development: the multiplication of international scientific organizations in the late nineteenth century.

It is sometimes claimed that science is by nature international, but science has not always been embodied in international institutions. In fact, many of these institutions only developed in the second half of the nineteenth century, especially with the rise of the scientific conference. When conferences in various branches of science and technology began to become recurrent phenomena, they began to be institutionalized. Permanent offices were installed to organize them, sometimes with a permanent seat in one or another country, and eventually for most sciences, international unions or associations were formed, complete with memberships, journals and subcommittees to carry out various tasks (Rasmussen 1990, 115–30).

Eijkman wanted to jump onto the bandwagon of *this* development, as is clear from his desire to locate the permanent offices of all international scientific organizations, most prominently the International Association of Academies, in his World Capital. But he did not present this accommodation as a separate function of his projected city—it was intimately associated with the aim of world peace. Eijkman regarded the ever increasing international organization of science (what he called "internationalism") as an inevitable process that was advancing with the necessity of natural law. Eijkman quoted Fried, who compared the process of internationalization to the natural fact of the rotation of the earth. Just as Galileo could not deny that fact, even when forced to by the inquisition, and had secretly whispered "and yet it moves," the current-day observer simply had to conclude about international cooperation: "and yet it organizes itself" (Eijkman 1908a, 4 and 12). The reason for this automatic process of internationalization was that scientists, by the nature of their work, simply *needed* international cooperation—research and the advancement of science more generally required international cooperation. Because of this, Eijkman believed that science was a much better guide to attaining international brotherhood than the pacifist movement. He repeatedly argued that one of the flaws of most forms of pacifism was that they were too sentimental. They were too much based on idealism and the hope that people would do good. International cooperation in science, by contrast, was not an ideal but a necessity, and hence science was a much more solid promoter of world peace than any pacifist ideology could ever hope to be. An exception was the pacifist pursuit of arbitration. Eijkman did not regard this as a sentimental endeavor, but as a pragmatic and rational development on a par with international cooperation in science (Eijkman 1910, 102–3). Both in their pragmatic processes and in their peaceful effects, science and arbitration were comparable in Eijkman's view, and this is one reason why scientific academies and the Peace Palace stood side by side in his capital plan.

The accommodation of international organizations, however, was just one scientific function of the World Capital. Another was to provide for the actual practice of research. Throughout his utopian city, Eijkman projected various kinds of facilities for performing experiments, testing theories, and trying out the applications of new scientific insights. The city's pedagogical Institute, for example, possessed an experimental boarding school where new methods in co-education, open-air classes, and sports instruction could be tested. The medical department included a hospital and sanatorium, small-sized but set up to the most modern standards where the latest medical advances could be tried out. And the economic institute disposed of an entire "Workers' Garden City," where scientists could experiment on "living test material" (that is the World Capital's service personnel who lived there). Here economists could try out innovations such as cooperative production, collective eating places, and joint laundry provisions. The Anthropological Academy would take advantage of the presence of representatives of the many nations present in the World Capital to study their different cultures (Eijkman 1905a, 868; anonymous 1906, 108–9; Netscher 1906, 184–6).

At no point did Eijkman discuss such disciplines as biology or physics. In fact, his definition of science seems to coincide to a large extent with the fields in which he was practicing himself: medical research, public health, and anthropology. The coincidence becomes even more striking if one considers the precise location where the future activities were supposed to take place: right next to his own house and clinic on the outskirts of The Hague. In fact the whole landscape of the World Capital, with freestanding estates in a park-like environment, was a very close replica of Eijkman's own neighborhood, the Van Stolkpark (Figures 12.2 and 12.3). From a cynical point of view then, we might say that the entire World Capital was nothing but a gigantic enlargement of Eijkman's own daily existence—he magnified his home and made it the center of the universe. Figure 12.3 below shows the relationship of Eijkman's house and elements of the World City. The Peace Palace is at A where the World Capital stretches out to the right,

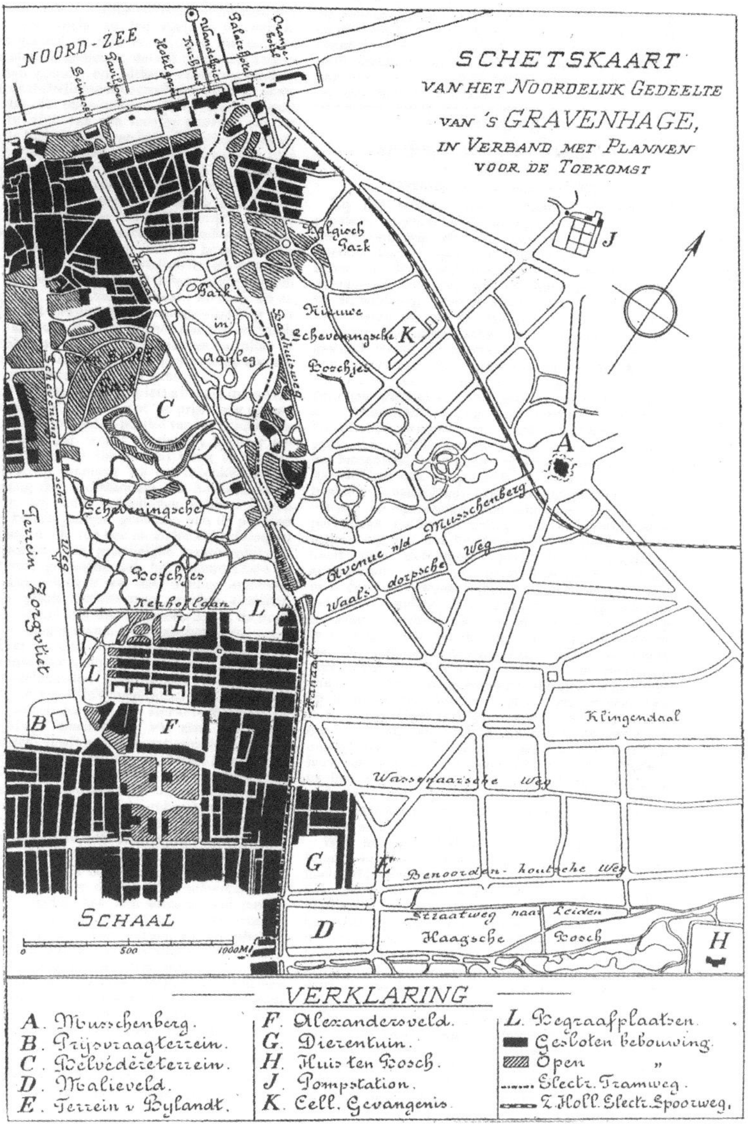

Figure 12.3 Sketch of part of the projected world capital with Eijkman's residence

Source: "Plannen van Internationalisme" *De Opmerker* 40 (1905), p. 309.

most of it not included. Eijkman's house and clinic are in the Van Stolkpark, left of point C.

It would seem then that Eijkman's perspective was not so grand after all and that he was actually no less narrow-minded than he presumed his fellow-countrymen to be. But his plans can also be read in a different way—as reflecting a conception of science and its role in the world that he followed in his own practice *and* tried to realize in the World Capital on a large scale. This conception was rooted in his position as a physician. According to Eijkman, the work of a doctor was not restricted to curing individual patients, or "spiriting off diseases with little prescriptions."[6] The modern practitioner had a wider social task that included public health measures such as combating alcoholism, providing vacation facilities, conducting food inspection, encouraging housing improvement. Eijkman saw this social task as the application of scientific insights to the welfare of society—a typical hygienist point of view that he had held from the beginning of his career. It had already informed his initiatives as a physician in Koog-Zaandijk, but it was also expressed in the physical lay-out of the World Capital. In his instructions to the young architect, De Bazel, Eijkman had made sure that the city would be spacious, with plenty of green and wide boulevards (up to 100 meters!), so that the sea breeze could provide fresh air everywhere. It would be "from the viewpoint of hygiene the healthiest terrain that there is" (Eijkman 1905a, 871; Eijkman and Horrix 1907, 4–5).[7]

While Eijkman's hygienism was expressed in the World Capital street plan, it was also incorporated into his wider view of science. For him, hygiene was not a branch of medicine, but the other way around. Hygiene embraced regular medicine in a much larger social task. In fulfilling this larger task, however, the hygienist physician quickly ran into problems that were of an even larger scope: issues for example of work organization, food production, poverty, and education. These issues belonged to such fields as economics and pedagogy—two sciences that Eijkman indeed brought together with hygiene into a single International Academy in his World Capital. But issues of social and physical welfare that informed his view of science went further. Although Eijkman did not elaborate on them as much, he envisioned all kinds of disciplines as being present in his World Capital—from architecture and anthropology to technology and traffic studies.[8] What these various fields had in common was that their work benefited mankind in one way or another. These benefits could come in the form of cultural products (Eijkman mentioned areas like fine arts and even "pure intellect"), but they could also be of a more practical nature, and these were clearly at the forefront of his mind (Netscher 1906, 186). In one of his articles he described how the International Tuberculosis Congress was currently devising ways to combat that disease by proposing hygienic measures. This led him to suggest that this

6   "Den Haag als Wereld-Hoofdstad (Intellektueel Wereld-Centrum)." De Hollandsche Revue 11, 1906, 107–10

7   I thank Heiner Fangerau for pointing out the hygienist features of the city plan.

8   "Den Haag als Wereld-Hoofdstad ... ," see note 6 above.

method of proceeding could be generalized to all sorts of problem areas: experts could meet on them, discuss possible solutions, and then try them out.[9] *This* was the kind of practice that the World Capital would institutionalize. The Academies would be "permanent conferences" (Eijkman literally called them that) and the Institutes would be testing grounds for the solutions that the Academies came up with (Eijkman 1905a, 864). As Eijkman declared in an interview, his entire plan of science and its social mission was an extension of his activity as a physician.

Global Government

The World Capital was more than a coherent whole that consisted of scientific organizations, research, and a Peace Palace. It also had a political dimension. In their beneficial work for humanity, scientists inevitably found themselves handling problems that were ordinarily understood to be the province of politics. In a sense, Eijkman stated, the scientific or medical practitioner "contributes to the great divisive issues of political life, but in *his* politics stands above the battles driven by party interests."[10] While science and medicine entered the political realm, they did so to deal with problems in ways that were beyond ordinary party struggles. They were less partisan, more disinterested, and much more efficient than regular politics. Because of this, Eijkman believed that science should be given a form of political authority. The Academies of various disciplines should not just devise solutions to practical problems, nor should they merely publish accounts of these solutions and promote their benefits (although the last was also an important task) (Eijkman and Horrix 1907, 2). They should above all be able to *decree* the solutions arrived at as policy measures that would be officially implemented. Eijkman's International Academies "should be granted particular rights and powers, by which they could have an official role in international government" (Eijkman 1905a, 868). Hence the World Capital, as a collection of Academies in all sorts of relevant fields, was really meant to be a *capital* in the sense of the seat of a global administration. It would not merely organize science but truly govern the world.

Eijkman's model was the new International Court of Arbitration. After all, its function was not limited merely to studying the law. It was required to issue verdicts on particular cases and these verdicts were supposed to be implemented globally. Eijkman saw close parallels between the Court and his Academies, both in their work and in how they were the products of the advancement of reason in public affairs. International relations had been handled before the Court's time in ways that were both unregulated and unreasonable, in Eijkman's view. The fact that some people believed that war was beneficial to the health of nations was not just wicked but actually absurd. Eijkman repeatedly tried to show how irrational opinions like this were. He observed, for example, that he had never seen those

---

9   "Den Haag als Wereld-Hoofdstad ... ," see note 6 above.
10   "Den Haag als Wereld-Hoofdstad ... ," see note 5 above (emphasis added).

who adhered to such a view accept its logical consequences and organize a brawl for their own personal benefit. He said that when an old professor told him that it was honorable to fight for one's nation, he replied that he hoped that the professor might to die on the battlefield—a deduction from the professor's position, he added, that did not receive the endorsement "which I would logically have expected" (Eijkman 1908a, 11–12). The fact that most countries spent thousands of times more money on weaponry than on art and science reminded Eijkman of the household choices of a drunk (Eijkman 1908a, 12–13). And from the point of view of a physician who was educated to save lives (always his favorite perspective), the slaughter of war was both an absurdity and a waste.[11]

But now, Eijkman claimed, reason had made its way into international relations. And the rational alternative to war was arbitration. Conflicts were no longer to be decided by violence and mistrust, but handled by legal experts at an international court. These experts represented precisely the sort of scientific attitude that Eijkman believed could advance all disciplines. This is also why the Peace Palace was a natural neighbor in his World Capital for the academies of such disciplines as economics and anthropology. Arbitration was to international relations what hygiene was to social policy.

The World Capital was entirely to comprise institutions that were comparable in the sense that they all advanced knowledge; they all served humanity; and they all issued decrees based on their expertise in the fashion of a true global administration. The only exception seemed to be what Eijkman called the "International Parliament" and its "two Chambers of World Government" (quoted in Reinink 1965, 114). These institutions seem to have been modeled on conventional political institutions. But what Eijkman probably had in mind was something like the Interparliamentary Union, which he hoped (and at one point claimed) would settle in The Hague (Eijkman 1908a, 40). It is true that the Interparliamentary Union was a gathering of ordinarily elected parliamentarians, but it was also an institution closely associated with the cause of arbitration. Moreover, in its projected congress hall, Eijkman's "International Parliament" would resemble the kind of permanent conferences that the Academies were supposed to be. As such it stood precisely at the conjunction of science and arbitration that was so central to Eijkman's World Capital.

Conclusion: Colorless Politics

Looking at the overall coherence of Eijkman's World Capital, we may finally ask the question: what kind of a utopia was it? It was certainly ambitious, like any utopia, but in its time it was not as outlandish as it might appear today. Eijkman's association of science with arbitration was not uncommon, and re-appeared, for example, in the work of the American diplomat Andrew Dickson White. White who had led the American delegation to the first Peace Conference, as mentioned above, was an

11   "Den Haag als Wereld-Hoofdstad … ," see note 5 above.

ardent promoter of arbitration. At the same time, he strongly believed in the progress of science versus unreason in fields of learning and public affairs. His famous *History of the Warfare of Science with Theology* (White 1896) was an account of exactly that struggle in various realms of human endeavor, including hygiene. Similar connections can be detected, for example, in the heritage of Alfred Nobel. Nobel designed his prizes to reward those who created major scientific benefits for humanity, but also "for the holding and promotion of peace congresses"—which was literally how the aim of the peace prize was stated in his will (quoted in Friedman 2001, 14). Things were different thirty years after Eijkman, when the influential writer H.G. Wells campaigned for a comparable kind of world government based on science (Somsen 2006). But Wells's socialist scheme left no place for separate nation states and hence there was no longer any need for arbitration. Eijkman's World Capital would not have fitted into such a socialist world.

This raises the question as to what kind of utopia the World Capital was *politically*. Eijkman himself presented his plans as a-political because they were based on science and reason and as were supposed to transcend political party divisions. But any claim to being a-political should be read as a political statement in its own right, and so the question about the politics embodied in the World Capital remains important. Eijkman inadvertently gave some indication of political relevance when he reported how his pleas for the World Capital were received by audiences in Germany. Here, he said, internationalism tended to have a negative ring as it was associated with what were basically the three *Reichsfeinde* of the Wilhelminian empire. The first fear was of the "red internationalism" of the social-democrats and their belief in a future classless world state. The second fear was of the "black internationalism" of the Catholic clergy and its global network under the leadership of Rome. And finally the third fear was that of the menace of a "gold internationalism" of mainly Jewish businessmen whose capitalist corporations would give them world power. Eijkman's answer to these fears was that he had nothing to do with any of the groups involved, and that what he promoted was a "*colorless* internationalism" (Eijkman 1908a, 1–2, original emphasis). But as he defined this colorlessness in contradistinction to other political positions, Eijkman inevitably revealed a political orientation despite himself.

Eijkman's internationalism was in no way a socialist ideology. The changes he envisioned were supposed to take place within the existing economic order. He never spoke of revolution as leading to the desired goals. Nor did he focus on an exploited proletariat. His attention was fixed on the educated upper class, the "aristocracy of the mind," who advanced society along the road of reason and internationalization (Eijkman 1905a, 871). As far as class structure was concerned, Eijkman did not seem to want to plan any change to what existed. But neither was he a conservative in the sense of confessional politics. He did not adhere to any organized religion. In his plans for future pedagogy, he proposed to keep religion out of education. Stead once characterized the World Capital as a "*Vatican laïque*," and on the whole that was accurate. It is true that Eijkman wanted to include the *study* of religion in his Academies. This could have been a reflection of the theosophical belief in the value and unity of all religions, a belief also shared by

218 *Information Beyond Borders*

his architect De Bazel (Netscher 1906, 185). But Eijkman said very little about this or about religious matters of any kind. He was much more outspoken about his faith in science. Finally, there is no anti-Semitism in his writings, no critiques of world-wide capitalism. On the contrary, he believed that in his plans the existing forms of trade and commerce would become better organized than they were, not controlled or abolished.

On the whole then, the World Capital appears to reflect the dream of a liberal elite who saw itself as fully facing the future with its back turned on the conservative forces of militarism and organized religion. In this view, progress was inevitable and its path predetermined, not by the laws of dialectic materialism but by the steady advancement of science. If anything, Eijkman was a positivist who saw science and reason dominating more and more areas—from public health and economics to the management of international conflict. Eijkman was perhaps lucky to die three months before the First World War.

References

De Bazel, K.P.C. 1906. "Stiftung für Internationalismus." *Der Städtebau*, 3: 36–9.
Björnson, Björnstjerne. 1907. "Een Grootsch Plan." *Revue voor Internationalisme*, 1: 6–8.
Brooshooft, P. 1906. "Schevingen Wereld-Centrum?" *De Beweging*, 2: 172–94.
Eijkman, P.H. 1896. *Koud Water voor Gezonden en Zieken*. Second edition. Leiden: Sijthoff.
Eijkman, P.H. 1902a. *Kanker en Röntgenstralen*. Haarlem: Bohn.
Eijkman, P.H. 1902b. *Krebs und Röntgenstrahlen*. Haarlem: Bohn.
Eijkman, P.H. 1902c. *Kurzer Inhalt des Vortrages über ein neues System fur die Anthropologie*. Den Haag: Haag.
Eijkman, P.H. 1902d. *Het vaccinatievraagstuk*. Amsterdam: Bladen voor Hygiënische Therapie.
Eijkman, P.H. 1903a. *Reformkleeding*. Amsterdam: Van Rossen.
Eijkman, P.H. 1903b. *Der Schlingact, dargestelt nach Bewegungsphotographien mittelst Röntgen-Strahlen*. Bonn: Emil Strauss.
Eijkman, P.H. 1905a. "Reorganisatie der Internationale Congressen." *Vragen van den Dag*, 20: 866–71.
Eijkman, P.H. 1905b. *Un Nouveau Système Graphique pour la Craniologie*. Lyon: Rey.
Eijkman, P.H. 1908a. *Over Internationalisme*. Den Haag: Voorbereidend Bureau der Stichting voor Internationalisme.
Eijkman, P.H. 1908b. *Bewegungsphotographie mittelst Röntgenstrahlen*. Amsterdam.
Eijkman, P.H. 1909. "Nieuwe Toepassingen der Stereoscopie." *Nederlandsch Tijdschrift voor Geneeskunde*, 53: 971–72.

Eijkman, P.H. 1910. *l'Internationalisme Médical*. La Haye: Bureau préliminaire de la Fondation pour l'Internationalisme.

Eijkman, P.H. 1911. *l'Internationalisme Scientifique (Sciences Pures et Lettres)*. La Haye: Bureau préliminaire de la Fondation pour l'Internationalisme.

Eijkman, P.H. and P. Horrix. 1907. *Internationalisme en de Wereld-Hoofdstad*. Den Haag: Voorbereidend Bureau der Stichting voor Internationalisme.

Eyffinger, A. 1988. *Het Vredespaleis, 1913–1988*. Amsterdam: Sijthoff.

Eyffinger, A. 1999. *The 1899 Hague Peace Conference. "The Parliament of Man, the Federation of the World."* The Hague: Kluwer Law International.

Fasseur, C. 1998. *Wilhelmina. De Jonge Koningin*. Amsterdam: Balans.

Friedman, R.M. 2001. *The Politics of Excellence. Behind the Nobel Prize in Science*. New York: W.H. Freeman, Times Books, Henry Holt and Company.

Horrix, M. 1956. "Wat Drie Generaties Opbouwden." *Jaarboek Die Haghe*, 44: 61–111.

Lorentz, H.A. 1913. "De Internationale Wetenschap Bevordert de Wereldvrede." *Vrede door Recht*, 14: 5–6.

Mijnhardt, W.W. 2004. "De Akademie in het Culturele Landschap rond 1900." In *De Akademie en de Tweede Gouden Eeuw*, edited by K. van Berkel. Amsterdam: Koninklijke Nederlandse Akademie van Wetenschappen, 15–41.

Netscher, Frans. 1906. "P.H. Eijkman, Arts." *De Hollandsche Revue*, 11: 179–89 and 249–64.

Rasmussen, A. 1990. "Jalons pour une Histoire des Congrès Internationaux au XIXᵉ Siècle: Régulation Scientifique et Propagande Intellectuelle." *Relations Internationales*, 62: 115–30.

Reinink, A.W. 1965. *K.P.C. de Bazel – Architect*. Leiden: Universitaire Pers.

Somsen, G.J. 2006. "De Metawetenschap van H.G. Wells." *Gewina*, 29: 293–305.

Somsen, G.J. 2009. "Science, Medicine and Arbitration: Pieter Eijkman's World Capital in The Hague." In *Utopianism and the Sciences, 1880–1930*, edited by Mary Kemperink and Leonieke Vermeer. Leuven: Peeters Publishing, 125–44.

Somsen, G.J. and Van Acker, W. 2012. "A Tale of Two World Capitals: The Internationalisms of Pieter Eijkman and Paul Otlet." In *Beyond Belgium: Transnational Social and Cultural Entanglements, 1900–1925*, edited by Daniel Laqua, Christophe Verbruggen, and Gita Deneckere. Special issue of *Revue belge de philologie et d'histoire / Belgisch tijdschrift voor filologie en geschiedenis*, 9: 1389–1409.

Stead, William. 1907. "La Haye, la Capitale des États-Unis du Monde." *Courrier de la Conférance de la Paix,* August 11, *supplément*.

Trapman, J. 1999. *Het Land van Erasmus*. Amsterdam: Balans.

Van der Vies, A.B. 1914. "Groote Dooden." *Vrede door Recht*, 15: 152–3.

Van Suchtelen, N. 1907. "De Hoofdstad der Wereld." *Revue voor Internationalisme*, 1: 9–19.

Van Vollenhoven, Cornelis. 1910. "Roeping van Holland." In *One Way or Another*, 74: 185–204.

White, A.D. 1896. *A History of the Warfare of Science with Theology in Christendom*. London: Macmillan.
White, A.D. 1905. *Autobiography*. London: Macmillan.

Chapter 13

Dynamics of Networks and of Decimal Classification Systems, 1905–35

Charles van den Heuvel

Classifications as codifications of knowledge and its dynamics are often explained by quoting texts of their designers and by analyzing the classes and notations of their systems. Such approaches, not seldom underpinned by philosophical arguments and statements about progress in science, do not take into account the often eclectic, inconsistent and sometimes contradictory world views of the classificationists themselves. We propose to complement the conventional approaches to classification by means of an exploration of the networks of individuals and institutions of the developers of the classifications. Such an exploration will reveal that classification dynamics are not regulated by their designers alone, but are shaped as well by the people in various networks around them. This chapter discusses the impact of aspects of the social networks of the protagonists of the Universal Decimal Classification (UDC) as developed from Melvil Dewey's Decimal Classification (DC) by Paul Otlet (1868–1944) and his colleagues in Belgium and the protagonists in the US of the Decimal System itself. First is the network that begins with a group of Dutch officials before the First World War and reaches out through one of them, Donker Duyvis (1894–1961) after the war, to a group of Americans, especially Godfrey Dewey, the son of Melvil Dewey (1887–1977). The two men, Donker Duyvis and Godfrey Dewey, worked through the organizations they dominated, the International Committee for the Decimal Classification of the International Institute of Bibliography (IBB) in Brussels and The Hague and the Lake Placid Education Foundation in the US. In particular we place a special emphasis on the role of Frits Donker Duyvis, who was chairman of the IIB Classification Committee essentially from its foundation in 1921 and who became in 1928 the third Secretary General of the Institute itself, in discussions in the period directly after World War I until just after World War II about how to develop a concordance between the DC and the UDC.

The second network we discuss was involved in continuing the daily work of the Mundaneum, the great international center that Paul Otlet and Henri La Fontaine had created in Brussels and had at first called the Palais Mondial. It had been set up in the huge space of the left wing of the Palais du Cinquantenaire in the Parc du Cinquantenaire in the center of Brussels. The Palais Mondial or Mundaneum was a combination of a universal bibliographic service, an international library, an international museum, a center for adult education activities and the headquarters of the Union of International Associations and a range of other international bodies affiliated with the UIA. A group of volunteers, "Les Amis du Palais Mondial," worked steadfastly with Otlet in the periods during which the Mundaneum was closed and reopened in the years from 1924 until Otlet's death in 1944. They continued their work in the Mundaneum after the Second World War during which it had been moved to new quarters in the Parc Léopold (Rayward 1975). We focus here especially on one of the Friends of the Palais Mondial, a strange autodidact, Walter Théophile Glineur, who experimented with developing a new decimal classification system inspired by the UDC as it had been elaborated by Paul Otlet and his colleagues: the Decimal Classification of Consciousness. (Classification Decimal de Conscience—CDC)

Frits Donker Duyvis and the Deweys: Networks at a Distance and Negotiated Classifications

Frits Donker Duyvis (1894–1961) was not yet born in 1892 when Paul Otlet and Henri La Fontaine began work that would lead to the creation in 1895 of the International Institute of Bibliography (IBB) and to discuss their first views on classification. This resulted in the publication in 1905 of the first edition of the Universal Decimal Classification (UDC) which was developed from the American Dewey Decimal classification. When Donker Duyvis met them in 1920 for the first time, Otlet was over 50 and La Fontaine almost 65. After working for almost 30 years on their mission to organize all of the knowledge of the world, Otlet and La Fontaine must have been very pleased with the arrival of a young, bright and eager chemical engineer who shared their interest in decimal classification. Donker Duyvis had been employed after his graduation in the offices of the United Patent Agencies (Verenigde Octrooibureaux) in The Hague. In 1919 he moved to the city of Deventer in the eastern part of the Netherlands to join the information office of the Governmental Industrial Service (Rijksnijverheidsdienst) where he remained until 1929 when the President of the Dutch Patent Office, Dr. J. Alingh Prins, invited him to join this organization. During his work for the Governmental Industrial Service, Donker Duyvis had become acquainted with the Universal Decimal Classification system and, even more importantly, with the application in Brussels of the system to the classification of abstracts of scientific and technical literature (Clason 1964, 5). The first evidence of contact between Otlet and Frits Donker Duyvis was in a letter of 18 January 1920 in which the Dutchman wrote:

"I am in the process of forming a library for the chemical department of the government service of industrial information. I frequently need bibliographical information. This is why I would like to remain in touch with your institute. Would it be possible to receive all the cards for chemical technology? Perhaps I could contribute to your work by having regularly sent to you the titles of the periodical articles etc. which appear in Holland."[1] The same letter reveals that it was the director of the "Centraal Normalisatie Bureau" (Central Office for Standardization) in Delft, Ernst Hijmans (1890–1987), who had brought the activities of the IIB and the decimal system to his attention. We do not know how Hijmans had learned of the IIB and the UDC. It is fascinating to think that it might have been in 1911 when Hijmans had gone to Brussels to work as a laborer in a steam machine factory. There were good reasons to remain in contact with Hijmans when Donker Duyvis moved to Deventer. Hijmans, like Donker Duyvis, had experience with patent development. Between 1916 and 1918 Hijmans had worked for Van Berkel's Patent Co. Van Berkel had invented and patented a mechanical meat slicer in 1898. In the period of the First World War, his enormously successful company diversified its manufacturing base to other machines such as lathes, engines and even airplanes for the government. In 1918 Hijmans became the director of the Central Office for Standardization and in 1920 he founded along with Vincent Willem van Gogh (nephew of the famous painter), the first Dutch private advice office for organizations, the Organisatie Advies Bureau, a company that did work similar to that of the Governmental Industrial Service, where Donker Duyvis was employed (Bloemen 1988 and 2004).

Donker Duyvis had become part of a small but dedicated network that had begun to advocate the implementation of the decimal system in Dutch administrative bodies in the first decade of the twentieth century (van den Heuvel 2012a). The leading member of this group was Johan Zaalberg (Ketelaar 2000). Zaalberg was the secretary of the municipality of Zaandam. He had been trying to find a comprehensive filing system [*registratuur*] for arranging the administrative decisions' registers and notes and so on for the municipality.[2] When the *Manuel du Répertoire Bibliographique Universel* (1905–1907), the first complete edition of the main and auxiliary tables of the UDC, was brought to his attention by H.J. Romeyn, registrar of the Senate of the Dutch parliament, Zaalberg recognized the classification's potential for administrative purposes and immediately wrote a letter to Brussels seeking a meeting with Otlet.[3] Zaalberg's interest had been triggered because the UDC promised a solution to a continuing problem in the

---

1   Mons, Mundaneum, Dossiers Numerotés-592: Donker Duyvis, Donker Duyvis to Paul Otlet, 18 January 1920.

2   The Hague, Archief van het Registratuurbureau van de Vereniging van Nederlandse Gemeenten, zoals dit berust bij het Algemeen Rijksarchief te Den Haag [Archives of the Register Office of The Society of Dutch Municipalities] Inv. 2.19.140– nr 1: "Geschiedenis van de invoering der administratieve documentatie volgens het decimale stelsel in Nederland, door het Nederlandsch Registratuurbureau. Door J.A. Zaalberg, 20 oktober 1930" [History of the implementation of administrative documentation in the Netherlands according to the decimal system by J.A. Zaalberg 20 October 1930, partly autobiographical]. Dates in this autobiographical account are not always reliable and differ from sources in the archives of the Mundaneum and FID.

3   Mundaneum, Mons Box PPPO 929, File 277: Zaalberg, Zaalberg to Otlet 28 March 1905.

management of the records of Dutch governmental administrations. A Royal Decree of 1823 prohibited an arrangement of these documents by subject. A subject arrangement was thought to create the possibility of bias and manipulation of information. Instead the Decree prescribed that filing should be chronological. The system did not work satisfactorily and Zaalberg wondered if Otlet's decimal order might be more suitable.

His meeting with Otlet in Brussels made an enormous impact on Zaalberg. Zaalberg described Otlet in his autobiographical history of the implementation of the decimal system in Dutch administrations as a man with comprehensive and profound ideas who influenced him for the rest of his life (Zaalberg 1908).[4] Otlet sketched for Zaalberg a future ideal of universality in the arrangement of all administrative records. Otlet stressed that to obtain such a goal a universal bibliographical index or decimal code was required. He agreed to a collaboration between the International Institute of Bibliography and Zaalberg on two conditions: 1) Zaalberg had to bring together prominent men in the Netherlands to achieve the ideals of the IIB; and 2) once this group had been formed, it had to organize itself into a Netherlands public limited company and contribute a considerable annual subsidy of 1000 Dutch guilders to the IIB. This was the origin of The Netherlands Filing Office [Het Nederlandse Registratuurbureau] that was incorporated by Royal Decree on 6 March 1909 with Zaalberg as its first director.

The collaboration between the International Institute of Bibliography and The Netherlands Filing Office was immediately successful. A model for a municipal filing register based on the decimal system was presented at the Premier congrès international des sciences administrative that was held in Brussels in 1910. After the organization of the records of the municipal administration of Zaandam on the basis of in the decimal system, other Dutch cities such as Utrecht and Zandvoort, soon followed. Zaalberg began to spend so much time on the implementation of the decimal system across the Netherlands that in 1912 he was gently requested to make a decision between his position as secretary of the municipality of Zaandam and that of director the Netherlands Filing Office. He chose the latter, "enabling him to support Mr. Otlet in fulfilling his ideals."[5] With the creation of the Netherlands Filing Office in 1909, more and more Dutch governmental and industrial bodies began to implement the decimal system so that by the end of 1925 Donker Duyvis could state that in the Netherlands, "thanks to the combined work of Otlet and Zaalberg [...] some 170 official institutions and commercial enterprises use the Classification Decimal for correspondence filing."[6]

---

4 Zaalberg's book (1908) on municipal administration has a preface by Otlet and the text of the book was in part based on the study that Otlet was then preparing, "La Documentation en matière administrative," *IIB Bulletin*, XII (1908), 342–8. Otlet announces in the preface the conference that would be held in 1910.

5 See note 2, "Geschiedenis," p. 9.

6 Royal Library The Hague, Archives of the FID-Box 96–c 001.2 DC:CD: "Overeenstemming Dewey Code en U.C.C. II"–C.C. 1148, Donker Duyvis provides the

We do not know exactly when Zaalberg and Donker Duyvis met for the first time. But 1920 was a crucial year in which two different applications of the UDC in the Netherlands would come together: the application of the decimal system in public administrations as promoted by Zaalberg since 1905 and a new area of application of the system, in which Donker Duyvis would play a central role, in private companies. The first meeting between Zaalberg and Donker Duyvis appears to have occurred during the first post war conference of the International Institute of Bibliography. This took place from the 5th through 20th September 1920 in Brussels as part of a range of celebratory post-war activities based in the Palais Mondial and called the Quinzaine Internationale. Zaalberg and Donker Duyvis were part of a large Dutch delegation that included H.J. Romeyn, mentioned above, who was now the director of the Dutch Council for Habitation ('s Rijkswoningraad); two board members of the Netherlands Filing Office, Ch. W.A. van Bergen and G.A.A. de Voogd; J. Alingh Prins, Director of the Office of Industrial Property (Bureau van de Industriële Eigendom), later director of the Dutch Patent Office and president of the International Institute of Bibliography; two board members of the Society of Dutch Municipalities; and two board members of the Dutch Postal and Telegraphy Office. At the conference Donker Duyvis and other members of the Dutch delegation were dismayed by what they found in the Palais Mondial. It seemed clear to them that the centralist aspirations of Otlet and La Fontaine to create a world center for documentation in Brussels were unrealistic. They became convinced that documentation should be organized in the first instance on a national basis.

In 1921 Donker Duyvis founded the Netherlands Institute for Documentation and Filing (Nederlandsch Instituut voor Documentatie en Registratuur-NIDER) of which he was the president from 1922 until 1929 before joining the Dutch Patent Office headed by Alingh Prins. H.J. Romeyn convinced Zaalberg to move the Netherlands Filing Office to the Hague into one of the rooms of the office of the Dutch Council for Habitation where the Netherlands Institute for Documentation and Filing was also housed. This move diminished the role of Zaalberg who had begun to be criticized for over-zealous implementation of the UDC in municipal archives and records management systems and for his dependence on the International Institute of Bibliography (Ketelaar 2000, 161). The activities of Zaalberg's Netherlands Filing Office in promoting the application of the decimal system to the filing systems of municipal governments were taken over by the Society of Dutch Municipalities (VNG). It assumed responsibility for over 52 contracts involving the Netherlands Filing Office and Zaalberg's decimal index system that had become known as "the yellow booklet" (Ketelaar 2000, 161–2). Moreover, Zaalberg's international role in promoting the Otletian idea of "documentation" was gradually taken over by Donker Duyvis's Netherlands Institute of Documentation and Filing. Although the headquarters of NIDER was

---

number in a general report of 24 December 1925 on proposals for concordance between DC and CD, typescript p. 3 (original in English).

officially in The Hague, Donker Duyvis operated from his house in Deventer (as he also did for the Dutch section for the International Commission for Decimal Classification).[7] When Otlet invited Zaalberg instead of Donker Duyvis to represent the Netherlands Institute of Documentation and Filing at the second International Congress of Administrative Sciences in Brussels in 1923, Donker Duyvis reacted with some annoyance on behalf of "the official Dutch center for documentation and filing." Otlet responded perhaps with a touch of sarcasm: "the Dutch like formalities" (Ketelaar 2000, 162).

But Donker Duyvis's importance in all matters concerning the International Institute of Bibliography and the UDC, would soon become paramount. He was already the active secretary of its most important committee and his influence would stretch beyond the Netherlands. In 1923 the Belgian government had announced to Otlet and his colleague La Fontaine that early in 1924 it would require the Palais Mondial or Mundaneum to vacate a major part of its locations in the Palais du Cinquantenaire so that they could be used to house a commercial rubber fair being supported by the Belgian Government. The Palais Mondial was to be closed for a period of several weeks and its work substantially disrupted by the removal or unavailability of its collections. On the initiative of Donker Duyvis, representatives of the International Institute of Bibliography from Germany, France, Belgium and The Netherlands met in The Hague for three days, from the 12th through the 14th of June, 1924, to consider the future of the IIB and to reorganize it under new statutes. Nine of the seventeen participants were from NIDER (Rayward 1975, 275). The IIB's statutes were revised to change its organizational structure to a federated grouping of national members. NIDER became in effect the first of the national members of this kind (Michailov 1964, 32). Especially important, the new statutes created a Classification Committee to develop appropriate procedures for controlling and updating the UDC (Rayward 1975, 276). The committee was to be "the guardian of the unity and integrity of the system, responsible for all of its developments and necessary principles of application, it acts as the sovereign authority in this matter … ." and it was intended to become "a liaison center for all who co-operate in the development of the tables of the Decimal Classification" (IIB Pub 140, 4, quoted from Rayward 1975, 277). Despite the focus in the new statutes on a confederation of national documentation centers, there was no loss of belief that the federation could only operate effectively if its members shared a single uniform system for classifying documentary items. Appointed secretary of the Classification Committee, Donker Duyvis was engaged for almost 40 years in adjusting the UDC to meet practical needs in organizing scientific knowledge and in developing concordances for interoperability with other classification systems, especially with the Decimal Classification in the US. Dealing with the last, however, turned out to be a process of tiresome negotiation in which tensions would arise between the European

---

7   Mons, Mundaneum, Dossiers Numerotés-592: Donker Duyvis, Donker Duyvis to Paul Otlet, 14 January 1922.

developers of the UDC and the Americans responsible for the DC with Donker Duyvis acting as mediator (van den Heuvel 2012a).

Melvil Dewey had given permission to Otlet to translate the Decimal Classification into French and other European languages in June 1895 and he and Paul Otlet were on good personal terms with each other. They shared the dream of disseminating the decimal system globally and very early on were concerned to keep as much concordance as possible between the two systems. However, in the first decade of the twentieth century there was a tumultuous period with "the Belgians and Americans in conflict" (Rayward 1975, 97–105). While differences continued, though slowing after the publication of the first full edition of the UDC in 1905, both sides never gave up hope of eventually achieving a detailed concordance between, and ultimately the unification of, the two classifications despite their awareness of the divergences that were continuing to take place.

This belief in eventual unification took an unusual expression when Otlet wrote to Dewey in 1919 asking for his support for the Palais Mondial. It was, he pointed out, a complex of international institutions in Brussels that functioned as a Congress building, a Palace of Documentation, a Library and a Museum all ordered on the principles of the decimal classification, or as Otlet described it to Dewey: "the decimal classification will serve as the architectural basis of the building […] the Decimal Palace! [Palais de la Décimale!]." Otlet suggested that an American committee be formed, perhaps under the patronage of the American Library Association and with the financial assistance from the Carnegie Foundation, to support the center.[8]

By 1923 as we have seen, the Belgian government's originally supportive attitude to Otlet and the Palais Mondial had changed. Otlet now appealed again to the Americans for support. Melvil Dewey's son, Godfrey, visited Brussels in September 1924 and was sympathetic to the plight that Otlet described to him. They agreed that Godfrey Dewey would seek what help he could from interested organizations in the US (Rayward 1975, 263–4). A few days later Godfrey Dewey presided over a formal meeting of the IIB in Geneva at which the statutes drawn up in the Hague were ratified. In attendance was not only Godfrey Dewey but Dorcas Fellows, editor of the DC. But despite this meeting, the tone of the correspondences between Duyvis and Otlet representing the collaborators of the International Institute of Bibliography on the UDC and Dorkas Fellows and Godfrey Dewey representing Melvil Dewey's Lake Placid Education Foundation in upstate New York, the publisher of the DC, would change considerably during the process of implementing the resolutions of the Geneva meeting. During the meeting, the Classification Committee had decided that there should be three versions of the Decimal Classification: an abridged version, a library version and a bibliographic version, the last being the European version. Furthermore, "the editions of the Decimal Classification Codes of Dewey and of the IIB should

---

8 Mons, Mundaneum, PPPO 0439– PM: PM-Courier Otlet, 1919–1933, Otlet to Dewey of 23 June 1919.

be unified" and so modified that "the best of both will be adopted, each party consenting to sacrifices" (Comaromi 1976, 300).

How difficult it would be to implement this decision became immediately clear in one of the letters that Dorcas Fellows, editor of the Decimal Classification, wrote to Donker Duyvis on 31 July 1924: "A difficulty has sometimes risen in our acceptance of the IIB tables, because of the signs used for form divisions, which are so extremely valuable for bibliographic work (such as is done by the Institute) but which do not seem practical for marking the backs of the books for shelf arrangement, which is one of the principal uses of the Classification in America."[9] However, the differences were not just conceptual. Donker Duyvis had used the UDC initially as a bibliographical tool for classifying the libraries of commercial and industrial enterprises and this had worked reasonably well. But, when he started to design new classification numbers for Sections 5 and 6 (pure and applied science) of the UDC, the implications for the unification of the DC and UDC for scientific specialization that was stressed by the compilers of the UDC began to become clear. At first Dorcas Fellows replied courteously to all requests of Donker Duyvis for changes in the tables of the Decimal Classification, but her tone changed rapidly: "and if the I.I.B. is not legally restrained because Dr. Dewey, with characteristic altruism, set aside the restraints, it seems to me that the obligations of I.I.B. are not less than those of the Germans [i.e. the editors of the German translation], to the effect in IIB's case, that, in developing C.D., care should be taken that it should not be developed in such a way as to work injury to the original system and to its development for its original purposes."[10] The conflict came to a climax in 1928 when Dorcas Fellows learned that Donker Duyvis wanted to alter the tables for botany, 580. Donker Duyvis wanted to use the modern Engler-Gilg botanical classification, while Dorcas Fellows wanted to keep the older Bentham-Hooker system.

The conflict lasted for years with the secretary of the Decimal Classification and the editor of the Universal Classification Decimal both mobilizing their networks. Godfrey Dewey in a statement about the DC and the UDC to the Lake Placid Club Education Foundation Trustees of July 1937 declared that 580 had been "the chief issue" that he had discussed in several meetings in Brussels and London in the years 1928, 29, 30 and 31. The 580 question was still mentioned in a historical overview by Donker Duyvis as late as 1948 (discussed below).[11] Dorcas Fellows had written to Godfrey Dewey, who in his turn wrote to Donker Duyvis to express his anger over "the apparent complete disregard of

---

9   The Royal Library The Hague – Archives of the FID: Box 96-C 0254x0012:DC–CD – Dorcas Fellows to Donker Duyvis, 31 July 1924. Underlinings by Dorcas Fellows.

10   Dorcas Fellows to Donker Duyvis, 3 December 1925, See note 9 above.

11   Godfrey Dewey to Lake Placid Club Foundation Trustees], July 1937 – typescript pp. 7–8. Royal Library The Hague – Archives of the FID, Box 96, *Outline of the DC and U.D.C. relations* historical overview by Donker Duyvis Box 96A-Dewey Classification, 13 December 1948 (see note 6 above).

our position."[12] In 1929 when Godfrey Dewey found Donker Duyvis's answers to Dorcas Fellows and to himself to be unsatisfactory, Godfrey Dewey wrote to Otlet of 580 that: "it would be probably the last straw which would break down and destroy any possibility of further efforts at cooperation and concordance between DC and CD" and he concluded: "Please do not underestimate the seriousness of the issue regarding 580."[13] [14]

Donker Duyvis countered on 9 December 1929 with a long letter in which he argued that in the UDC "1000 to 1500 numbers were altered to avoid collisions with the American code" while the Americans had changed no more than 10 in favor of the IIB. He explained the modification of the botany class as follows: "It is for the IIB that disposes over staff of competent scientific collaborators very painful to produce work of less quality, just because they have to fit into a general schema that is made in a less scientific and sometimes amateurish way."[15] The atmosphere was so bad that the 80-year old Melvil Dewey two years before his death felt the need to intervene with his son and Dorcas Fellows (Rayward 1975, 316). In Europe, Otlet expressed his embarrassment to Godfrey Dewey that he could not change the date of the IIB conference in Zürich in 1930 to allow the participation of the American visitor. He suggested that instead Godfrey should meet Donker Duyvis and himself in Brussels (Rayward 1975, 317). In Brussels, Dewey and Otlet came to some sort of arrangements and both were again hopeful that a reconciliation between the two classifications might be achieved. During the Zurich conference, however, it became clear that they were gradually losing control of the Classification to new influential figures in the IIB, especially the Englishmen Bradford and Pollard, who in 1927 had founded the British Society for International Bibliography to be the English national member of the IIB and, of course, Donker Duyvis.

After the 1930 Zürich conference, Otlet complained to Dewey that he had defended their project but had preached in the desert and that the Dutch and English had declared that it was impossible for them to commit themselves to it: "Mr. Donker Duyvis should have written to you about what the situation is now. The English had proposed to call him 'Dictator' of the classification" (quoted by Rayward 1975, 318). The situation would change once again. In 1931 the International Institute of Bibliography changed its name to the International Institute of Documentation and its

---

12   Godfrey Dewey to [Donker Duyvis], 30 May 1929, Royal Library The Hague – Archives of the FID, Box 96, see note 6 above.

13   Donker Duyvis to Godfrey Dewey, 5 November 1929, Royal Library The Hague – Archives of the FID, Box 96, see note 6 above

14   Mons, Mundaneum, Dossiers Numerotés 591: Donker Duyvis, Dewey to Otlet 14 November 1929. Otlet sent a copy to Donker Duyvis with the heading Classification Decimale Botany 58 (Royal Library The Hague: FID Archives Box 96, see note 6 above).

15   Mons, Mundaneum, Numerotés 591:Donker Duyvis, Donker Duyvis to Samuel Bradford (Chief Librarian of the Science Library), Blondin, La Fontaine, von Hanffstengel, Otlet, Alan Pollard (President of the IIB from 1928–31) and Carl Walther (Chairman of the Classification Sub Committee of the German Standards Institute) 9 December 1929.

president at the time, Allan Pollard, was replaced by Alingh Prins. The former Institut International de Bibliographie now the Institut International de Documentation had a Dutch president and a Dutch Secretary-General who also chaired the institute's most important committee, the International Committee for the Decimal Classification. Otlet had used strong words to describe the power of Donker Duyvis and would soon find out what the presidency of Alingh Prins implied for the management of the old "Brussels Institute" when the latter wrote on the 20th November 1931 to the co-founder of the IIB: "We very sincerely request that your refrain from any intervention in the management of the finances of the IID (not the IIB)" (quoted from Rayward 1975, 326). It seemed the period was now over in which the grandfathers of the American and European decimal classification systems could resolve the squabbling between those involved with developing the two systems by referring to old contracts and gentlemen's agreements. The character of administration and tone of communication of the actors in these two arenas had changed drastically over time. Letters of understanding had been changed into directives.

Achieving concordance between the Decimal Classification and the UDC in the inter-War period seemed even more remote than during the crisis in the relationships between the Americans and the Belgians before World War I. Had nothing changed? Donker Duyvis had defended himself in the letter of 9 December 1929, mentioned above, against the allegations of Dorcas Fellows and Godfrey Dewey that he was ignoring the needs of the Decimal Classification by pointing out that he had adopted 1500 classes of the DC for the UDC. When the idea of publishing an English translation of the UDC (the complete edition was in French) had been raised in 1931, Godfrey Dewey had officially protested. He wanted to discuss the underlying problem of the concordance between the two classifications (Comaromi 1976, 310). In 1933 when the Americans finally granted permission for the publication of the translation, they did so on the condition that a "substantial concordance on 1000 heads be reached" (quoted by Rayward 1975, 321; Comaromi 1976, 306–7). It was at this point that the British Standards Institution agreed to sponsor a translation of the UDC as a British standard.

It can be questioned whether the number of classes for which concordance was achieved would have been the best yardstick to use in assessing the extent of the reconciliation between the two classifications. References to large amounts of added classes seemed like arguments offered to the outside world to show goodwill. Perhaps they could be used to gain some time for what was needed for real concordance to be achieved. However, internally both camps seemed to be convinced that for real concordance radical changes were necessary in their own systems. Godfrey Dewey, in his statement in 1937 to the Lake Placid Education Foundation Trustees on the DC and CD, concluded dramatically his lengthy historical exposé of previous efforts toward concordance: "Without concordance, it is quite possible that DC and CD alike contain within themselves the seeds of their own destruction."[16]

---

16   Godfrey Dewey to Lake Placid Club Foundation Trustees, July 1937–typescript p. 11. The Royal Library The Hague – Archives of the FID, Box 96, see note 6 above.

At a meeting in 1946 in the offices of the British Standard Institution to discuss the state of affairs of the new English UDC edition, F. Steggerda, who had replaced Donker Duyvis as chair of the classification committee, again stressed the importance of concordance and referred to Godfrey Dewey's statement of 1937. Steggerda reported that Donker Duyvis had suggested "that the U.D.C. should move as far as possible toward the D.C. in Classes 1, 2, 4, 7, 8 and 9 and that in case of Classes 3, 5, and 6 the D.C. should move as far as possible towards the U.D.C."[17] In 1948, in a historical overview, *Outline of the DC and U.D.C. relations*, Donker Duyvis described how his attempt to bring concordance between the UDC and the D.C. had met a "resistance movement" by European users. But his view was that it had to be carried out anyway as promised. He concluded that "a close co-operation with the American colleagues can only improve the scientific contents and increase the practical value of the classification."[18]

While Donker Duyvis argued that promises about concordance made over the years should be met, at the same time he realized that this would only be possible following a serious revision of the UDC. "If fundamental modifications are realized, I think that the time is ripe, they should be realized now in complete agreement with the American colleagues."[19] He suggested not trying to realize too much at once but to prepare what he called "some schemes" and to carry them out later.

In fact, Donker Duyvis had already made plans for revision which he had presented in an address, probably in 1933 to the British Society for International Bibliography, as a "Tentative scheme of complete revision of C.D.U. (1940 or later)" which he introduced as follows: "Allow me to develop before you some suggestions of a radical character with a view to the standardized system of classification we have built up with so much care. This radicalism is moreover rather harmless since it deals with a revolution announced at least 7 years before."[20] Donker Duyvis suggested that the current method of evolutionary revision should be continued but at the same time he claimed that the progress of

---

17   Godfrey Dewey: "Notes of a meeting held at the offices of the British Standards Institution on Tuesday 19th February, 1946"– typescript p. 3. Present were F. Steggerda (Chair), Godfrey Dewey, S.C. Bradford, J.F. Stanley and D.M. Roberts. The Royal Library The Hague – Archives of the FID, Box 96, see note 6 above.

18   Donker Duyvis, "Outline of the DC–U.D.C. relations," p. 2. The Royal Library The Hague – Archives of the FID, Box 96, see note 6 above.

19   See note 18 above.

20   The Hague – Archives of the FID, Box 95-c.004.65 File: "Algeheele revisie der U.D.C." The presentation has no signature, but was definitely by Donker Duyvis. It has the title "Address to the B.S.I.B. on March 31st," but unfortunately no year. In the text he refers to the latest edition of the DC which came out in 1932. Probably the presentation is of 1933 or of the years directly following since Donker Duyvis refers to this revision as a project. Twenty-five years later, G.A. Lloyd of the British Standards Institution later Head of the FID Classification Department and Secretary to what was then known as the Central Classification Committee referred to the plan again in a letter of November 3, 1958 with the title: "On the Future of the Dewey DC and UDC," Box 96A-"Dewey Classification" (see

modern knowledge and the mutual relations between the sciences would require a complete reconstruction of the UDC at certain intervals, perhaps of 30 or 50 years. He considered that such a major revision was justified "also for giving guidance and direction to our present work" and that it would "be useful to have an ideal classification behind the screen in order to follow a well pondered policy in developing the old scheme." In fact, Donker Duyvis suggested that he had special reasons for wanting to renew the UDC in its current form, "Since we want to come to concordance with our American friends." Such a renewal was necessary all the sooner to "respect the free numbers blocked by them." His plan entailed leaving three of the ten classes open for "renewal and starvation" by erasing the distinction between certain categories such as that between the pure and applied sciences. [21]

Donker Duyvis did not believe in a static system. What he called an ideal scheme was important in serving in the background to give direction to continuous development of the existing classification that would change in a more fundamental way only twice or three times in a century. With this ideal background scheme in place, editors would be well prepared to act as soon as a serious request for revision arrived. He stressed that the preparation of the scheme he was proposing was intended to "help us to 'evolutionarise' our present U.D.C." What we see here is that Donker Duyvis is developing a strategy for constant, systematic revision of the classification by international cooperation. In 1948 he even proposed that what was now called the International Federation of Documentation (the new name give to the IID in 1937) establish a committee for comparative classification (with Ranganathan as its chair) (Donker Duyvis 1953, 33; Ranganathan 1961, 98). Perhaps Donker Duyvis's best explanation of his views on flexibility and the dynamics of classification occur in his article: "The UDC: What it is and What it is not" of 1951. Here he stated that despite the eagerness of Otlet to give full credit to Dewey, Otlet had changed the fundamental basis of the decimal classification system (Donker Duyvis 1951, 99). Otlet's solutions were not dissimilar to those of Ranganathan: "In fact from the moment that Otlet started his 'colon classification' the Dewey scheme was dethroned as [the] basic scheme" (Donker Duyvis 1951, 100; see also van den Heuvel 2012b). The flexibility of the UDC that resulted from the changes that Otlet had proposed to the DC implied that no philosophical or scientific system could be recognized as underpinning the UDC (Donker Duyvis 1951, 100). The consequence was that the UDC would not become an inflexible scientific system. "On the contrary the classification should reflect the reality of the development, of the contradictions and even of the errors in science. In a purely scientific system there is no place for an erroneous conception. In a practical classification a place should be allotted to any scientific error as long as publications on such error are available" (Donker Duyvis 1951, 101). Despite

---

note 6 above): G.A. Lloyd to Donker Duyvis with copies to several colleagues, November 3, 1958; see also C van den Heuvel (2012a).

21   The Hague – Archives of the FID, Box 95–c.004.65 File: "Algeheele revisie der U.D.C." The quotations are from pp. 3, 4–5, 5, 12 and 13. See note 20 above.

the fact that Donker Duyvis tried to underline the differences between the UDC and "Dewey's philosophical conception of the classification of knowledge," we can recognize in his plea for a practical classification of publications instead of a classification based on a scientific system the same arguments that Dewey and his supporters had made (Donker Duyvis 1951, 101). In fact Donker Duyvis had ignored the fact that the classification of knowledge by Otlet and La Fontaine, perhaps more than that of Dewey, had been based on philosophical notions (coherent or not) of universalist thought about the unity of the sciences. However, Donker Duyvis' pragmatic approach (in Otlet's words a "dictatorial" approach) in the seemingly never-ending negotiations about revision and about concordance between the two decimal classification systems resulted in a more dynamic, continuously and collectively revised UDC than had previously been the case.

An Unlikely Friend of the UDC in the Network of les Amis Du Palais Mondial and his Alternative Decimal Classification

Whereas Otlet and La Fontaine seemed to have gradually lost control of the UDC's development to Donker Duyvis, they would be confronted at the end of their careers with a person who certainly had remained loyal to their ideals, but who would follow them in an almost obsessive way in creating his own alternative classification. This person was the self-declared autodidact scientist, Esperantist, anarchist, shoemaker and proud owner of the Universal Shoeshop, "Cordonnerie Universelle" in Uccle near Brussels, Walter Théodore Glineur (1887–?).

Glineur was a man on a mission. This becomes clear from an announcement in 1959 in the lengthy title of his, "*The Sense of Life. Universal Unity. Know Yourself, know life: microcosmic and macrocosmic through the Ten Senses and the 5th Dimension by Wal-Theo-Gli-Jongh.*" The author's name is a combination of his father's and mother's surname we learn from the first two and a half pages of this ten page-long document. The self-taught scholar explains that he had created "his own university" by attending the lectures and discussions at the Palais Mondial in Brussels during a period of more than 30 years and by being a member of several committees related to its work.[22] This had resulted in his plans to write in the period between 1949 and 1951 a comprehensive document of about 400 pages with the title: *Testamento*. A random selection of topics from the preliminary table of contents of the work that was still in preparation in 1959 is revealing: Biography 20 pages, Secrets of Colors 10 pages, Secret doctrines (theosophical) 46 pages, New Testament 19 pages, Liberalism 7 pages; Socialism 6 pages, Anarchism 22 pages, and Is God a mathematician?, The Writings of Khrishnamurt etc. etc. However, apart from his views on political and philosophical systems of the

---

22   Mons, Mudaneum, RUD-BIO 268: Glineur W.T.– The Sense of Life.etc., Printed cover typescript 10 pages, p. 3.

Figure 13.1   Portrait of Walter Théodore Glineur © Mons, Mundaneum

Western and Eastern World, he also compiled an *Encyclopedia of Shoes* consisting of 611 pages, organized three conferences on the topic and planned five films on the history of the shoe of which three were realized.[23] He chaired various local and national syndicates and guilds of shoemakers. In 1929 Glineur was one of the cofounders of the grandiosely named, but essentially local, Committee of World Rights for Workers and co-creator of the Constitution of World Rights for Manual and Intellectual Workers. He was a member of the Belgian Syndicate of Artisans and Shopkeepers of Shoes: Uccle and Surroundings Section, was President of the Masters of the Shoemakers guild and was a member of the Cooperative of the Professional Union of Shopkeepers (Shoemakers Section).[24]

---

23   Mons, Mundaneum, RUD-BIO 268: Glineur W.T.—schema film cinéscopie-chaussure.

24   Mons, Mundaneum, RUD-TH (858): 331.88: 69, Glineur.

Although Glineur sometimes refers to the works of the forefathers of communism, whose ideology he would follow after World War II, he was known to Otlet and La Fontaine as an anarchist. Glineur donated a large library on the Anarchist Movement to the Mundaneum and like Marcel Dieu (better known as Hemday) and Georges Lophèvre, both of whom also donated similar material to the Mundaneum, he actively participated in the activities of the "Amis du Palais Mondial" (Fueg 1995 and 1999). As has been mentioned, this was a group of devout followers that Otlet gathered around him when the Palais Mondial had to close its doors for a period in 1924. Glineur indicated that he had been "general treasurer and member since 1926" of the Amis du Palais Mondial when he excused himself to Paul Otlet for not being able to attend a meeting on the 6th June 1937 "on the occasion of the third year of the closure and sequestration of the documentation of the Palais Mondial, Museum, I.I.B etc." to express his continuing support. [25] The Belgian government had finally closed the Mundaneum in 1934 in the Palais du Cinqantenaire and it was not to open again in that location, though during the War, as mentioned, it was moved to new quarters in the Parc Léopold.

Glineur must have been one of the most enthusiast members of the Friend's group, perhaps an over-zealous friend who in 1959, 15 years after the death of Otlet, still felt the need to fine-tune a dialogue he had written with (or for) Otlet in 1933.[26] He was also a "friend" who donated a beautiful plaque to the "Citizen of the World Henri Lafontaine" with the title: "Federal Organization of the World in zone 1,296 in the year 134773v," that is 1931.[27] La Fontaine and Otlet and their UDC were a source of inspiration for Glineur's own attempt to order his knowledge of the world. This becomes clear from an alternative decimal classification system in typescript signed by Glineur (and a certain Reinoud Welvaert about whom nothing is known) kept in the archives of the Mundaneum. Glineur first called it the Décimal Classification of the Human Senses (Classification Decimale d'après les Sens Humains) but changed it later the Decimal Classification of Consciousness (Classification Décimale de la Conscience—CDC) to distinguish it from Otlet's UCD which Glineur described as the Classification Décimale d'àpres des Sciences (CDS). In French "conscience" has a wider meaning than in English. It has both the idea of conscience but also it can mean consciousness or awareness as well. Since Glineur's classification focuses on the senses we prefer the term consciousness. However it is also clear that for Glineur classification also carried a moral signification. It was an instrument of social action for the liberation of the mind. Liberation of the mind expressed in knowledge classes went hand in hand with the liberation of the oppressed social classes.

Glineur presented his CDC in two letters partly in French and partly in Esperanto dated 6 April 1935, copies of which were sent to Paul Otlet, to the Permanent Council for the Study and Diffusion of the World Plan (a section of the

---

25   Mons, Mundaneum, PPPO-0442, APM OP 131, Glineur to Paul Otlet, 2 juni 1937.
26   Mons, Mundaneum, RUD-BIO 268:Glineur W. T.
27   Mons, Mundameum, RUD-TH (Thematique (858) Box 3G 02, Notes –3601–5.

Association of the Rights of Workers [Association pour le Droit de Travailleurs], of which he was the secretary, and to Krishnamurti. [28] The first letter was in the form of an announcement ("Avis"). It was 11 pages in length and was to serve, says Glineur, as an introduction to his publication that related the Ten Human Senses to the Plan Mondial, which was probably a reference to Otlet's book, *Monde* (Otlet 1935). The "Avis" suggested that the project he was introducing was an initiative of the Permanent Council for the Study of the World Plan. This reference to the World Plan is of importance since Glineur aimed at connecting his classification to his conception of the world just as Otlet had attempted to connect the UDC and *Monde* (Ducheyne 2009; van den Heuvel and Smiraglia 2010; Kouw, van den Heuvel and Scharnhorst 2013). This introductory missive was accompanied by a 16-page letter that described the classification system. Both letters arose out of a sense of grievance. In the "Avis" Glineur stated that Otlet had co-signed with Glineur and others a resolution that was intended for the Permanent Commission of the World Plan and that would be shared for study and "emancipatory action," but Otlet had "strictly, (jealously) [jalousement]" kept the resolution to himself. And, continued Glineur, "am I not boycotted by André Colet [the longtime associate of Otlet, who had helped create and presided for many years over Les Amis du Plan Mondial], who it seems was now opposed to "the subject of 'my' conference-debate of 6 January 1935?" Glineur is presumably referring to his participation in one of the discussion meetings, lectures and debates that were regularly sponsored by Les Amis du Palais Mondial and that were continued in the quarters of the Institut des Hautes Etudes and elsewhere in Brussels after the closing of the Palais Mondial. Believing he had been silenced by those he could not convince, Glineur now attempted to clarify both his intentions in presenting his classification and the classification itself.

Glineur described and illustrated his Decimal Classification of Consciousness as a step-by-step transformation and sublimation of Otlet's UDC, or as he calls it, the Classification Décimale des Sciences (CDS) (see Fig. 13.2). The point of departure is the transformation of the ten main classes of the UDC (or CDS), consisting of nine subject classes represented by the branches of the tree of knowledge and a class 0 or Generalities, the stem of the tree. From branch 1 a key is hanging that symbolizes the *index-scientia* that provides access to the knowledge of the world via the decimal classification. He then suggests, at the next level, that these classes are related to each other in various ways and presents what he calls a "rough sketch of a partial rectification" of these relationships. He combines classes 1 and 2 of the UDC (philosophy and religion), 3 and 6 (social sciences and applied sciences), 4 and 8 (philology and literature), 5 and 9 (natural sciences and history) and leaves the number 7 (Arts) and 0 (Generalities) unchanged.

---

28   Mons, Mundaneum:Box PPPO-0442: Courrier APM III 1934–3–5–Courrier Otlet 1935 [Two typescripts]: "Avis Préliminaire" 11 pages (referred to below as "Avis") and Letter to P. Otlet, 6 April 1935 with explanation of the classification (referred to below as "Classification").

After this re-ordering and re-grouping of Otlet's classification of the sciences, Glineur provides a complex representation of his Decimal Classification of Consciousness (Classification Décimale de la Conscience—CDC). He gives special meanings to the ten classes of the Otletian tree of the Classification Décimale des Sciences (CDS). He adds an extra branch with an eleventh class 0 for generalities. This class however is given a far more important meaning in the CDC than in the CDS. It symbolizes a continuous process of science and of life or in Glineur's cryptic words: "Eternity consumes the 0 in eternal becoming." Glineur's tree represents evolution. The tree contains a spiral. This expresses the dynamics of the sciences. It is also a star tree that represents what he calls the "flames" of the human senses and fundamental biological functions that produce sensation along with their related sciences (i.e. the branches of the Classification Décimale de la Conscience—CDC). These "flames" are represented by Arabic numerals and are ten in number (unlike the nine substantive classes of the UDC or Classification Décimale des Sciences—CDS whose combinations as presented in the middle of the figure are now given Roman numerals). These flames of the senses which Glineur describes inconsistently in nouns and verbs and with spelling errors are the branches (sciences or disciplines) of the Star Tree associated with them are: 1 Taste (*goûter*)—biology; 2 Smell (*odorat*)—Hygiene; 3 Touch (*toucher*)—sociology; 4 Sight (*vue*)—encyclopedia; 5 Hearing (*ouïe*)—science of rhythm; 6 Physical motility (*motilité physique*)—physical motility, locomotion, 7 Psychic motility (*motilité psychique*)—psychic motility, intuition, psychology; 8 Respiration (*respire*)—phonology (sic); 9 Emotionality (*emotivité*)—cosmology, 10 Procreation (*procreation*)—geneaology. (Table 13.1 summarizes the relations between the classes of sciences-branches of the CDS and CDC and of the senses-flames).

The dynamic interactions and relationships revealed by the spiral in this complex figure allow in Glineur's words for the "harmonious transmutation of universal energeticism. Peace and happiness are not then an illusion but a fact." This process is represented at the bottom of the figure in the form of an eye. The pupil of this eye consist of the 9 substantive classes of the UDC or Classification Décimale des Sciences (CDS) with class 0 right at its centre. The pupil is surrounded by an iris consisting of the 10 senses (flames) of the Classification Décimale de la Conscience—CDC, now represented by letters. This iris is not static but seems to circle and expand as indicated by the dotted lines in order to symbolize "the perpetual re-equilibrium of life" in the "eye of clairvoyance." A horizontal line through the eye divides the sciences from senses related to nature (upper part) and those related to the divine (lower part). The final transition from the UDC (or Classification Décimale des Sciences–CDS) to the Classification Décimale de la Conscience—CDC involves "Life acting through and across Being" and results in "The Eye of Clairvoyance."[29]

---

29   Mons, Mundaneum, PPPO-0442. See note 28 above. The quotations are from "Classification" pp 3–6.

A l'arbre de l'Index-Scientia de la C.D.S. pend une olé,dit"passe-partout"par la-
quelle on pénètre dans "l'arbre de la Vie". Après reflexions il se fixa l'erreur
de cette image séparant   l'Esprit de la Vie.

Cherchant les mots les 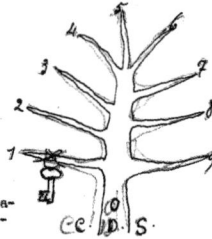 plus interprétatifs
& de ce qui est le + agissant danschaque
chose & être du micro- cosmique au macrocosmique,
il surgissait a la X les mots& une figure des:
SIGNES,le SIGNO GENERA (en espéranto)
= signe générique se fit JOUR & iln'est
pas ce qui entre dans la Vie mais ce
qui le Meut,ce qui en Est laMani-
festations festations
mots : +═ Croiser, ═Croître,X═
Multiplier & G═ *Réaliser furent
RE-ENSEIGNES au MONDE .
C'est ce qui précisa fonda- mentalement mieux le SENS
de la Classification Déci- male de la Conscience,de
cette Réalité agissante que ns sommes,ainsi que TOUT.

Aucune continuité réelle,créatrice où le discernement quand au plan général.
Des branches sont + ou - interverties & certaines coincées; exemple: le 3 = Science-
sociale,entre la Réligion & la Philologie;5 = Science-naturelle,entre la Philologie & entre
la Science-appliquée; etc.
Ebauchant une rectification partielle voici le troé qui en résulte :
I/Philosophie & 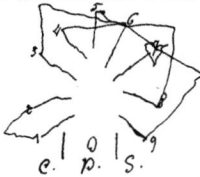 5/Science-naturelle &
2/Réligion 9/Histoire.

3/Science-sociale & 7/Beaux-arts reste
6/Science-appliquée. + ou - isolé (selon..)

4/Philologie & 0/d'où l'ensemble,ou
8/Littérature. Généralité limitée.

Maintenant,VOICI l'arbre étoile a 10 flammes de la C.D.C.où ns avons placées les IX
branches de la C.D.S. & le 0 : (les chiffres arabes ns concernent,les romains le tiers).

La I° Gouter ou Biologie 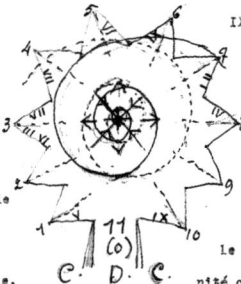 Comme il est visible: les

" 2° Odorat " Hygiène IX parties ou sciences

" 3° Toucher " Sociologie sont indiquées du côté

" 4° Vue   " Encyclopédie A ou Z selon qu'elles

" 5° Ouie   " Rythmologie sont Naturelles ou

" 6° Motilité physique artificielles dans le SENS
" " Locamotion
" 7° Motilité psychique (les sens de la vie huma-
"Intuition,psychologie ne) & de la Vie Réelle donc.
" 8° Respir " Phonologie.
le II° SENS: Présentement Éter-
9° Emotivité " Cosmologie. nité consumme le 0 par l'incessant
Devenir.
" 10° Procréation " Généalogie.

La polarité en action Uniconsciemment & Unisconsciemment,conçu & compris,rééquilibre ou
perpétue l'équilibre,fait que ns vivons "transmutant"harmonieusement l'énergétisme Uni-
versel.La Paix le Bonheur ne sont alors pas une illusion mais un Fait.

La Vie agissante physique au travers de l'être
ou l'OEUIL de la CLAIRVOYANCE

**Figure 13.2   Transition from the universal classification of science (i.e., U.D.C.) to
the universal classification of conscience and the eye of clairvoyance**

Source: PPPO–0442 –Courrier Otlet 1935, Mons, Mundaneum ©

Although Glineur's explanations are elaborate, they are often incomprehensible and what he is aiming at can only be guessed. Table 13.1 represents Glineur's attempts to provide a concordance between the major categories of the UDC (Or Classification des Sciences–CDS) and the two aspects of his Classification Décimale de Conscience–CDC: the senses or the biological functions that create sensation (flames) and the sciences or disciplinary areas (branches) associated with them.

Table 13. 1   Comparison of the Decimal Classification of the Sciences (i.e., UDC) and Decimal Classification of Conscience: main categories, senses (flames), and disciplines and disciplinary areas (branches)

Decimal Classification of the Sciences (compare UDC)	Decimal Classification of Consciousness–Flames (Flammes)	Decimal Classification of Consciousness–Branches (Branches)
0 – General (*Généralité*)	0 (11) Eternity (Eternité)	0 (11) – Clairvoyance
1 – Philosophy (*Philosophie*)	1 – Taste (*Gouter*)	1 – Biology (*Biologie*)
2 – Religion (*Religion*)	2 – Smell (*Odorat*)	2 – Hygiene (*Hygiène*)
3 – Social sciences (*Sciences Sociales*)	3 – Touch (*Toucher*)	3 – Sociology (*Sociologie*)
4 – Philology (*Philologie*)	4 – Sight (*Vue*)	4 – Encyclopedia (*Encyclopédie*)
5 – Natural Sciences (*Science Naturelles*)	5 – Hearing (*Ouie*)	5 – Knowledge of Rhythm (*Rythmologie*)
6 – Applied Sciences (*Science Appliquées*)	6 – Physical Motility (*Motilité Physique*)	6 – Locomotion (*Locomotion*)
7 – Fine Arts (*Beaux Arts*)	7 – Psychic Motility (*Motilité Psychique*)	7 – Intuition, Psychology (*Intuition, Psychologie*)
8 – Literature (*Littérature*)	8 – Respiration (*Respir*)	8 – Phonology (*Phonologie*)
9 – History (*Histoire*)	9 – Emotivity (*Emotivité*)	9 – Cosmology (*Cosmologie*)
	10 – Procreation (*Procreation*)	10 – Genealogy (*Généalogie*)

Dynamics in Networks and in Classifications: Epilogue

Glineur's classification, although inspired by the UDC, presented a complete, new, alternative system to the UDC. It seems that Glineur had intended to overcome the shortcomings of alphabetical classifications in Esperanto by using a decimal system similar to the UDC (but with 11 Main Classes) and at the same time to improve the decimal classification of the sciences by including the senses and consciousness.

Despite the fact that Glineur, as an active member and former treasurer of the "Amis du Palais Mondial," had been close to Otlet and had presented his ideas at the Mundaneum before he and Otlet apparently fell out with each other, his views on classification had no impact on the UDC. Paradoxically it seems that the more distant, European and trans-Atlantic networks centered on Frits Donkers Duyvis and the American developers of the DC, despite all the rhetoric's of resistance, had a gradual but more steady impact on the UDC and the renewal of its classes, especially those for the sciences. The change and dynamics of development of both of the decimal classification systems, the UDC (and to a lesser degree of the DC), were not only the result of the flexibility in the system enabling it to accommodate new scientific insights, but also of the effectiveness of negotiations through networks.

Acknowledgements

I am indebted to Sofia Kapnisi (Office manager of the UDCC) for providing access to the closed archives of the FID in the Dutch National Library. Magda Bouwens (KB-Dutch National Library) provided me important inventory lists. Furthermore, I would like to thank Boyd Rayward who brought the first letter from Donker Duyvis to Paul Otlet to my attention. As always he had important suggestions for the text.

References

Bloemen, E. 1988. *Scientific Management in Nederland 1900–1930*. Amsterdam: NEHA.

Bloemen, E. 2004. *From Industry to Services at a Personal Level. Life and Work of Ernst Hijmans, Consultant (1890–1987)* (http://www.econ.upf.edu/ebha2004/papers/1D1.doc).

Clason, W.E. 1964. 'The Life of Frits Donker Duyvis', in *F. Donker Duyvis. His Life and Work*. The Hague: NIDER publication series 2 no. 45, 5–8.

Comaromi, John Philip. 1976. *The Eighteen Editions of the Dewey Decimal Classification. A Revision of the Author's Thesis. University of Michigan*. Albany (NY): Forest Press.

Donker Duyvis, Frits. 1951. "The UDC: What it is and What it is Not," *Revue de la Documentation*, 18 (2): 99–105.

Donker Duyvis, Frits. 1953. "On the Future Work of Comparison of Classifications," in K. Chandrasekharan (ed.) *Library Science in India. Silver Jubilee Volume*. Madras: Madras Library Association, London: Blunt & Sons Ltd., 32–37.

Ducheyne, Steffen. 2009. "To Treat the World. Paul Otlet's Ontology and Epistemology and the Circle of Knowledge," *Journal of Documentation*, 65 (2): 223–44.

Fueg, Jean-François. 1995. "Des sources pour l'histoire du movement anarchiste," in *Cent Ans de l'Office International de Bibliographie*. Mons: Editions Mundaneum, 337–68.

Fueg, Jean-François. 1999. *Aperçu des collections; collection des inventaires*, No. 4. Mons: Editions Mundaneum.

Ketelaar, E. 2000. "Zaalberg en Otlet. Een hoofdstuk uit de geschiedenis van Belgisch-Nederlandse innovatie op archiefgebied," in G. Janssens, G. Maréchal and F. Scheelings (eds) *Door de archivistiek gestrikt. Liber amicorum prof. Dr. Juul Verhelst, Archiefinitiatie(f) [4]*. Brussels: VUB press, 157–64.

Kouw, Matthijs, van den Heuvel, Charles and Scharnhorst Andrea 2013. "Exploring Uncertainty in Knowledge Representations: Classifications, Simulations and Models of the World," in P. Wouters, A. Beaulieu, A. Scharnhorst and S. Wyatt (eds) *Virtual Knowledge: Experimenting in the Humanities and the Social Sciences*. Cambridge (Mass.):MIT Press, 89–125.

Michailov, A.I. 1964. "Donker Duyvis Contribution to the Progress of Scientific Information and Documentation," in *F. Donker Duyvis. His Life and Work*. The Hague: NIDER publication series 2 no. 45, 30–38.

Otlet, P. 1990 (i.e., 1891–92). "Something about Bibliography," in *International Organisation and Dissemination of Knowledge: Selected Essays of Paul Otlet*, translated and edited by W. Boyd Rayward. Amsterdam, New York: Elsevier, 11–24.

Otlet, P. 1935. *Monde, essai d'universalisme: connaissance du monde, sentiment du monde, action organisée et plan du monde*. Editiones Mundaneum; Bruxelles: D.van Keerberghen & Fils.

Ranganathan, Shiyali R. 1961. "Sayers and Donker Duyvis. Theory and Maintenance of Library Classification," *Annals of Library Science*, 8: 85–99.

Rayward, W. Boyd. 1975. *The Universe of Information: The work of Paul Otlet for Documentation and International Organisation*. Moscow: Published for International Federation for Documentation (FID) by All-Union Institute of Scientific and Technical Information (VINITI).

Van den Heuvel, Charles and P. Smiraglia Richard 2010. "Concepts as Particles: Metaphors for the Universe of Knowledge," in C. Gnoli and F. Mazzocchi (eds) *Paradigms and Conceptual Systems in Knowledge Organization: Proceedings of the Eleventh International ISKO Conference, 23–26 February 2010 Rome Italy*. Würzburg: Ergon-Verlag, 50–56.

Van den Heuvel, Charles. 2012a. "The Dutch Connection: Donker Duyvis and Perceptions of American and European Decimal Classification Systems in the First Half of the Twentieth century," in T. Carbo and T. Bellardo Hahn (eds) *International Perspectives on the History of Information Science and Technology. Proceedings of the ASIS&T 2012 Preconference on the History of ASIS&T and Information Science and Technology*. ASIST Monograph Series: Medford (NJ): Information Today Inc, 174–86.

Van den Heuvel, Charles. 2012b. "Multidimensional Classifications: Past and Future Conceptualizations and Visualizations," *Knowledge Organization*, 39 (6): 446–60.

Zaalberg, J.A. 1908. *Het Nieuwe Registratuur-Stelsel bij de Gemeente-Administratieën*. Amsterdam: L.J. Veen.

Chapter 14

The Great Classification Battle of 1910: A Tale of "Blunders and Bizzareries" at the Melbourne Public Library

Mary Carroll and Sue Reynolds

Australia in the late nineteenth and early twentieth centuries may seem geographically far removed from the turbulent social, political and artistic milieu of the old European centres of culture and from the emerging influence of the United States. Yet Australia as a modern nation was fashioned during this period. The flow of ideas and the complex forces at work, both nationally and internationally, were to be fundamental to the construction of Australia's nationhood and its identity. Rather than being isolated from the wider international context, or in the thrall of its colonial beginnings as it is often characterized, Australia was a crucible of competing influences and ideas which contributed at times to conflicting and contradictory cultural aspirations. As such, Australia shared many of the characteristics that define the European Belle Époque. Among the forces at work shaping the new nation were an aggressive and self-conscious nationalism, the emergent influence of the United States, and continuing allegiance to Western European—particularly British—institutions and beliefs. Against the turbulent backdrop of the times, strands of nationalism, modernism, imperialism and colonialism were distilled and transformed in the new nation's quest for cohesion and identity. Tensions in Australian society between the conflicting influences of the new and old worlds were to prove ongoing. Surprisingly, such tensions and influences are exemplified by a series of events at one of the nation's great cultural institutions, the Melbourne Public Library, with librarians E. Morris Miller and Amos Brazier as central figures in this narrative.

The Emergence of a New Nation

In 1901, after almost a decade of inter-colonial debate, the independent colonies of Australia had united as a federation of autonomous dependencies known as the Commonwealth of Australia. The formation of the Commonwealth had been preceded by an intense period of nationalistic growth typified by the assertion of a uniquely Australian character, the flowering of an Australian perspective in literature and the arts, and the active promotion of a cohesive national identity. The capital of the new Commonwealth was to be located in "marvelous Melbourne," a phrase coined by well-known English journalist, George Sala (1885), during a visit to this "pre-eminent colonial city of the 1880s" (McIntyre 1977, 76). Melbourne was the premier city of the previously independent Colony of Victoria and in the late nineteenth century, fuelled by the discovery of gold in the 1850s, it displayed wealth and confidence to the world. Contrary to popular characterizations of Australia as an isolated frontier society in the tradition of the American West, this affluence endowed Melbourne, in particular, with a rich architectural, intellectual and cultural legacy. It helped to create a quintessential Victorian-era city which was "a shining example of the product of Empire, with its culture, prosperity, *noblesse oblige,* and religion in full evidence"(MacIntyre 1997, 81). Federation Melbourne had a population of almost half a million people, making it one of the largest cities in the British Empire (Blainey 2009, 96). It was home to an influential network of "secular, humanist, scientific intellectuals, engaged with the world and its issues" (Clark 1962, 17). Melbourne was to remain the capital of the Australian Commonwealth, accommodating both Victorian state and Australian federal parliaments, until the first national parliament house was opened in the purpose-built city of Canberra in 1927.

Crucially, "marvellous Melbourne," with its rich and diverse influences, its goldrush legacy of cultural institutions, and as the home, at least temporarily, to the nation's leaders and thinkers, provided the opportunity for both nationalism and internationalism to intersect in a way which denies common assertions of cultural isolation. Rather, the city and its citizens in this Australian Federation period bear all the hallmarks which define the Belle Époque in Europe, including social and political unrest, emerging modernisation in its social and political organisations, and changes in artistic and literary life. Despite this, the story of the new Commonwealth's quest for cohesion, international recognition, and national identity has often been framed exclusively in terms of the impact of remoteness on the formation of its national character and identity without consideration being given to the impact on its intellectual life of an often extensive international engagement.

Isolation and Geography

The Australian Commonwealth was formed, some commentators contend, not through idealism but, in contrast to the origins of the United States, "in the shadow of nineteenth-century utilitarianism" (Brady 1981, 1), a fear of invasion fuelled by Australia's proximity to Asia (McQueen 1986, 9), and at a time of unprecedented change internationally. Belief in a character shaped by isolation and the remote pastoral interior became central to the country's emerging nationalism and were among a number of themes that were to dominate its quest for identity. These included an emphasis on the "uniqueness" of the white Australian character, identification with a new and modern world, and a vision for the country of a utopian and classless society. The construction of nationhood and the definition of national character were considered to be products of landscape and geography and focussed on that strand of Australian life which celebrated isolation and the "outback spirit." The emphasis on isolation was framed in terms of the impact of Australia's remoteness on its character and identity and on the development of ideas—Geoffrey Blainey's "tyranny of distance" (1966).

The emerging narrative valorised traits such as anti-imperialism, egalitarianism, anti-intellectualism, radicalism, and the collectivism of "mateship." After a visit to Melbourne in 1885, American philosopher, Josiah Royce, saw such traits as being different from the frontier individualism of the United States. According to historian Russell Ward, who agreed with this assessment, in Australia, unlike in the United States, both "geography and legislation made it impossible for the small man to succeed or for any man to live in isolation" (quoted in McQueen 1986, 3). Thus in Australia, the national character was seen not to be defined by the actions of the lone individual pitted against adversity but by the loyalty of "mates" acting in concert against the vicissitudes of nature. Accordingly, concepts such as "mateship," citizenship and duty were seen as fundamental to defining the Australian character. The State was seen as vital in this endeavour because it provided access to knowledge and education to "the masses" and so ensured that the community was fully cognisant of its responsibilities and duties as citizens. In this tradition, "the goal of the institutional provision of education" was pursued "with missionary fervour … At all levels, access to knowledge was a concomitant of the extension of democracy" (Boucher 1997, xxvii). Federation Melbourne was heavily influenced by such views largely through the work of a network of influential idealist philosophers and politicians.

The themes and ideas associated with the emergence of a sense of national identity were promulgated by a radical and populist nationalist press, particularly during the 1890s, and the emergence of a cultural movement which self-consciously celebrated the Australian landscape and condition. The popular construction of national identity can be seen in the artistic endeavours of a group of now celebrated nationalistic poets, writers and painters, many of whom, ironically, were influenced by cultural movements emerging in Europe at this time. Emblematic in this respect was the work of a group of painters who had what has been called a "blue-gold vision" of Australia (Serle 1973, 73). These painters were part of an Australian art movement in the late 1880s known as the "Heidelberg school" because of their location in a rural suburb of Melbourne called Heidelberg. Here, the artists painted in the impressionist open or *plein-air* tradition. They drew inspiration consciously from the light and colours of the Australian landscape and in the depiction of Australian rural life. Their "blue-gold vision" focused on the depiction of the nation's pastoral interior and the celebration of the unique character of the "Australian native"—conceived of as a locally born white Australian—as framed within the rural landscape. A major, controversial and now landmark exhibition of their work revealed their nationalist aspirations, their sense of iconoclastic modernism and their awareness of being part of an international movement. This exhibition entitled the "9 by 5 Impressionism Exhibition," was opened in Melbourne in 1889 and was so called because most of the impressionist-influenced works were painted on wooden 9 by 5 inch cigar box lids.

Utopianism

Also holding sway in the national imagination was an identification with the "new world," progressivism, the values of democracy and equality as espoused by the United States, and a vision of a new utopian world order free from the shackles of the old world. Fed by the development of the radical press, this utopianism was to resonate through much of the popular writing of the period. According to historian Geoffrey Serle, "the utopian assumption of Australia's destiny as another United States, peopled by a chosen white race, superior to the Old World and free from its vices, at this time held sway as never since" (Serle 1973, 60).

A reflection of this identification with the new world, together with the underlying elements of fear of invasion from Asia and the racism that this displayed, interwoven with the new Australian nationalism can be found in the poetry of Melbourne's Bernard O'Dowd. O'Dowd corresponded regularly with the American poet Walt Whitman to discuss the issue of Australia's Federation and his poetry was admired by many Australians as strongly nationalist in their orientation. O'Dowd wrote:

> From Northern strife and Eastern sloth removed
> Australia and her herald gods invite
> A chosen race, the sternest ordeals proved,
> To guard the future from exotic blight.
> (1944, "Our Land," lines 1–4)

Such visions were in contrast to an often passionate attachment to the "old world" as represented by Europe with its accompanying sense of colonial inferiority.

The Urban Reality

The emphasis on isolation and the tyranny of distance does not reflect the diverse cultural and intellectual urban life which existed in cities such as Melbourne, or the stream of ideas and people from all corners of the earth who were drawn to the former colonies by the prospect of wealth, adventure, or land. The story of the new nation is more complex and multifaceted than the isolated frontier pastoral myth allows. Such a narrative, for example, does not take into account the almost obligatory travel of the "educated" classes from the antipodes to the intellectual centres of Europe. As commentator Humphrey McQueen contends in his examination of the Australian Labor movement, the "frontier thesis of Australia is more accurately framed in terms, not of isolation from ideas, but of the new nation as being a frontier of white capitalism." In this thesis "only by relocating Australia in the mainstream of world development, will it be possible to understand the nature of our radicalism or of our nationalism" (1986, 3).

Rather than being shaped solely by the nationalist pastoral tradition as one narrative suggests, Australian society, according to a more complex and convincing narrative, emerged in the context of a turbulent turn-of-the-century period characterised by political and social unrest and the exposure of its leaders to new international movements and philosophical ideas flowing from Europe and the United States. These new interests led to the establishment in 1881 of the first chair of philosophy at the University of Melbourne to which Henry Laurie was appointed. Laurie was a proponent of the Scottish idealists and he was to influence many of the intellectual leaders in Melbourne in this period, including the librarians E. Morris Miller and Amos Brazier. The effect of these international influences caused a dichotomy in the cultural paradigm which saw the new nation looking both intensely inward to establish a national identity, and optimistically (and some would say aggressively) outward as it attempted to assert the presence of that identity on the world stage.

Melbourne, as the newly federated nation's temporary capital after 1901 and the focus of much of its nationalist cultural activity, was central to a merging of these national and international influences. Perhaps no other city in Australia better illustrates the complexity of Australian life as the country moved into the twentieth century. At this time of new nationhood, as the writers, artists, and politicians strove to define the uniqueness of the Australian vision in terms of the frontier spirit and pastoral interior, Melbourne was in fact a large, complex and multifaceted urban centre. Melbourne was a city in which "the modern commercial office, the mechanised factory, the large speculator-builder, the militant professional association ... together represented a new stage in the subordination of urban life to market structures and industrial technology ... we may fairly call 'metropolitan'" (Davison 1978, 130).

Federation Melbourne was a city of literary clubs and scientific and philosophical societies, an established university, and a large world class public library. Underlining its complexity was the fact that it was concurrently a centre for Australian socialism, feminism and unionism and home to private schools and private clubs in the elite British tradition. Melbourne had hosted two major international exhibitions or World's Fairs at the height of its wealth in the 1880s. As well as the previously mentioned visits of Josiah Royce and Henry Sala, the city attracted visitors as diverse as Rudyard Kipling, Sarah Bernhardt, Anthony Trollope, Mark Twain and Sydney and Beatrice Webb. The inward flow of these international visitors was paralleled by the flow outward of citizens seeking inspiration and education beyond Australian shores.

Among those in Melbourne who participated in this flow of influence were many of the artists of the "pastoralist" Heidelberg School mentioned above, such as Tom Roberts, Arthur Streeton, Charles Conder and Frederick McCubbin, for whom the city and aspects of city life were also subjects that attracted their attention. In Melbourne during its "golden age" they benefitted from the arrival of a number of European artists. Important among these was George Folingsby, an Irishman broadly educated in art as a young man in the US and Europe. A portraitist and historical painter, he became director of the National Gallery of Victoria and director of its School of Painting. Other notable arrivals in Melbourne of the time were: the Swiss Landscape painter Abram Buvelot, English *plein-air* artist A.J. Daplyn, whose stay in Melbourne in the early 1880s was brief, Portuguese artist, Arthur Jose De Souza Loureiro, who had exhibited at several of the Paris salons and was equally at home with realistic domestic as well as "impressionist" landscape and occasional symbolist techniques and the Italian Girolamo Pieri Ballati Nerli who had been influenced by the radical *plein-air* Macchiaioli school of painters of Tuscany. These artists taught in Melbourne's schools of art and design, were members of various associations of artists, and contributed European perspectives, techniques and ideas into Melbourne's artistic cauldron. In turn, young Australian artists such as Conder, Roberts and Streeton travelled overseas. They, and others like them often spent extensive periods of time living in Europe and working in its artist mileau. Ultimately these artists and their art typified the fusion of old and new, national and international, which was to shape Australia. Geoffrey Serle reflecting on this believes that while "Much of the painters' inspiration and what they had to say was clearly a reflection and product of the ferment of nationalistic idealism of the day" (1973, 71), rather than creating something new, their work applied what they learned in Europe to the visual and social conditions present in Australia. Their experience was also mirrored by that of contemporary academics, writers and politicians who also travelled extensively, re-shaping their experiences to reflect the Australian condition. Their efforts to shape "other into own" can be seen in all aspects of Australian cultural, social and intellectual life.

These intertwined forces are encapsulated in Melbourne during 1909 and 1910 in a battle over the classification system to be used at the Public Library. A battle between librarians in conflict over the old traditional way of doing things and the new; between a proven Australian method and a "modern" international system considered by some to be more appropriate for the times. The "great classification battle" at the Melbourne Public Library embodies the maelstrom of social, cultural and political life in Federation Australia, mirroring the Belle Époque in Europe, and also engaging at its periphery many of the leading characters of the era.

The Network of Influence

In 1909, pre-eminent Melbourne politician, Alfred Deakin, was in his third term as Australia's Prime Minister. At the Melbourne Public Library, Deakin's protégé, Morris Miller, had returned from a European tour to the position of cataloguer in the Lending Library (a collection of popular books), where he was under the direction of Robert Boys. The Melbourne Public Library's lending collection had been successfully classified by Boys and his staff by the American Dewey Decimal Classification system in 1900 and Morris Miller reported in his memoirs that its introduction in the Lending Library had been accomplished "with comparative ease" (Miller 1951, 55).

Amos Brazier, the head cataloguer at the library, was in charge of the Reference Library, which held a more serious and intellectual collection than that of the Lending Library. The Reference Library was housed in Queen's Hall, a rectangular room in the European tradition with bays down each side and shelving at each end. Books were arranged by subject category. Downstairs were theology, philosophy and literature, literary periodicals, foreign literatures and British history, the sciences and technology. Upstairs were law, political science, and bound periodicals. The ground floor of Barry Hall, which adjoined the south end of Queen's Hall, held periodicals, history, genealogy, travel and description, and biography; and on its second floor were art and medical books, the latter subjects being only accessible to readers with passes.

Each book in the Reference Library was given a shelf number for its place on a particular shelf of a particular bookcase. Knowledge of which books were held and where required reference to the card catalogue which indexed the subject matter of each volume. Morris Miller described how Brazier "enjoyed the art of classifying by means of shelf inspection" (Miller 1951, 65). He carried "books around, almost lovingly in his arms and fitted them into shelf positions, where each book would be at home with the fellows he chose for it. He worked in daily contact with the shelves, the contents of which were within his mental grasp" (Miller 1951, 65). Thus he was "not an arm-chair classifier, at the mercy of a system rigidly set in book form" (Miller 1951, 65). Brazier argued strongly against use of the Dewey schedules which were prescriptive and thus removed the possibility of individual, idiosyncratic classification.

In October 1900, in Adelaide at the third conference of the Library Association of Australasia, Brazier had detailed his robust objections to the Dewey system. In his paper, "The Principles and Practice of Library Classification," he not only expressed his dissatisfaction with the Dewey system but also his contempt for "modern inventions" generally. He declared at the start of his presentation a clear preference for "that venerable compilation the British Museum Catalogue" over the "distracting yet ingenious methods of some of the American experts." He urged his audience to resist the veneration of such experts as Melvil Dewey and Charles Cutter (Brazier 1900, 4). Brazier embodied the established, traditional method for providing access to information at the Melbourne Public library and was resistant to the new methods coming from the United States of America. His position represented a microcosm of what was happening in Australia at large.

While Brazier's 1900 paper presented a lengthy case against Dewey's decimal system, it did not include an argument for any of the approximately 30 known classification schemes in existence at that time. His objection to any classification method, other than one he employed in the Reference Library in Melbourne, was based on his opinion that the Dewey method and the other existing classification schemes merely allowed for items to be organised physically on shelves. The Dewey scheme, with one classification number representing all of the content of a book, did not adequately classify the information within books to make it more readily accessible for scholarly readers. Brazier considered the Dewey system to be impossible to apply, unnecessary, and aesthetically unappealing with books in public view of all sizes—folios, duo-decimos, quartos, sexto-decimos, imperial, royal and demy octavos—being kept together by subject rather than appearance. This led Brazier to reject Dewey's idea of relative location and Charles Cutter's "expansive system," which allowed books related to each other by subject content to be moved around depending on what items were added to or removed from the collection. Brazier's fixed location system, with books numbered by shelf location rather than subject, did not efficiently accommodate a growing collection but he insisted that the only expansive system required was a "good substantial charge of dynamite" since the need for space was the issue. This could only be resolved by removing books from one fixed location to another location altogether, not moving them around in the space available (Brazier 1900, 7).

However, by the end of his paper, Brazier seems to have talked himself around and he concluded that the Dewey system (or any similar system), *without* the decimals—they "can go; we don't want them" (Brazier 1900, 9)—would be appropriate for classification *within* a card catalogue, where unlimited expansion of individual cards was possible (that is, a classified catalogue). Brazier also set out a blueprint for assigning subject headings that would be understood today. This involved entries for co-ordinate, sub-ordinate and super-ordinate subjects, up to a dozen, applied to any book, with cross references being made between the headings. His final words on the topic of library classification enthusiastically endorsed the operation of both a dictionary catalogue in alphabetical order by subject and a classified catalogue using Dewey's classification scheme. This would allow for both intellectual access to the content of each book in a library and provide an effective finding aid for locating books on the shelves: "What more does any library want? No library has ever asked for more; no library has ever expected so much. And yet this is possible ... by taking what is best and by applying it in the best way" (Brazier 1900, 11).

He had in effect come full circle from sentimental attachment to the "venerable" catalogue of the British Library to implementing what has "been done for us by our fellow-librarians ... in America, where the conditions and purposes of libraries generally are more like our own than are those of the old country," although he did maintain that "Australian librarians ... must make modifications to suit their own conditions" (Brazier 1900, 10, 11). Brazier's paper appears to be something of a "stream of consciousness" composition, conceived as a negative response to the introduction of the Dewey system into the Melbourne Lending Library but ultimately providing a rationale for how this system could be best used in a library's catalogue. One may see Brazier's arguments as expressing the conflict between old and new, European and American, being felt around the world and as demonstrating the confusion of the times.

In particular, Brazier's arguments are similar in tone to the resistance that was expressed in late nineteenth century Europe to Paul Otlet and Henri La Fontaine's idea of using the Dewey Decimal Classification (DDC) system as the basis from which to develop their Universal Decimal Classification (UDC) scheme. This was intended to be used to organise the world's knowledge in what became the International Institute of Bibliography. European librarians argued against the usefulness of the modern American system as did Brazier in the early part of his Adelaide paper. Louis Polain objected to decimal classification because "far from helping searching, [it] rebuffs the reader" (Polain in Rayward 1975). Leopold Delisle, librarian of the Bibliothèque Nationale in Paris, noted the heavy American bias in the schedules. Ferdinand van der Haeghen, chief librarian of Ghent University, described the decimal system as "ridiculous" and declared that it "should not have passed the American border." Frantz Funck-Bretano, sub-librarian at the Bibliothèque de l'Arsenal in Paris, objected to Otlet and La Fontaine's version of the decimal system as being "a fixed, rigid, artificial and inadequate framework" (Uyttenhove and Van Peteghem 2008, 99). According to Rayward, the general objection in France to the proposed Universal Decimal Classification system was that "it seemed to imply that older classifications should be thrown out, and with them should go the old bibliographical order itself" (1975, 62). These objections to, and criticism of the decimal system as proposed by Otlet and La Fontaine are particularly interesting in view of Brazier's later support for UDC over DDC.

William Ifould from Adelaide used the Library Association of Australasia Conference in Melbourne in 1902 to take on Brazier's "bold and original suggestions" from two years earlier (Ifould 1902, 24). In his paper Ifould both disagrees and agrees with Brazier's earlier arguments, taking each point in turn: the dictionary versus classified catalogue, the possibility or impossibility of close classification on the shelves, the Dewey system as a classification system versus the Dewey system as notation, and moveable versus fixed location. Ifould claimed that he wished to consider Brazier's point of view in "the searching light of critical discussion" (Ifould 1902, 23). Amos Brazier's response to Ifould, via an "Editor's Note" in the conference proceedings, dismissed Ifould's suggestion of a discussion by describing his comments as "splendidly irrelevant—cometic, if there be such a word" (Brazier 1902, 32). Brazier, ever convinced of his own rightness, expressed the hope that he would be able to convince Ifould to agree with him in the future and proposed to give a paper on the subject at the next conference. Unfortunately, there were no more conferences as the Library Association of Australasia and its journal, *The Library Record of Australasia*, collapsed.

Modernisation, Change and Disaffection at the Melbourne Public Library

In 1908 Morris Miller took leave from his position in the lending section of the Melbourne Public Library to travel abroad. His doing so caused embarrassment for the Chief Librarian, Edmund la Touche Armstrong, who initially refused permission but was forced to accede after Miller used his personal influence with the Chief Secretary of the State to have the decision overturned. Miller proposed to study philosophy in Scotland and to visit some of the world's leading libraries of the time. His visits also included the International Institute of Bibliography in Brussels. While Miller was travelling, Brazier wrote to him to inform him that Armstrong himself was taking leave to travel and to inspect overseas libraries in order to inform himself of best practice prior to the construction of a new reference library building in Melbourne. Miller met up with Armstrong in Edinburgh and reported having had a pleasant time with him although they had argued over the shape of the new reading room for Melbourne, with Armstrong inspired by the round, domed buildings at the Library of Congress and the British Museum and Miller favouring the rectangular shape of European research libraries.

On his return to Melbourne in 1908, and on Brazier's suggestion, Miller wrote a report of his library visits. Miller described the International Institute of Bibliography as "an attractive place ... [which] looks like one huge array of catalogue cases" (Miller 1951, 75). It is likely that his admiration of the Institute's card catalogue—which left him "spellbound"—was readily absorbed by Amos Brazier when it was described to him on Miller's return. Edmund la Touche Armstrong had also returned from his travels and was determined to proceed with building a monumental reading room in the Melbourne Public Library similar to the great reading rooms of the Library of Congress and the British Museum Library. The question of how to catalogue the books to be transferred into the proposed reading room on its completion began to loom large for the protagonists of the different systems. Their conflicts rapidly developed into what was called "disaffection" at the library.

In 1909 Edmund Armstrong and his ally, Robert Boys, moved to have the Reference Library reclassified according to the Dewey Decimal Classification system without consulting Amos Brazier, the librarian in charge of the Reference Library and also the library's chief cataloguer. It is likely that the failure to discuss the reference collection's reclassification with Brazier was to avoid conflict because, as discussed above, it was well known that he strongly disapproved of aspects of the Dewey system. Brazier was given a written direction by Armstrong to commence reclassification. Always reactive, he responded with a memo in which he vehemently objected to being treated inappropriately with respect to his position as the second in command of the library and its chief cataloguer. Miller reported that Brazier was "adept at invective, and his first draft was caustic to a degree, the language going beyond the bounds of administrative relations" (Miller 1951, 79). Although Brazier asked for Miller's opinion of what he had written and accepted some judicious editing, he refused to remove the strongest condemnations of Armstrong's actions.

Armstrong was outraged by the memo and reported Brazier to the Library Trustees who in turn referred the matter to the Under-Secretary who was responsible for public library employees under the Public Service Act. At about the same time Armstrong became ill and took sick leave and the Chairman of the Public Library Trustees, Gyles Turner, directed Boys to take charge of the library instead of Brazier, the Sub-Librarian. Brazier, slighted again, also objected to this decision and informed Turner and Boys that he was in charge. He telephoned the Under-Secretary to report the incident and in response the Board of Trustees established a committee to investigate the developing conflict. Brazier and Miller agreed that they would go before the "Library Staff Disaffection Committee" if requested to do so, but would only respond to matters occurring at the library and not related to disciplinary action since they were employees of the State rather than of the Library Trustees.

The transcript of the subsequent committee hearing (Report 1910) shows that Brazier was belligerent, querulous, and disagreed with much of the evidence put forward by Armstrong. Brazier refused to answer questions about his behavior—as opposed to consideration of library classification systems—and frequently queried the intent of the Committee's line of questioning before providing a response ... or not. Brazier expressed his disdain for the American decimal classification system, writing, "It was designed especially for the United States, and it is a pity it did not stay there" (Brazier 1912, 13).

Armstrong in his turn was arrogant and belittled Brazier. He told the Committee that he did not consider Brazier to be an expert in classification despite agreeing that Brazier had all of the qualifications required for his position. Armstrong also said that Brazier did not have any practical experience with the Dewey system. Brazier, on the other hand, claimed that Armstrong "was considerably impressed by the views I expressed as an expert in Adelaide" (Brazier 1910). He also claimed that Armstrong had implied to him that various classification systems were under consideration for the Reference Library. Yet it now appeared that Armstrong had already decided on a particular system, the Dewey Decimal System, and had begun to apply it without Brazier's knowledge.

Brazier now expressed a preference for Universal Decimal Classification, as developed by the Belgians, Paul Otlet and Henri La Fontaine, and in use at the International Institute of Bibliography in Brussels. As mentioned above, Morris Miller had brought the International Institute of Bibliography to Brazier's attention after his visit there in 1908. Brazier said that he had wished to consult the "Brussels Manual" (Manuel 1907) and accused Armstrong of having kept the library's copy from being publicly available by not putting it "through the books" or on the library shelves in the usual way, thus making impossible a comparison of the UDC with the Dewey system (Report 1910, 7). The significance of this claim was that the "Manual," over 2,000 pages in length, contained not only the fully developed tables of the first edition of the Universal Decimal Classification (in French), but the rationale behind its development and lengthy instructions about how to use its complex system of number compounding and assignment (Rayward 1975, ch. V).

Following the Committee's questioning of Brazier and Armstrong and publication of the "Report of the Library Staff Disaffection Committee," the Trustees made it known to the Under-Secretary that they intended to transfer Brazier out of the Reference Library. They indicated that their preference was that another position altogether be found for him outside the library. Brazier was officially demoted to a position in the Lending Library and Robert Boys promoted to head of the Reference Library with the title, "Assistant Librarian" rather than Sub-Librarian, the title of Brazier's former position.

The upshot of all of this was that at the Melbourne Public Library Melvil Dewey's American system had won out over the Brussels system and the DDC was successfully introduced into the Reference Library in 1913 when it moved into Armstrong's newly constructed Reading Room. This fulfilled Armstrong's dream for the library and enabled him to boast of a reading room that had the largest reinforced concrete dome in the world at the time (Armstrong and Boys, 1932, 21–2).

The Brussels Influence

Amos Brazier continued to object to the Dewey Decimal Classification system and wrote a further justification for his objections in 1912. He pointed out that Melvil Dewey did not invent classification, as Brazier claimed many people seemed to think. He agreed with Julius Kaiser, who was writing about classification at this time, that "a library cannot emancipate itself from local influences … and … would be the first to suffer by the adoption of a universal classification; it would tend to destroy them" (Kaiser in Brazier 1912, 3). Brazier asserted that the Dewey system of classification was full of "blunders and bizarreries" and that "Australia ought to be able, by this time, to look after herself in library matters" (Brazier 1912, 13).

Amos Brazier believed that Dewey had confused the classification *of* books with the classification of knowledge *in* books and had used his decimal numbers for subjects as well as for shelf notation, "forgetting that he cannot handle books on the shelves as he can information on paper" (Brazier 1912, 9). Brazier suggested that "in a word, the 'Dewey system,' as usually practised, is a confusion in principles, a confusion that is quite unnecessary. Use the principles properly and the confusion disappears, and cosmos will come again" (Brazier 1912, 11). He claimed that the International Institute of Bibliography at Brussels had "come to the rescue" with its *Manual*, which had taken what was good and right with Dewey and improved and extended it so that it provided for classifying information minutely on cards by the methods of "documentation," a neologism coined by Otlet. Brazier proposed that a "new" Dewey should be developed and asserted that "the glory that was America has departed … Brussels is the seat of a new method of classing information, not books, that must supersede the old 'Dewey,' if, indeed, it has not already done so. It is an 'improved Dewey'; all fashionable librarians should at once see to their wardrobes" (Brazier 1912, 7). It was also true in Europe that by this time the Universal Decimal Classification system "seemed to be flourishing," and that strong opposition to it had disappeared (Uttenhove and Van Peteghem 2008).

Brazier's final claim was that the *Manual* could be used to "class anything under the sun. All that would be wanted would be capable librarianship—not navvies, but directive ability" (Brazier 1912, 18). He recommended that all libraries should employ: a classification scheme with headings under which to record information from books (that is, to classify the information); a card catalogue arranged by a subject classification, with an alphabetical index to the shelf notation; the arrangement of books on the shelves by the shelf notation; and a record of both shelf and subject notations on the cards, "thus completing the connection between the information in books and the books themselves" (Brazier 1912, 18).

After the "Library Staff Disaffection Committee" hearings and his demotion to the Lending Library, Amos Brazier spent his remaining years at the Melbourne Public Library reading books and concentrating on his private printing activities at home. He never interacted with Armstrong again. His principle publication was the philosophical and metaphysical *The Terms of Grammar and Creation*, privately published in 1919 (1919).

E. Morris Miller's career at the Public Library (and his mental health) also suffered from his friendship with, and allegiance to, Brazier and from his own part in creating staff disaffection at the library. Miller himself noted that the Chief Librarian, Armstrong, considered that he was *persona non grata* because of his relationship with Brazier and believed that he had too many outside interests to be suited to library "drudgery" (Miller 1951, 74). Armstrong also disapproved of Miller's activities in the Australian National Party and Imperial Federation League and of his friendships with powerful men such as the Prime Minister, Alfred Deakin. Armstrong refused to promote Miller and actively recommended his transfer out of the library. Miller was convinced that a librarian should be active both within the library and outside in the community and was disappointed that the Melbourne Public Library demonstrated such "indifference ... to the educative side of a library's service." He reports that he was further disciplined by Armstrong in 1912 for keeping a library book on his desk as a reference for writing his treatise on *Libraries and Education*, perhaps the first Australian book on librarianship. With any chance of promotion at the library lost, Miller left Melbourne in 1913. He took up a lectureship in "mental and moral science" (psychology and philosophy) at the University of Tasmania where he later became vice-chancellor. He was also the university librarian and to this day the Dewey Decimal Classification system is not used in the university's Morris Miller Library.

Conclusion

In the conflict over the introduction of Dewey Decimal Classification at the Melbourne Public Library we can see a complex distillation of both internationalism and parochialism, a collision of eras as the old world came face-to-face with the new, and as emerging philosophical and social paradigms competed with older and more entrenched ideas. The librarians, Brazier, Miller, Armstrong and Boys, like the contemporary artists, poets and writers, were attempting to negotiate, redefine and reconcile competing and often contradictory influences emerging from the United States and Europe for application to a unique set of geographic, social and cultural conditions. As suggested at the start of this chapter, the events at the public library highlight these competing influences. They illustrate Australia's increasing engagement with international movements and events and the impact of emerging modernisation on Australian society which were to shape the long-term development of the new Commonwealth and its institutions. Little attention has been given to the impact of these influences in Australia but they can be seen in the experiences of many of the leaders of the new Commonwealth, amongst the educated elite and, at a micro level, they are made transparent in the 1910 battle over the future of the library in "Marvellous Melbourne."

References

Armstrong, Edmund La Touche and Boys Robert Douglass. 1932. *The Book of the Public Library, Museums, and National Gallery of Victoria, 1906–1931.* Melbourne: Trustees of the Public Library, Museums, and National Gallery of Victoria.
Blainey, G. 1966. *The Tyranny of Distance: How Distance Shaped Australia's History*. London: Macmillan.
Blainey, G. 2009. *A Shorter History of Australia*. Sydney: Random House.
Boucher, D. 1997. "Introduction." In *The British Idealists*, edited by David Boucher. Cambridge: Cambridge University Press. viii–xxxiii
Brady, V. 1981. *A Crucible of Prophets: Australians and the Question of God.* Sydney: Theological Explorations.
Brazier, Amos. 1902. "Editor's Note." *Transactions and Proceedings of the Third General Meeting held at Melbourne, April 1902, of the Library Association of Australasia.* Melbourne: McCarron Baird & Co., 31–2.
Brazier, Amos. 1912. *Libraries and Librarianship by A Mere Librarian.* Melbourne: [A.W. Brazier].
Brazier, Amos. 1900. The Principles and Practice of Library Classification: A Paper Read at the Adelaide Conference of the Library Association of Australasia on Thursday October 11, 1900. Adelaide: Government Printer.
Brazier, Amos. 1919. *The Terms and the Grammar of Creation: A New Method by An Old Student.* Melbourne: Privately Printed.
Bryce, James. 1888. *The American Commonwealth*. London: Macmillan.
Clark, M. 1962. "Melbourne: An Intellectual Tradition." *Melbourne Historical Journal*, 2: 17.
Davison, G. 1978. *The Rise and Fall of Marvellous Melbourne.* Carlton: Melbourne University Press.
Ifould, W.H. 1902. "Library Classification." *Transactions and Proceedings of the Third General Meeting held at Melbourne, April 1902, of the Library Association of Australasia.* Melbourne: McCarron Baird & Co., 23–31.

McIntyre, P. 1997. "From Canvas Town to Marvellous Melbourne: Melbourne in Colonial Children's Novels." *LaTrobe Journal*, 60 (Spring 1997): 74–83.

Macintyre, S. 2003. "Alfred Deakin: A Centenary Tribute." Senate Occasional Lecture Series; Canberra: Parliament of Australia.

McQueen, H. 1986. *A New Britannica*. Ringwood, Vic.: Penguin.

Manuel du Repertoire Bibliographique Universel. 1907. Bruxelles: Institute International de Bibliographie.

Miller, E. Morris. 1912. *Libraries and Librarianship*. Melbourne: George Robertson.

Miller, E. Morris. 1913. *Kant's Doctrine of Freedom*. Melbourne: George Robertson.

Miller, E. Morris. 1911. *Moral Action and Natural Law in Kant: And Some developments*. Melbourne: George Robertson.

Miller, E. Morris. 1985. (i.e., 1951). "Some Public Library Memories, 1900–1913." (With an introduction by Derek Drinkwater). *La Trobe Journal*, 35: 49–88.

O'Dowd, B. 1944. "Our Land." *The Poems of Bernard O'Dowd* (with an introduction by Walter Murdoch, 257–60). Melbourne: Lothian.

Rayward, W. Boyd. 1975. *The Universe of Information: The Work of Paul Otlet for Documentation and International Organisation*. FID 520; Moscow: VINITI.

"Report by the Library Staff Disaffection Committee to be considered by the Trustees at the meeting to be held on Thursday, 17th February, at 8 P.M." 1910. MS6875 Box 360/8 State Library of Victoria, 4–7.

Roe, M. 1984. "Edmond Morris Miller 1881–1864." in *Nine Australian Progressives: Vitalism in Bourgeois Social Thought 1890–1960*. St Lucia, Qld: Queensland University Press.

Sala, George Augustus. 1885. "The Land of the Golden Fleece VII: Marvellous Melbourne." *The Argus*, August 8, 5a.

Serle, G. 1973. *From Deserts the Prophets Come: The Creative Spirit in Australia 1788–1972*. Melbourne: William Heinemann.

Uyttenhove, Pieter and Van Peteghem Sylvia. 2008. "Ferdinand van der Haeghen's Shadow on Otlet: European Resistance to the Americanized Modernism of the *Office International de Bibliographie*." In *European Modernism and the Information Society: Informing the Present, Understanding the Past*, edited by W. Boyd Rayward. Aldershot, England: Ashgate, 89–104.

Chapter 15

From Display to Data: the Commercial Museum and the Beginnings of Business Information, 1870–1914

Dave Muddiman

In 1886 the Secretary of the London Chamber of Commerce, Kenric B. Murray, embarked on a fact finding tour of continental Europe to investigate a new kind of public institution: the commercial museum. Murray's enquiries were prompted by a number of economic concerns: the Great Depression of 1873–1895 and its effects on British trade and industry; worries about rising overseas challenges to British commercial pre-eminence; and, intriguingly, an interest in a new phenomenon—commercial intelligence or information—which seemed to be emerging as a dimension of international trade. In part because of these concerns, a report in the *Manchester Guardian* of July 31, 1885 had resulted in a flurry of interest in the new museums. At the Antwerp Universal Exhibition of 1885 they were hailed as a "new weapon" in the international and commercial "struggle for supremacy." The British government, desperate for any ideas which might stimulate trade, dispatched its officials to Belgium; they returned in November 1885 urging caution, as officials do (Kennedy and Bateman 1885). Murray himself, however, was more enthusiastic. He reported his findings in late 1886. In travels to Austria–Hungary, Belgium, France, Germany, Holland, Italy and Switzerland, he had visited a total of 56 commercial museums. Although some of the museums, especially in France, were still at the "initiatory stage" and many in Germany were little more than local "export pattern depots," Murray admired especially the larger, multipurpose museums like those he had seen in Vienna, Brussels, and Milan. These, he warned, threatened to "sap the approaches to our export trade" by providing commercial intelligence, technical education, and exhibitions of industrial samples. It would appear, his report concluded, "astonishing to an outsider" that Great Britain—the first industrial nation—"has not already a commercial museum" (Murray 1886, 2–3). He urged its speedy establishment in the City of London.

To twenty-first century readers the concept of a commercial *museum* might appear unusual, even a contradiction in terms. However, before the twentieth century, museums commonly had much wider objectives than their latter-day focus on preservation, heritage and high culture (Bennett 1995, 75–86). Such purposes included a general principle of "leçons de choses": the display of objects on any theme with an educative or informative purpose (Mairesse 2010,

141). A museum hence seemed in the mid-nineteenth century a not unnatural location for a storehouse of materials and information relevant to business and commerce. According to the Director of the United States National Museum (1887–1896) and author of *The Principles of Museum Administration* (1895), George Brown Goode, commercial museums could therefore become centers for the display of "saleable crude materials and manufactures" and for information about "markets, means of distribution, prices and the demand and supply of trade." They would aim to enlighten "home producers about the character and location of foreign markets and foreign buyers about the location and products of the home producer." Essential to their success would be a "bureau of information, through which practical knowledge concerning prices, shipments and quality of products might be obtained … and samples distributed for use in experiment and competition" (Goode 1896, 158). Murray himself concurred in this view. His 1886 report proposed a multifunctional business institution for London which classified and displayed both raw materials and manufactured goods (both old and new); provided information on prices, tariffs, duties, transportation charges and markets; incorporated a laboratory for analysis of samples and a commercial library; and offered courses in commercial and technical education. It would equip the nation, Murray claimed, with "method in business, as there is in the sciences" (Murray 1886, 6, 3).

   Over the next quarter century, commercial museums became a common feature of international networks of business and trade. Eleven years later, in 1897, Murray presented another report, this time to the United Kingdom (UK) government Board of Trade *Enquiry into The Dissemination of Commercial Information*. By then, Britain had indeed developed its own commercial museum under the umbrella of the Imperial Institute, South Kensington, London, and Murray in an updated survey was able to list 153 commercial museums worldwide. Some of these were located in far-flung trading cities such as Santiago, Philadelphia, Tokyo and Sydney, although the majority, especially the smaller institutions, were clustered in continental Europe (Murray 1898). Many of these museums, significantly, had developed the "bureaux of information" that Goode had recommended, responding to a rising demand for information about new products and machinery, transport systems and arrangements, market conditions and demands, tariffs and regulations, raw materials and supplies, and competitor intelligence. Some museums, like those described later in this chapter in Brussels, London, and Philadelphia, were for a time large and complex with relatively generous funding, numerous visitors, a wide range of objectives, and multiple operations. Others were small and provincial. However, for most, decline was swift: by 1918 almost all had ceased to exist in their original form. A good number in Europe were of course ravaged by war. Many more, even if unaffected by such dramatic events, found it difficult to simultaneously remain a "museum" and also adapt post-1918 to an age of incipient mass media and mechanical reproduction. India, as in many other spheres of human activity, confounds such generalizations. As far as is known the Kolkata Commercial Museum, founded in the late nineteenth century (Anstey 1929: 210), is still functioning and open for business.[1] The Philadelphia Commercial Museum (PCM), also one of the longest survivors, became essentially a convention venue and educational centre after 1926.

In the remainder of this chapter I examine the origins, functions and significance of the commercial museum during the years of the European *fin de siècle*. Utilizing contemporary sources, I focus upon the activities of three of the major museums noted above: the Musée Commercial in Brussels founded in 1882; the Imperial Institute in London opened in 1892; and the Commercial Museum of Philadelphia dating from 1894. Broadly, I identify two initially complementary functions engaged in by all of the museums: those associated with, respectively, *display* and *data*. For each museum I examine how these functions gradually evolved into separate practices: display becoming gradually divorced from commerce and enmeshed with propagandist spectacles of the age (often, but not always, colonialism and imperialism); data leading into the foundation of usually separate state sponsored commercial information bureaux. I then explore the significance of these developments in the context of what is a relatively undeveloped dimension of the information history of the late nineteenth century—the beginnings of commercial information and intelligence services. Unlike the static systems of knowledge embodied in the classified displays of the museum, these new "data" services, I suggest, dealt in knowledge that was quantifiable, reproducible, adaptable to individual needs, changing constantly over time and mobilized over a new global space. They created, I argue in conclusion, a new type of knowledge—modern business *information*—which emerged to service the needs of capitalist modernity: in 1900 what was already becoming, according to contemporary observer W. Colgrove Betts (1900, 222), "one big mart, that of the world."

Beginnings: The "Great Boom," International Exhibitions, and Vienna

Any history of the commercial museum must commence by noting the economic context of its development: in particular the huge expansion, after 1850, of international commerce and trade. Broadly, the advent of the museums coincided with two phases of global economic development: the "great boom" of 1848–c.1875 and its successor, a period of "organised capitalism" between c.1875 and 1914 (Hobsbawm 1987, 44). During the first of these phases the foundations of modern market economies were established: "this was the period," according to Hobsbawm, "when the world became capitalist" (1975, 29). Between 1850 and 1870, world trade increased by 260 percent, compared with a mere doubling in previous half century. Facilitated by new transport and communication technologies—steamships, cable and especially railways—previously dispersed producers, consumers, raw materials and commodities were networked into an emerging world economic system. World railway mileage rose from 23,600 in 1840 to 228,600 in 1880 including the first transcontinental track across North

---

1 See www.travel-westbengal.com/museums.htm for its current incarnation.

America in 1869. The Suez canal opened in 1872. Such links brought gold from California and tin from South America to the European metropole. Conversely, the advent of consumerism in South America, Africa and Asia fuelled the expansion of traditional industries (British cotton and Belgian iron exports more than doubled between 1850 and 1860) and the foundation of new industries such as German chemicals and machine tools and British iron ships and railways. All of this was underpinned, so free market theorists observed, by the advance, beginning in Britain, of the doctrines of economic liberalism and "free trade" and by the gradual reduction and, in the British case, abolition of trade restrictions and import/export tariffs. Even the British "imperialism" of this period was arguably driven by trade, dedicated, as it was, to the often coercive negotiation of free trade treaties which opened up overseas markets to cheap British goods (Lynn 1999).

By around 1875, however, the forces underpinning this global commercial revolution began to shift. From 1870 to 1890 a series of economic depressions, due in part to overproduction in European industries, intensified the competition for commercial hegemony. An emerging group of industrial powers—France, the Low Countries and (especially) Germany and the United States—began to challenge and supplant British commercial dominance. These pressures fuelled a return to protectionism in the years after 1875, and, more dramatically, a rush to colonialism (or "new imperialism") in which the European powers, and eventually others such as the United States, Russia and Japan, annexed territory in Africa, Asia and America in an attempt to secure captive markets and supplies of raw materials. According to later theorists such as J.A. Hobson and V.I. Lenin, all of this amounted to "a new phase of capitalism" (Hobsbawm 1987, 60–61), linking nationalist and colonialist expansion. It was paralleled by the emergence of large industrial conglomerates, such as Standard Oil (United States), Armstrong Whitworth (shipbuilding and armaments, Britain) and BASF (dyes and chemicals, Germany). Fuelled by these developments, world production recovered after 1890, more than doubling before 1914. Advanced industrial economies diversified, prompting the massification of advertising, the popularization of a wide range of consumer goods, the artistic and intellectual movements and the cultural excesses of the *belle époque*, and, more menacingly, the growth of a science-based military/industrial nexus, especially in Britain and Germany. This last development especially cemented the role of the state in commercial affairs. Such involvement of nation states, as we shall see, was to have significant consequences for the forms and institutions of business information as they emerged in the years prior to 1914.

One inevitable result of this quantum expansion of commerce was an accompanying leap in demand among manufacturers, traders and retailers alike for information and "intelligence" about new products, services, markets, transport facilities and regulations.[2] Firms had traditionally employed commercial

---

2   On the origins of commercial "intelligence" see, in particular, Alistair Black and Christopher Murphy: "Information, Intelligence and Trade: The Library and Commercial

agents to provide this knowedge but their effectiveness naturally varied widely and, as the complexities of trade increased after 1850, the benefits of collective information provision were increasingly recognized. In particular local chambers of commerce came to represent an important conduit of both formal and formal information. Dating from as early as 1599 in Marseilles, and from 1774 in Britain (when the Committee for the Protection and Encouragement of Trade was formed in Manchester), chambers gradually assumed a quasi-official place in nineteenth century trading networks. National associations of these chambers, such as that established in Britain in 1858, began to collect and disseminate "statistics bearing upon the staple trade of respective districts" (Ilersic and Liddle 1960, 15). Trading nations gradually established chambers abroad—for example Belgium in New York in 1867; Austria in Constantinople in 1872; Britain in Paris in 1872; Italy in Alexandria in 1874—and these gradually became centers of information provision and exchange about bilateral trading links (Bairoch 1989, 97–8). Some of the chambers were heavily sponsored by the state (by 1913 France had 36 chambers of commerce overseas). In Britain and the United States, however, they remained independent of government. Instead, in Britain, official involvement in commercial intelligence was originally channelled through government ministries: the Board of Trade, the Foreign Office (FO) and the Colonial Office. The FO established a commercial sub-department in 1864. Under its able and energetic head, C.M. Kennedy, it began gathering and disseminating overseas commercial information throughout Britain. In 1880 it started to appoint commercial attachés within embassies located in the countries of major trading partners. These attachés were required to submit regular consular reports on trading conditions (Ilersic and Liddle 1960, 22–6). The reports, together with a wide range of other commercial information, were eventually gathered and disseminated after 1901 by a centralized "commercial intelligence bureau" established under the auspices of the British Board of Trade, an initiative arising from a government enquiry into commercial information in 1897 to 1898 (Board of Trade Commercial Intelligence Committee 1898b).

In addition to this institutional activity, and arguably of much greater significance in the realm of *public* knowledge about trade and commerce, were the 60 or so spectacular international exhibitions of "products of trade and industry" which became a defining feature of the internationalization of trade in the years between 1850 and 1914. Inaugurated by the Great Exhibition in London in 1851, these "giant new rituals of self-congratulation" (Hobsbawm 1975, 32) attracted thousands of exhibitors and millions of visitors to the metropolises of Europe, North America and their overseas empires: by 1900 the Universal Exposition in Paris could boast 48 million visitors compared with the (mere) 6 million at the Great Exhibition in London in 1851(Alwood, 2001, 206–7). Initially, the promotion of trade and commerce constituted a core utilitarian rationale for such

Intelligence Branch of the British Board of Trade, 1834–1914" (forthcoming). My thanks to the authors for letting me see a draft of this paper.

exhibitions. They were, in effect, an early mass medium of consumerism and display designed to gather together "the commercial travellers of the universal world, side by side with their employers and customers, with a showroom for their goods … such as the world has never before beheld" (Greenhalgh 1988, 23). Moreover, the exhibitions, according to their proponents, brought "nations together … enabling them to compare each others methods of work, mechanical appliances, artistic ideals and scientific progress" (Barwick, quoted in Geppert 2010, 203). However, as the exhibitions became increasingly grandiose, their focus on commerce and manufactures became diluted by other agendas: colonialism and national competition; urban and architectural boosterism; and, pre-eminently, popular entertainment and spectacle. The emphemerality and unreliability of such exhibitions as a source of information and data also gradually became apparent, leading to calls for the formation of "permanent institutions" which might provide a more enduring and reliable display of the "arts, manufactures and commerce" of the new global marketplace (Imperial Institute 1894, 803).

Some of the early commercial museums emerged directly out of these exhibitions. Probably the most important example of this phenomenon was the Oriental Museum in Vienna, founded in 1873. Set up initially to place on permanent display many of the samples and exhibits of the Vienna International Exposition of 1872, the museum was supported by the Austro-Hungarian government in order to encourage trade with the East, in particular the Ottoman Empire (Bogert 1976). It operated partly on the basis of subscriptions from banks, railway companies, commercial firms and private individuals, and partly on the basis of state subsidy. It lost some of its commercial focus in the early 1880s as a result of changes in government trade policy, and it drifted towards an emphasis on the fine art aspects of its oriental exhibits (Law 1898). From 1886, however, when it was renamed simply the Austrian Trade [Handels] Museum, its commercial information functions in particular were revived, coinciding with a new push for Austrian trade with the Orient. By the 1890s, the museum was reported by British visitors to have had a complement of about 15 to 20 staff members and to be operating a commercial information and enquiry office, a library, and a publication programme. Its journal, *Das Handels Museum*, became one of the leading commercial journals of the Habsburg Empire and the museum also provided Austrian industry with a specialist information service related to foreign customs duties, transport rates and creditworthiness (Murray 1886, 12–13; Law 1898). By 1907, these commercial information functions seem to have been recognised as the museum's primary concern and the majority of its collection of oriental samples and exhibits were transferred to the Vienna Museum of Art and Industry (MAK 2011). As we shall see, such a separation of the informational and curatorial activities of the commercial museum was to become increasingly common as the twentieth century progressed.

Display, Data and Trade: The Major Museums and their Bureaux of Information, 1880–1914

The Musée Commercial, Brussels 1882–1897

After his tour of European Museums in 1886, Kenric Murray concluded that although the Vienna Museum was "very fine and well arranged," it hardly attained "more than half the importance of the Brussels Museum … the largest and best organised in Europe" (Murray 1886, 9). According to Murray, the source of the pre-eminence of Brussels was its sponsorship by the Belgian state which gave it resources, stability and legitimacy. Like Vienna, the Brussels Musée Commercial had originally been established in the wake of an international exhibition—in its case the 1880 "cinquantenaire" exhibition celebrating 50 years of Belgian independence. However, even more than its counterpart in Vienna, it seems to have been viewed as an instrument of a government commercial policy of industrial expansion into export markets based on free trade and overseas investment (Aberloos 2008, 108–11). The museum's sponsors included powerful figures in the Belgian establishment such as Auguste Lambermont, the head of the commercial branch of the Belgian Foreign Office and one of the architects of the policy of state support for commerce (Lee 1898, 159). Opening its doors in 1882, by the mid-1880s the museum, located in a "large and handsome" building in the commercial district of Brussels (in the Rue des Augustins, near the Bourse), could boast exhibition space of over 3500 square yards (Murray 1886, 9). Ten years later it had a complement of 12 administrative staff plus porters costing some 400,000F (c.£1600) per annum in salaries, which were paid by the state. In addition, it received an annual grant from the government of 20,000F for purchases and running expenses (Lee 1898, 159).

The objectives of the Musée Commercial initially centred upon its collection of industrial and commercial samples that had grown to 25,000 in number by 1897 (Lee 1898, 158). The samples fell into two main classes: overseas raw materials which might by usefully imported by Belgian industry and foreign (mainly European) manufactured goods which were available in overseas markets. In the second case especially the museum effectively aimed to supply Belgian merchants and manufacturers with a form of "competitor intelligence" whereby they could assess the quality and value of foreign competition. To this end, Belgian consuls overseas were instructed specifically by the Ministry of Foreign Affairs to remit suitable samples to the museum. Museum staff then organised the specimens into 44 industrial/material groupings and "about 400 specific classes" (Kennedy and Bateman 1886, 156). All of the samples were catalogued and users could request detailed information and further particulars (for example blueprints, patterns, prices, if available) of articles which took their interest. New acquisitions were listed in a weekly bulletin, initially a simple supplement to the museum catalogue.

Although this system seemed at first to have functioned well enough, by the 1890s serious deficiencies had begun to emerge. The sheer number and physical bulk of some samples led to instructions to overseas consuls to be more discriminating about what they selected. Gradually, the collection seems to have become dated, because of a lack of clarity about when to discard or relegate objects to storage. The cost of samples, too, affected the quality of the collection. Consuls often remitted freely available samples and goods which did not always represent the keenest competition (Lee 1898, 158–9).

Because of these problems of availability and currency of material artefacts, by the 1890s visitors, such as British Paris-based commercial attaché H. Austin Lee, were reporting a refocusing of activity in the Museum around the provision of information. By then, the museum had established a Bureau de Renseignements [Information Bureau] which collected and tabulated "every kind of information with regard to foreign trade," including copies of trade intelligence reports from Belgian consuls and agents; prospectuses and price lists; statistics; transport rates and freight charges; customs tariffs; and contracts open to public tender. A library and reading room contained trade directories and a large collection of international commercial journals. The museum's own bulletin had also by this time expanded to become a general weekly business journal offering what was in effect a current awareness service. In the 1890s, levels of use of all of these services were claimed to be growing and, if statistics are to be believed, they justified Murray's initial praise of the museum. In 1898, Lee reported that there was an average of 123 visitors to the museum per day, a 200 percent increase over the early 1880s when the average daily figure was 45. The Museum's *Bulletin*, in the same year, claimed to have 700 paying subscribers. However, Lee was sceptical of some of these figures, noting that many Belgian traders of his acquaintance responded to his enquiries about the Musée Commercial with "a shrug of the shoulders." In particular, he observed, its sample galleries were usually deserted. The Musée was, he concluded, basically a "succès d'estime": a prestige driven project which functioned as an emblem of Belgian commitment to overseas trade (Lee 1898, 158–9).

However, as the century turned, the wider current of events in Belgium seems to have resulted in the marginalization and diminution of the status of the Musée Commercial. In *fin de siècle* Brussels, hard commercial realism was increasingly replaced, in both the popular and intellectual zeitgeist, by grand narratives about colonialism and (paradoxically) pacifist internationalism. "Grands projets," such as the huge 1910 International Exposition, and even grander plans to establish Brussels as a "world city" or global capital became omnipresent in public discourse and museums were naturally expected to play a part in these (De Groof 2008). Two projects in particular seem to have adversely affected the public profile of the Musée Commercial. The first, the Palais des Colonies, in effect superseded the museum's function as a showcase of colonial trade. Developed by Leopold II as a temporary construction in the suburb of Tervuren for the Brussels International Exhibition of 1897, the Palais over the next ten years gradually became a permanent "Musée du Congo." After 1908, when it adopted that name, it assumed the role of the official government centre for information about the colony and the display of Congolese products and raw materials (Stanard 2011, 130). The second project, a vast scheme proposed in 1905/6 for an "acropolis of Brussels" in the district of the Mont des Arts, similarly ended any possibility of the Musée Commercial evolving into a multipurpose "international" museum along the lines of that in Philadelphia (De Groof 2008, 100). The Musée Commercial appears to have been strangely unrepresented at the International Congress on World Economic Expansion, held in Mons in 1905, a major meeting sponsored by the King with a strong Belgian colonialist orientation where many of these national plans were originally discussed. Most important Belgian organizations were represented at the Congress together with, among many international representatives, those from the Commercial Museum of Vienna. Although more detailed research in Belgium would be required to confirm this, the suspicion must be that the Musée Commercial was by this time being excluded or bypassed in these policy debates (Congrès International d'Expansion Economique Mondiale 1905, CXV111).

Although imperfectly realized, the utopian scheme of the Mont des Arts included proposals for both a "world museum" and a "social museum," concepts which, in modified form, eventually materialized between 1911–14 and 1918–22 in the shape of Otlet and Lafontaine's Palais Mondial (Rayward 1975, 172–96). This was an international museum sponsored by the Belgian government, which functioned as a headquarters of the Union of International Organizations, an internationalist library, and a museum organizing and exhibiting materials donated from exhibitions worldwide.

Partly as a result of these developing trends, it is clear that as the new century progressed the Musée Commercial's focus shifted ever more towards its business information work. In 1898, Austin Lee's report had observed that perhaps, for the Brussels Musée Commercial, the name "museum" was a mistake "as it gave a wrong idea of the collection" and its focus on contemporary commercial activity and information. In 1896 the name of its weekly journal had been changed from the *Bulletin du Musée Commercial* to the more generic *Bulletin Commercial* (Lee 1898, 158–9) This publication was to remain the main Belgian government's trade journal and was published by the Office Belge du Commerce until the 1960s. By 1911, in *Encyclopaedia Britannica*, the Musée Commercial is simply described as a branch of the Belgian Ministry of Commerce aiming to "centralize commercial intelligence and facilitate its dissemination" (Trade Organization 1911, 139). By then, it seems, its public profile as a museum was greatly diminished and its transition to government commercial information bureau substantially complete.

The Imperial Institute, London 1886–1903

Kenric Murray, as we have seen, visited the Vienna and Brussels Museums in 1886. Impressed by their successes, he campaigned for a similar museum in the commercial heartland of London to be operated, he hoped, by the Chamber of Commerce with state support (Murray 1886, 1–7). However, his arguments seem to have made headway with neither a sceptical UK government nor the Chamber's powerful "free trade" lobby. Instead, in a decade of "new aggressive imperialism" (Mackenzie 1984, 97), the proposal for a British commercial museum was in fact hijacked by an imperial establishment keen to "draw closer the bonds which unite the empire" and operationalize the idea of a self-sufficient "Greater Britain" (Baden-Powell 1886, 442). After the first uniquely "imperial" exhibition in Britain—the Colonial and Indian Exhibition of 1886—a steering group led by Edward, Prince of Wales (the future Edward VII), was established to examine the potential of a "permanent" exhibition which might reflect the imperial zeitgeist. This quickly assumed the shape of a multipurpose institute in South Kensington with a commercial museum at its heart. Planned to commemorate Queen Victoria's Golden Jubilee in 1887, the funds for the project were raised largely by colonialist arm twisting. It obtained support from most overseas empire administrators, although not a sceptical British parliament. Lord Salisbury, the then Prime Minister, who was by no means an anti-imperialist, advised Queen Victoria against royal patronage of a scheme that he argued "might mean anything from a lecture room to a tea garden" (Salisbury, in Sheppard 1975, 221). After numerous construction delays, the "Imperial Institute," as it was called, was finally opened by Queen Victoria in 1892. Its primary objective according to its charter was economic: the "formation and exhibition of collections representing the important raw materials and manufactured products of empire, so maintained as to illustrate the development of agricultural, commercial and industrial progress" (Imperial Institute 1893, 44–5). It aimed also to operate a "commercial intelligence service," an emigration bureau, a technical education service and a network of satellite commercial museums throughout Britain and the Empire. *The Times*, in an editorial on the 23rd April 1887, grandly hoped all of this would make it a kind of "brain centre" of empire. Others however, like its pragmatic first director, military scientist Sir Frederick Abel, saw it rather more prosaically as an attempt to bolster British trade and industry against the rising economic power of Germany and America (Abel 1887).

From its inception, despite royal patronage, the Institute functioned only with difficulty. As we have seen, elements of the British government were sceptical and, unlike Belgium, no state support was forthcoming for running costs. In the 1890s the Institute survived on less than £2000 per annum in the form of subscriptions from fellows and a mere £4000 per annum in interest from investments (Abel 1898, 2). As a result, its activities as a commercial museum were inevitably constrained. It never managed to persuade British companies to display products for export and in 1898 its United Kingdom galleries were converted to accommodation for London University. Its mooted network of sister museums and corresponding agents in the empire never bore fruition—although a few commercial museums developed in, for example, India; one being founded in Calcutta at the end of the nineteenth century (Abel 1897, 4–5; Anstey 1929, 210). Some collections were successfully established at the Institute. The best were the Indian and Ceylon "courts," fitted out in the popular orientalist style. These were largely financed by the Government of India and comprised artefacts inherited from the Colonial and Indian Exhibition of 1886. Other display sections were also established with materials from New South Wales, South Africa, Canada and the West Indies prominent. By 1897, however, it was clear that these collections were proving difficult to update and maintain and were operating primarily as an attraction for a curious public rather than a commercial or industrial resource (Abel 1898, 3).

To provide a resource for business it was necessary for the Institute to diversify: into science (a scientific and technical department was established in 1894) and information. Its Department of Commercial Intelligence (as it became known) was the brainchild of E.F. Law, an advisor to Frederick Abel and a commercial attaché at the British Foreign Office who had visited the Commercial Museums in Vienna and Brussels. In 1887 he persuaded Abel to incorporate a Department of Commercial Intelligence (DCI) into the Imperial Institute blueprint and by 1892 it had commenced operation. Initially it focused on establishing a library and reading room and on producing high quality publications: the *Imperial Institute Yearbook* (1892–94) and from 1895 the *Imperial Institute Journal*. The latter was actually an innovative and impressive monthly broadsheet which functioned as a current awareness bulletin containing up to date commercial data and topical articles and reports (Muddiman 2010, 114). In 1895, when the DCI attained a full complement of seven staff, it opened an "enquiry and correspondence service" which by 1897 was receiving between 500–600 enquiries per annum, equally split between visitors and correspondents. The latter came from the length and breadth of the UK and the empire and, whilst some were straightforward requests for data, others clearly sought "intelligence" and required detailed research by staff or the advice of experts or colonial "correspondents" (Abel 1898, 1–4). In June 1899 the enquiry service was extended when the Institute opened a "City Branch" in the commercial heartland of London's East End. Interestingly, by this stage, a charge of ten shillings (a not insignificant amount of money in 1900) was being levied for some complex enquiries (Muddiman 2010, 115).

All in all, by the late 1890s, the Department of Commercial Intelligence had turned out to be a limited success, demonstrating the demand for, and usefulness of, the new commodity of information. Ironically, however, within five years this very success would lead to its demise. In 1897, the UK government Board of Trade (BOT), finally recognizing the potential of commercial information, and set up an enquiry into "the dissemination of commercial information" (Friedberg 1988, 45–51). Reporting in 1898, the enquiry recommended the establishment of a state funded "Commercial Intelligence Branch" of the Board of Trade, an office which opened in the City of London in 1899 (Board of Trade Commercial Intelligence Committee, 1898b). With a worldwide, rather than a narrowly imperial remit, and led by President of the Board of Trade and later Chancellor Charles Ritchie, the BOT office to some degree represented an ideological victory for the "free trade" lobby in the British political and commercial establishment (Ridley, 2004). Whatever the politics, in purely practical terms the new service clearly rendered the Imperial Institute's DCI surplus to requirements. In 1902, when the whole Imperial Institute (in dire financial straits by then) was brought under state control, the DCI's activities were transferred to the Board of Trade. By 1908, the new BOT office claimed to be receiving 11,267 enquiries annually (Advisory Committee to the Board of Trade 1909, 4).

After 1902, although the Imperial Institute as an entity survived, any prospect of its continuance as a "commercial museum" was dropped. The 1897 *Enquiry into the Dissemination of Commercial Information* had, it is true, recommended that some system of exhibiting commercial samples in UK centres of trade and industry be co-ordinated by the new Board of Trade Commercial Intelligence Branch. However, this failed to materialize. The government provided no financial backing for it and local chambers of commerce were less than proactive in taking up the idea (Board of Trade Commercial Intelligence Committee 1898b, xiv–xvi). Meanwhile, the Institute itself thrived for a time as an empire scientific centre, especially between 1904 and 1918 (Muddiman 2010, 116–21). Latterly, until 1955, it became mainly a propagandist educational museum and cinema, especially during the imperial "Indian summer" of the 1920s and 1930s. There was little, however, that was commercial about its audiences and visitors. The majority were school parties, scouts and guides, armed forces in training and other groups still receptive to the fading allure of empire (Mackenzie 1984, 132–42). Eventually, in the 1950s, even this existence became anachronistic and the Imperial Institute was finally closed and its buildings demolished (except for its bell tower which remains at the centre of "Imperial College," which took over the site). A newly formed Commonwealth Institute took on its cultural and educational work, albeit working within a rather more liberal and multicultural ideological frame. In contrast, little legacy remained of the Institute's commercial beginnings and of the links between empire, information and economic power they had once embodied.

The Philadelphia Commercial Museum 1893–1926

In one respect at least the pre-eminent commercial museum in the United States—that at Philadelphia—resembled its cousins in Vienna, Brussels, and London. It was established, like them, out of the remnants of a major international exhibition. In 1893, its founder, University of Pennsylvania botanist William P. Wilson, conceived the idea of purchasing surplus exhibits from the Chicago World's Fair of that year to form "a permanent collection representative of the world's industrial products" (Branford 1902, 243). He successfully persuaded his home municipality, then a rising new-world industrial powerhouse, to vote a sum of $4000 to complete the purchase. He set up the exhibits in downtown offices vacated by the Pennsylvania Railroad. Thereafter, the growth of what became the Philadelphia Commercial Museum (PCM) was rapid and spectacular. Wilson successfully courted state and municipal patronage and by 1902 had raised an annual operating grant of $115,000. In addition by then it had some 850 commercial subscribers to its services (at $100 per annum each). This provided an additional income of $85000, giving the museum an annual operating budget of about $200,000 (Branford 1902, 250). Two years later it moved to impressive, purpose built premises in West Philadelphia which covered 16 acres of grounds. By 1900 it was able to house some 200,000 samples of foreign manufactured goods (Conn 1998a, 120). Perhaps more impressively still, by 1902 its Bureau of Information employed about 125 clerks who co-ordinated a global network of 20,000 regular and 60,000 occasional correspondents on business matters (Branford 1902, 247). In effect, the Philadelphia Museum had quickly become, as its President, William Pepper (1897) claimed, "national in its scope and importance," a status underlined by its formal inauguration in 1897 by US President McKinley. Its growth reflected the spectacular entry of the United States into world trading systems during the late 19th century. By 1913 exports were estimated to be worth some $356 million, or 13 percent of the world total, compared with just 3 percent a century earlier (Lipsey 2000, 685).

Like European commercial museums, the PCM seems to have operated on two levels. On the one hand it was straightforwardly business-focused and pragmatic. On the other its purposes were complicated (and to some extent compromised) by attempts to adopt and inculcate a wider range of discourse linked to contemporary notions especially of progress but also of civilization and empire. From the beginning Wilson and his colleagues appear to have recognized the indispensability of information to commerce. The museum, according to William Pepper, existed "for the purpose of promoting trade with foreign countries and particularly with the object of obtaining information relative to the world's markets and trade conditions in foreign lands" (Pepper 1897). A Bureau of Information was planned from the very inception of the museum and this functioned by and large on empirical lines in response to demand. Organized into sub-sections comprising a commercial library, an enquiry and intelligence bureau and publication and translation services, it quickly became a model for a new brand of commercial intelligence service worldwide. An interesting case in point was the attempt by Julius Otto Kaiser, a leading figure in the establishment of the PCM's information bureau, to attempt to set up a bureau in London along PCM lines under the brand name "The Commercial Intelligence Bureau Ltd (1899–1902)." Its viability seems to have been short-lived, partly because, unlike the contemporaneous Board of Trade Office established by the UK government, it had to charge at full cost for its services (Dousa 2011, 170).

The scope of the Philadelphia Commercial Museum's information gathering operation was global. As well as intelligence received from the large number of correspondents already noted, it collated trade reports from US and foreign consuls and indexed and cross-referenced information from the 900 plus US and overseas periodicals received in its commercial library. Importantly, the publications and correspondence that it received were not simply passively stored and catalogued but were analysed, dissected, reformatted and mobilized for active use. It was at the PCM between 1896 and 1899 that Kaiser first developed his methods of "systematic indexing." These generated a series of card files which allowed information to be retrieved through techniques of faceted classification, enabling queries based on materials, ("concretes" in Kaiser's terminology), processes, or geographical locations (Dousa 2011). The PCM enquiry service was able, as a result, to construct detailed intelligence reports in response to special enquiries. UK visitor, Wyndham Dunstan (1898, 173–4), for example, reproduced one of these reports, entitled "Report on the Bicycle Trade in Sydney, New South Wales," which was around 2000 words in length. During his period of study at the PCM in October–November 1899, another British visitor, Victor Branford, reported that 295 such enquiries were processed during a period of 19 days, 180 from the United States and 115 from abroad (Branford 1902, 250). In addition, the Bureau produced 94 "general reports" for public consumption during the same period. As the PCM consolidated its information role, these reports evolved into a general publications programme. This comprised pamphlets focusing on US exports markets (Conn 1998a, 134) which eventually transitioned into a monthly current awareness journal entitled *Commercial America* highlighting US products for export (Wolfrom 2010, 2). More innovative was its "rush news" service, a prototypical experiment in the selective dissemination of information. This alerted subscribing US export firms to important breaking world news based on newly received intelligence being processed by the PCM relevant to their target markets and products (Betts 1900, 231–2). Through this and similar experimental services, such as the use of new techniques of photography to supplement remote requests, the Bureau of Information can be seen to be responding to the dynamic and rapidly changing nature of global commerce at the end of the nineteenth century.

In contrast, there seems to have been little in the way of a pragmatic focus on commercial facts in the other main branch of the PCM's activities: its operation as a kind of "permanent world's fair" and exhibition centre. Over time its displays of more than 200,000 artefacts, products and samples would become a veritable battleground of contemporary ideas and value systems. Initially the museum claimed to be above politics and to operate, in the words of President William McKinley in his speech at the official opening, on "broad and progressive lines" (McKinley in Conn 1998a, 133). Aspiring to be an "Institute of Economic Science" it aimed through its displays to reveal "the conditions of international commerce" through "systematic study" and "scientific methods" (Branford 1902, 240). To this end its samples were organised in a pseudo-scientific classification of species and classes in parallel with a section of geographically ordered displays: a system designed by Gustavo Neiderlein, Wilson's deputy at the museum and like him a former botanist. These permanent displays were supplemented by periodic exhibitions (like the National Export Exhibition of 1899 which attracted 1.25 million visitors) and conventions such as the "First International Commercial Congress" of the same year which promoted a utopian world of "commerce and civilization, economic prosperity and world peace" (Conn 1998a, 135).

However, gradually, it seems that this initial rhetoric of open markets and liberal values dissolved, being replaced by what Stephen Conn (1999b) has described as an "epistemology for empire." To some extent, this imperialist turn was undoubtedly linked to economic realities. Displays of samples often related to US trade interests and therefore focused upon its developing commercial hegemony in South America, the Pacific and other parts of the global economic periphery (Dunstan 1898, 171). However, especially after the US invasion of the Philippines in 1901, the museum's imperialism became more overt and took on cultural and colonial connotations comparable to those evident in Europe. These were particularly marked in a permanent Philippines exhibition brought to the museum after the St Louis Louisiana Purchase Exhibition of 1904 for which it had been organized by none other than William P. Wilson. In culturally colonialist terms, this displayed the Philippines as a "savage" land benefiting from the "impelling power of modern civilisation" (Neiderlein in Rydell 1984, 179). It also veered away from commerce towards ethnology, spuriously ranking various Philipino tribes according to degrees of intelligence and savagery. One exhibit in particular of a "negrito" male achieved notoriety and popularity under the epithet of "the missing link" (between man and the apes) (Rydell 1984, 175). More generally, Wilson and Neiderlein seem to have gradually become more and more preoccupied with such colonialist spectacles. Not only did they spend nearly two years in 1902–3 gathering exhibits for the St Louis Exhibition and setting up a (temporary) commercial museum in Manila (Rydell 1984: 167–178), but in 1906 Wilson took a three year leave of absence to put together a Philippines section for the 1909 World Exposition in Nancy, France. "This massive exhibit featured 1,200 living Filipinos as a 'living diorama'; inanimate portions of the exhibit were brought back to the [Philadelphia] Commercial Museum as a display" (Wolfrom, 2010: 4).

Such ethnological spectacles were, of course, popular with the white American public. Over 19 million people were estimated to have attended the St Louis Centennial Exhibition and the PCM no doubt hoped to benefit in its own attendances from this popularity. In contrast, however, expert visitors were often unimpressed by the museum's displays of commercial samples. Future Imperial Institute Director, Wyndham Dunstan, in 1898 regarded them as dull and outdated (Dunstan 1898, 171). In 1901 the City of Philadelphia itself temporarily withheld funding from the exhibition halls of the museum. The city's Republican caucus considered them to be a burden on the public purse and many tons of exhibits shipped from the Paris Centennial Exhibition (1900) had apparently to be stored away in cases until the conflict was resolved (A Philadelphia Squabble, 1901). No such problems, however, afflicted the Bureau of Information which was immune to the City's spending freeze because of its national popularity among manufacturers and exporters and the financial viability that resulted. By the early years of the twentieth century, the consensus amongst business commentators was that the Bureau had "soon surpassed in magnitude the original 'Museum' … [and] began to work in the commercial world changes similar to those which the introduction of steam produced in the industrial world" (Betts 1900, 226). By 1902, the Bureau was taking on a quasi-official role as a national centre for commercial information. "It is commonly said," reported Victor Branford in that year, "that US consuls are in the process of habituating themselves to report to Philadelphia rather than Washington" (Branford 1902, 248).

Nevertheless, in this bifurcation of the PCM's functions between display and data lay the seeds of decline. The Bureau of Information eventually became a victim of its own success—rather like the London Imperial Institute's Department of Commercial Information but on a larger scale. As the progressive era in America gathered momentum, an increasingly active federal state began to see it as its business "to provide American exporters and importers with information" (Collings in Conn 1998a, 146). In 1903 it established the US Department of Commerce which, by the 1920s in the guise of Herbert Hoover's International Trade Administration, had effectively usurped the role of the Philadelphia Bureau of Information whilst utilizing many of its methods (Wolfrom 2010, 3). Meanwhile, between 1910 and 1925 the museum itself attempted to expand its educational work, but with the coming of mass media and the waning of the "world's fair" concept the link between contemporary commerce and the idea of a museum became increasingly difficult to sustain. After Wilson's death in 1926, the PCM was downsized and took on a civic role as an educational resource and convention center. Its contents were gradually dispersed as the twentieth century progressed, some apparently finally being destroyed in 2010.[3]

Conclusion: The Commercial Museum, the Sense of the Present, and the Rise of Information

Although some historians, such as Arno Mayer (1981), have argued for the persistence of an "old regime," for most the 30 years or so before 1914 represent a "great acceleration" in global development and change—indeed the "birth of the modern world" (Bayly 2004, 451–87). For Stephen Kern in *The Culture of Time and Space 1880–1918*, this was an epoch when "a series of sweeping changes in technology and culture created distinctive new modes of thinking about and experiencing time and space" (Kern 1983, 1). According to Kern, the pervasive impact of these changes lay in their creation of a ubiquitous "sense of the present" made possible, for example, by daily newspapers, cinema newsreels and world standard time. Harnessed to advances in global communications (railways; steamships; telegraph and, more latterly, automobiles, radio and telephone), this sense of the present "expanded spatially to create the vast shared experience of simultaneity" that is such an important characteristic of modern life. The present, argues Kern, "was no longer limited to one event in one place sandwiched tightly between past and future and limited to local surroundings." It had become dynamic, kaleidoscopic, and global (Kern 1983, 314).

The rise and decline of the commercial museum, as described in this chapter is almost exactly contemporaneous with the period covered by Kern's book. The

---

3 See the Museums project "Commercial America," organized by the Slought Foundation of Philadelphia, which focuses on former PCM exhibits, http://slought.org/content/11448/ (Accessed December 2011).

museums too, we might observe, attempted to encapsulate and represent this new, modern world of the present. Their collections, according to their founders and sponsors, were supposed to exhibit "in an exhaustive manner, the products of the hour, because [these] are commercially the most valuable." More than that, they aspired to create order out of the ever-changing chaos of the market: to create "method in business, as there is in the sciences" (Murray, 1886: 3). For William P. Wilson in Philadelphia, in particular, this comprised an attempt to design an "intellectual framework for international commerce" (Conn 1998a, 116). In an early apologia for the science of marketing published in *The Forum* (September 1899, 116), Wilson argued that by studying the classified samples in the PCM, "the manufacturer may learn how styles may be modified to suit the tastes of [diverse] peoples" (quoted in Conn 1998a, 121). Objects, systematically arranged, were thereby claimed to generate useful knowledge and competitive advantage.

By 1900, however, the flaws in applying this object-based epistemology (and pedagogy) to the fast moving world of trade and commerce were becoming apparent. In practical terms, from the very beginning the London Imperial Institute, as we have seen, had encountered difficulties in gathering together suitably current and comprehensive displays of commercial materials. Similar problems were experienced in Vienna and Brussels in the 1890s where constantly *maintaining* currency became an ongoing battle. Another limitation of the spatially anchored museum was access. The notion of perpetual relevance was difficult to maintain when users were limited by distance and cost in the frequency of their visits. Brussels (situated at the centre of the city's commercial district) seems to have experienced this problem less than the Imperial Institute whose South Kensington location became a constant source of aggravation for commercial users based in the City of London (Muddiman 2010, 112). Philadelphia, of course, was another matter altogether. Most of its potential users had to travel vast distances to see its collections. Ever innovative, the PCM tried to overcome the "limits to its descriptive powers" by supplying photographs of objects to remote enquirers when requested (Betts 1900, 232). In the end, however this was not enough. Gradually, initially enthusiastic commentators began to recognize the contradictions between the static displays of samples in a single location and the dynamic and universal milieu of trade and commerce. In January 1902, an article in *Imperial Argus* suggested that the Imperial Institute's exhibits were of little use to the "practical" man and were "emblems of an extinct greatness" which failed to reflect economic and industrial change (Norman 1902, 393). Reporting from Brussels in 1898, H. Austin Lee concluded as we have seen that "the name 'museum' was thought to be a mistake, as it gave a wrong idea of the objects of the collection" (Lee 1898, 159). In similar vein, Paul Cherrington, writing about the PCM in *World Today* (May 14th 1908) concluded that "the very name Commercial Museum is a demand for reconciliation between two diametrically opposed ideas … what can a musty collection of specimens have to do with commerce?" (quoted in Conn 1998a, 120).

One response to these difficulties was for the museums to try to enhance the purposes, scope and popular appeal of their exhibits. This is what most commercial museums eventually did, although in the process they probably inadvertently distanced their displays from the world of commerce. In Philadelphia after 1905, as we have observed, Wilson and Neiderlein's neo-imperialist Philipino display, together with numerous congresses and conventions, assumed center stage. The Imperial Institute, after an attempt to carve out a niche as an empire scientific institute between 1902 and 1914, finally abandoned its commercial goals in the 1920s. It then tried to re-invent itself as an "empire under one roof," attempting to draw the crowds with Indian "courts" and other exotic and orientalist displays. In 1927 it set up a cinema to show empire focused documentaries and other propaganda films (Mackenzie 1984, 132–7). The museums, it seems as the twentieth century progressed, were pinning their hopes for survival upon spectacle, following a similar recipe to that of the numerous international exhibitions which peppered the United States and Europe between 1900 and 1939. "Display" hence moved away from the complex and often specialist world of commerce and industry. The target audience became the general public, rather than exporters, industrialists and merchants. Success began to be measured by attendance. Entertainment spiced with public "education" was the goal. In terms of content this involved a shift to eclecticism in that as well as commerce, displays mixed ethnology, travel, history, politics—very often blended in a colonialist and imperialist brew (Mackenzie 1984; Rydell 1984; Stanard 2011).

All of this is not to suggest, however, that the museums prematurely severed their commercial roots. On the contrary, before 1914 it is clear that they largely reconceptualized their mission as one which addressed the demands of business through the provision of information. As detailed in previous sections, the museums' bureaux of information in Brussels, London, and Philadelphia during the late 1880s and 1890s had all greatly increased both the variety and scale of their activities. In Brussels by 1897 the volume of information enquiries had tripled since the late 1880s and its widely circulated *Bulletin Commercial* ensured that the Musée had become much more than a repository of samples (Austin Lee 1898, 158–9). The publications of the Imperial Institute were similarly impressive and despite funding difficulties, even its commercial enquiry service finally began to expand in the late 1890s (Muddiman 2010, 114). The Philadelphia Bureau was, however, by the early twentieth century, generally considered by most observers to be *the* model information service: "senior of all … in maturity of development" (Branford 1902, 243). In part this was because of the size of its operation and the range of its services; but in part also because of its innovations and efficiency. According to W. Colgrove Betts, this service surpassed the museum itself in importance if not in size. It had become a national service essential to "the general extension of the foreign interchange and commerce of the United States" (Betts 1900, 223).

Eventually, then, all of the bureaux of information examined in this chapter evolved into services which were to outgrow and outlast their parent museums. Their techno-bureaucratic importance lay in their focus upon a new kind of knowledge—business information—which was geared to the fast changing demands of the commercial world. Beginning in the 1880s, the bureaux deployed an innovative series of techniques and technologies in order to assemble, organize and disseminate this information. Information gathering started to become a proactive endeavour which went beyond the acquisition of published material. Each of the bureaux examined in this chapter attempted, not always successfully, to develop a global network of "correspondents" from whom it could glean intelligence on market trends and developments. New types of data (much of it made possible by the standardization of concepts such as time and exchange rates) were constantly acquired, arranged, analysed and amended: commercial tariffs, timetables, carriage rates, economic statistics and so on. Information was organized, stored, and indexed in new physical and intellectual formats so that it could be easily updated, retrieved to answer individual enquiries, and utilized for publication and distribution. Kaiser's card index system at Philadelphia was an example of innovation in "information technology" (Dousa 2011). New and varied experiments in dissemination were tried. These included the weekly or monthly current awareness bulletin (developed by all bureaux examined in this chapter); the statistical *Year Book*; and pamphlets targeting specific market sectors and geographical locations. Selective dissemination techniques satisfying off-site enquiries such as standard format competitive intelligence reports and the PCM's "rush news" service were also notable innovations. All of these techniques benefitted from the contemporaneous revolution in office technologies. Typewriters, vertical filing, card indexes, duplicating and tabulating machines enabled the new bureaux to process, store and amend data rapidly and systematically compared with earlier manual techniques (Yates 2000). Cable and telegraph and more latterly the telephone supplemented carriage of information by improving surface mail and transformed the processes of data gathering and dealing with distant users. In theory, at least, it was now possible to gather and disseminate information throughout the connected world.

To claim, however, that the commercial museums and their bureaux were symptomatic of a seamless transition to a world of "information beyond borders" would nevertheless be to seriously oversimplify their role. It is undoubtedly true that they were created as a response to the "great boom" in world trade between c1850 and 1875 and that, by 1914, they in turn had helped to create new knowledge structures and systems for global commerce. Nevertheless, it is also clear that such developments took place in the context of what Cain and Hopkins (2002, 663) identify as the epoch of "modern" globalization (1850–1950), an era which paradoxically "strengthened the nation state and led to the formation of new or expanded empires." Hence, as we have highlighted in Belgium, Great Britain, and the United States, the internationalism of the commercial museums was sporadic and uneven. The informational frameworks they created for worldwide interconnectedness were skewed towards the "existing preference systems of empires and their spheres of influence" (Aberloos 2008, 107). Perhaps even more significantly in the long run almost all of their information bureaux were converted, as the twentieth century gathered pace, into state controlled business information services operating in the national, as opposed to the international or global, interest. In Belgium this had more or less always been the case, the museum after 1900 evolving as we have seen into a department of the Office Belge du Commerce. In Great Britain state provision by the Board of Trade replaced the Imperial Institute's Commercial Intelligence Department in 1902. Even in the free market-minded United States, by the 1920s the Federal Department of Commerce had effectively superseded the information services of the PCM.

In the final analysis, therefore, Commercial Museums were essentially hybrid and transitional institutions, but none the less important for that. Their initial project of displaying and categorizing commercial knowledge through the collection and arrangement of samples was at root an attempt to create an economic science based on the static kinds of taxonomy and order that had prevailed in the past. Inevitably, this was doomed to fail in an accelerating modern age of the present where knowledge was fast becoming mutable, dynamic and omnipresent. Much more in touch with their times, however, were the museums' information bureaux. These, according to W. Colgrove Betts, extended the scope and functions of the commercial museum "far beyond those usually associated with the name" and turned these museums into a "centre of intelligence on all matters appertaining to international commerce" (Betts 1900, 223). In hindsight, we can perhaps justifiably add that these bureaux were in effect key innovating institutions of what it is now possible to see as a late nineteenth century information revolution. They initiated the beginnings of twentieth-century business *information* by mobilizing knowledge in response to economic modernization, trade internationalization and imperialist competition. By engaging in "commercial intelligence" they helped develop, often for the first time, many of the concepts, forms and techniques of modern commercial information systems. And although in the early twentieth century most of the bureaux were replaced by state funded services, these (together with the information sections of large commercial enterprises especially after 1918) by and large extended and expanded, rather than overturned, the systems of information developed in the bureaux of the commercial museums. Commercial museums and their information bureaux might with some justice be said to have pioneered between 1870 and 1914 an era of business information based on documents, data, an international economy, and a new modernity with its distinctive "sense of the present." This era would endure for a century, more or less until the digital transformations of a later, electronic, information age.

References

"A Philadelphia Squabble." 1901. *New York Times*, July 13.

Abel, Frederick. 1887. "Sir F. Abel on the Imperial Institute." *The Times*, April 23.

Abel, Frederick. 1898a. "Memorandum on the Nature and Progress of Work Carried Out at the Imperial Institute." In *Minutes of Evidence Taken before the*

Departmental Committee Appointed by the Board of Trade to Inquire into and Report Upon the Dissemination of Commercial Information and the Collection and Exhibition of Patterns and Samples. Board of Trade, Commercial Intelligence Committee. London: HMSO, 65–8.

Abel, Frederick. 1898b. "Illustrations of Practical Results Attained … from Information Supplied by the Commercial and Industrial Intelligence Office and the Scientific and Technical Department of the Imperial Institute." In *Minutes of Evidence Taken before the Departmental Committee Appointed by the Board of Trade to Inquire into and Report Upon the Dissemination of Commercial Information and the Collection and Exhibition of Patterns and Samples*. Board of Trade, Commercial Intelligence Committee. London: HMSO, 102–3.

Aberloos, Jan-Frederik. 2008. "Belgium's Expansionist History between 1870 and 1930: Imperialism and the Globalisation of Belgian Business." In *Europe and its Empires*, edited by Csaba Levai. Pisa: Plus, 105–27.

Advisory Committee to the Board of Trade on Commercial Intelligence. 1909. *Report of the Proceedings, August 1905–August 1909*. London: HMSO.

Allwood, John. 2001. *The Great Exhibitions: 150 Years*. London: ECL.

Anstey, Vera. 1929. *The Economic Development of India*. London: Longmans.

Baden-Powell, George. 1886. "An Empire Institute." *National Review*, 46: 433–42.

Bairoch, Paul. 1989. "European Trade Policy 1815–1914." In *The Cambridge Economic History of Europe,* Vol. VIII, "The Industrial Economies: The Development of Economic and Social Policies," edited by Peter Mathias and Sidney Pollard. Cambridge: Cambridge University Press, 1–160.

Bayly, C.A. 2004. *The Birth of the Modern World 1780–1914*. Oxford: Blackwell.

Bennett, Tony. 1995. *The Birth of the Museum*. London: Routledge.

Betts, W. Colgrove. 1900. "The Philadelphia Commercial Museum." *The Journal of Political Economy*, 8: 222–33.

Board of Trade Commercial Intelligence Committee. 1898a. *Minutes of Evidence Taken before the Departmental Committee Appointed by the Board of Trade to Inquire into and Report Upon the Dissemination of Commercial Information and the Collection and Exhibition of Patterns and Samples*. London: HMSO.

Board of Trade Commercial Intelligence Committee. 1898b. *Report of the Departmental Committee Appointed by the Board of Trade to Inquire and Report Upon the Dissemination of Commercial Information*; Cmd. 8962. London: HMSO.

Bogert, Elvira. 1976. "Austro-Hungarian Maritime Trade with the Ottoman Empire." PhD diss, Tufts University.

Branford, Victor. 1902. "The Philadelphia Commercial Museum." *Scottish Geographical Magazine*, 18: 243–52.

Cain, P.J. and Hopkins, A.G. 2002. *British Imperialism 1688–2000*. 2nd ed. London: Longman.

Congrès International d'Expansion Economique Mondiale. 1905. *Documents Préliminaires et Compte Rendu des Séances*. Brussels: Goemaere.

Conn, Stephen. 1998a. *Museums and American Intellectual Life, 1876–1926*. Chicago: University of Chicago Press.

Conn, Stephen. 1998b. "An Epistemology for Empire: The Philadelphia Commercial Museum 1893–1926." *Diplomatic History*, 22: 533–63.

De Groof, Roel. 2008. "Promoting Brussels as a Political World Capital: From the National Jubilee of 1905 to Expo 58." In *Brussels and Europe*, edited by Roel De Groof. Brussels: Academic and Scientific Publishers, 97–126.

Dousa, Thomas M. 2011. "Concretes, Countries and Processes in Julius O. Kaiser's Theory of Systematic Indexing: A Case Study in the Definition of General Categories." In *Proceedings from North American Symposium on Knowledge Organization*, Vol. 3, edited by P. Smiraglia. Toronto, Canada, 160–73.

Dunstan, Wyndham R. 1898. "Memorandum by Professor Wyndham R. Dunstan FRS … on the Philadelphia Commercial Museums." In *Minutes of Evidence Taken before the Departmental Committee Appointed by the Board of Trade to Inquire into and Report Upon the Dissemination of Commercial Information and the Collection and Exhibition of Patterns and Samples*. Board of Trade, Commercial Intelligence Committee. London: HMSO, 170–76.

Friedberg, Aaron. 1988. *The Weary Titan: Britain and the Experience of Relative Decline 1895–1905*. Princeton NJ: Princeton University Press.

Geppert, Alexander. 2010. *Floating Cities: Imperial Expositions in Fin de Siècle Europe*. London: Palgrave MacMillan.

Greenhalgh, Paul. 1988. Ephemeral Vistas: *A History of the Expositions Universelles, the Great Exhibitions and the World's Fairs 1851–1939*. Manchester: Manchester University Press.

Goode, George B. 1896. "On the Classification of Museums." *Science*, New Series 3 (57): 154–61.

Hobsbawm, Eric. 1975. *The Age of Capital 1848–75*. London: Weidenfield and Nicholson.

Hobsbawm, Eric. 1987. *The Age of Empire 1875–1914*. London: Weidenfield and Nicholson.

Ilersic, Alfred and Patricia Liddle. 1960. *Parliament of Commerce: a History of the Association of Chambers of Commerce*. London: Association of British Chambers of Commerce and Newman Neame.

Imperial Institute. 1893. *Annual Report for 1893 and Appendices*. London: Imperial Institute.

Imperial Institute. 1894. *The Year Book of the Imperial Institute*. 3rd edn. London: John Murray.

Kennedy, C.M. and Bateman A.E. 1885. "Memorandum Respecting the Commercial Museums at Antwerp and Brussels." In *First Report of the Royal Commission on the Depression of Trade and Industry*, Board of Trade. Parliamentary Paper C4621, Session 1886. London: HMSO, 207–9.

Kern, Stephen. 1983. *The Culture of Time and Space 1880–1918*. Cambridge MA: Harvard University Press.

Law, Edward Fitzgerald. 1898. "Memorandum on the Objects, Organisation and Working of the Vienna Handels Museum." In *Minutes of Evidence Taken before the Departmental Committee Appointed by the Board of Trade to Inquire into and Report Upon the Dissemination of Commercial Information and the Collection and Exhibition of Patterns and Samples*. Board of Trade, Commercial Intelligence Committee. London: HMSO, 161–2.

Lee, H. Austin. 1898. "Copy of Memorandum by H.M. Commercial Attaché at Paris on the Brussels Commercial Museum." In *Minutes of Evidence Taken before the Departmental Committee Appointed by the Board of Trade to Inquire into and Report Upon the Dissemination of Commercial Information and the Collection and Exhibition of Patterns and Samples*. Board of Trade, Commercial Intelligence Committee. London: HMSO, 158–9.

Lipsey, Robert. 2000. "US Foreign Trade and the Balance of Payments 1800–1913." *In Cambridge Economic History of the United States*, Vol. 2 "The Long Nineteenth Century," edited by Stanley Engerman and Robert Galman. Cambridge: Cambridge University Press, 685–732.

Lynn, Martin. 1999. "British Policy, Trade and Informal Empire in the Mid 19th Century." In *The Oxford History of the British Empire: The Nineteenth Century*, edited by Andrew Porter. Oxford: Oxford University Press, 101–21.

Mackenzie, John M. 1984. *Propaganda and Empire*. Manchester: Manchester University Press.

Mairesse, François. 2010. "Paul Otlet, Apprenti Muséologue." In *Paul Otlet, Fondateur du Mundaneum (1868–1944): Architecte du Savoir, Artisan de Paix*, edited by Jacques Gillen. Mons: Les Impressions Nouvelles, 137–48.

MAK [Österreichisches Museum für angewandte Kunst / Gegenwartskunst-Austrian Museum of Applied Art/Contemporary Art]. 2011. "Die Geschichte des österreichischen Museums für angewandte Kunst / Gegenwartskunst [History of the MAK]." http://www.mak.at/das_mak/geschichte (accessed January 2013).

Mayer, Arno. 1981. *The Persistence of the Old Regime: Europe to the Great War*. London: Verso.

Muddiman, Dave. 2010. "Information and Empire: the Information and Intelligence Bureaux of the Imperial Institute, London, 1887–1949." In *Information History in the Modern World*, edited by Toni Weller. London: Palgrave Macmillan, 108–29.

Murray, Kenric B. 1886. "Special Report on Commercial Museums." *Chamber of Commerce Journal*, 5(56): 1–19 [Supplement].

Murray, Kenric B. 1898. "Appendix No.2. List of Commercial Museums and Special Agencies." In *Minutes of Evidence Taken before the Departmental Committee Appointed by the Board of Trade to Inquire into and Report Upon the Dissemination of Commercial Information and the Collection and Exhibition of Patterns and Samples*. Board of Trade, Commercial Intelligence Committee. London: HMSO, 104–9.

Norman, A. 1902. "The Imperial Institute, its Genesis, History and Possibilities." *Imperial Argus*, 1 (8): 393–407.

Pepper, William. 1897. "The Commercial Museum." *New York Times*, June 3.

Rayward, W. Boyd. 1975. *The Universe of Information. The Work of Paul Otlet for Documentation and International Organisation*. FID 520; Moscow: VINITI.

Ridley, Jane. 2004. "Ritchie, Charles Thomson." In *The Oxford Dictionary of National Biography*, Vol. 47, edited by Colin Matthew and Brian Harrison. Oxford: Oxford University Press, 16–19.

Rydell, Robert. 1984. *All the World's a Fair. Visions of Empire at American International Expositions, 1876–1916*. Chicago: University of Chicago Press.

Sheppard, F.H.W. (ed.) 1975. *Survey of London: Volume 38: South Kensington Museums Area*. http://www.british-history.ac.uk/source.aspx?pubid=364 (accessed January 5, 2013).

Stanard, Matthew. 2011. "Learning to Love Leopold: Belgian Popular Imperialism 1830–1960." In *European Empires and the People*, edited by John M. MacKenzie. Manchester: Manchester University Press, 124–57.

"Trade Organization." 1911. In Encyclopaedia Britannica, 13th edn, Vol. XXVII. Cambridge: Cambridge University Press, 135–40.

Wolfrom, Kathleen. 2010. "The Rise and Fall of the Philadelphia Commercial Museum. How a Forgotten Museum Forever Altered American History." http://www.phillyseaport.org/web_exhibits/mini_exhibits/philadelphia_commercial_museum/index.shtml (accessed January 4, 2013).

Yates, JoAnne. 2000. "Business Use of Information and Technology During the Industrial Age." In *A Nation Transformed by Information*, edited by Alfred D. Chandler and James W. Cortada. Oxford: Oxford University Press, 107–36.

Chapter 16

An Information Management Tool for Dismantling Barriers in Early Multinational Corporations: The Staff Magazine in Britain Before World War I

Alistair Black

The Staff Magazine: Definition, Early History, and Roles

Staff magazines are in-house periodicals produced for employees of an organization and potentially anyone associated with it. They are normally produced with the approval of the organization, and are often subsidized by it. They are to be distinguished from trade magazines, which focus on developments across sectors and industries.

The first staff magazines—synonyms include multiple permutations of the adjective-noun house/plant/factory/works combined with the noun organ/paper/newsletter/journal—emerged in post-Civil War America (Riley 1992). In Britain their development got underway two decades later. Research to date (Black 2011) has revealed that the first staff magazine in Britain was *Hazell's Magazine* (1887), the in-house organ of the printing and packaging company Hazell, Watson, and Viney, based in both Aylesbury and London. After 1900, the number of staff magazine increased rapidly. Although there is evidence of publication in non-commercial organizations—such as hospitals, public utilities, and libraries[1]—it would appear that the vast majority occurred in large private enterprises.

Staff magazines were founded in enterprises across the economy but they appear to have been particularly common in sectors that were supplying new mass-consumer goods and services, such as in the electrical, chemical, motor vehicle, pharmaceutical, banking, insurance, and food and drink sectors. One of the factors that distinguished these sectors from the earlier staple industries of

---

1   For example, *The Guyoscope*, magazine of the staff of Guy's Hospital, London, which was first published in 1905. Some magazines in public utilities are discussed in Black (2011): for example, *The Aquarius*, organ of the employees of the London Metropolitan Water Board, first published in 1905, was produced by the organization's Staff Association. Black (2012) includes a discussion of the *Croydon Crank*, the staff magazine of Croydon Public Libraries, first published in 1908.

coal, engineering, shipbuilding, and textiles was their expertise in, and exploitation of, new techniques of marketing, encompassing market research, branding, and advertising (Fitzgerald 1995, 5). To a degree the staff magazine was an important part of this turn towards systematic corporate promotion and intelligence.

The format and length of early staff magazines varied considerably. At one extreme some had the appearance of a flimsy newsletter; at the other extreme there are examples that compared favorably with the high standards of good commercial periodicals. In the setting of the private enterprise, as one would expect, editorial control was far from independent. Especially in companies that adopted modern management ideas, the room for manoeuvre that editors had in terms of content and direction was limited. Like format and length, content also varied from one magazine to the next. However, a number of frequently occurring ingredients can be identified: news about staff (for example, promotions, births, deaths, marriages, retirements, and new recruits), the company and the sector in which it operated, and social activities (for example, annual dinners, sporting and social clubs); histories of the company; games, puzzles, and quizzes; fictional stories; instructions and protocols; and what one might term motivational messages from managers. Most magazines aimed at a high degree of internal specialization. On its front cover *Hazell's Magazine*, for example, styled itself as "a monthly journal of literary effort, notes, news and gossip." In short, the aim was to ensure that there was something for everyone.

What little research has been conducted to date on the early history of staff magazines in Britain has focused on their function as instruments of management control in the workplace (Griffiths 1999) and, in the case of my own research, as tools for the management of information and knowledge-creation (including professional training) in the context of the rise of the science of management (Black 2011; Black 2012). These studies have emphasized the contribution a staff magazine could make to corporate prestige, identity, and heritage; to the incorporation of the workforce and the fashioning of a more conciliatory regime of industrial relations; to the construction of an *esprit de corps*; and to corporate cohesiveness at a time when enterprises and production units were enlarging to a degree that was unforeseen in earlier phases of industrialization.

However, none of this work has yet addressed the early history of staff magazines against the backdrop of the international activities of corporations that in decades approaching World War I were becoming multinational in character by virtue of establishing subsidiaries abroad and/or purchasing foreign properties, concerns, and concessions to secure their own sources of raw material. This chapter illustrates how staff magazines in a selection of the first truly multinational corporations—Lever Brothers (*Progress*), Rowntree and Company (*Cocoa Works Magazine*), Ford Motor Company (Britain) (*Ford Times, Britain*), and British Westinghouse (*British Westinghouse Gazette*)—reflected, explained, and justified their newly fashioned international activities and strategies; and how these magazines served as sources of information that sought to bind together a workforce and an array of operations that were increasingly dispersed not only within individual countries but also around the globe.

The Rise of the Multinational Corporation

In the decades immediately preceding World War I, across the advanced industrial economies of Europe and North America, private enterprises expanded sharply in both scale and scope (Chandler 1990). The period witnessed the emergence of large corporations which gradually began to extend their reach beyond national boundaries. This extended reach entailed not just huge growth in the exporting of goods and services to foreign markets but also a physical embedding in the manufacturing, service-industry, and distribution infrastructures of other countries.

Multinational companies were not a product of the post-Second World War economy, when their challenge to the sovereign power of the nation-state became apparent (Tugenhadt 1971). Indeed, it is possible to trace the multinational company back as far as the interregional and intercontinental trading companies of ancient times (Wilkins 2005, 47). In the late middle ages, the Florentine Medici Bank had branches in a number of European cities. In the golden era of merchant capitalism in the seventeenth century, powerful monopolies such as the British and Dutch East India Companies and the British Hudson Bay Company opened up vast new markets to satisfy an increasingly prosperous consumerism in Europe (Carlos and Nicholas 1988). It was not until the late nineteenth century, however, that the modern multinational company as we recognize it today—capable of contributing to a globalized culture, its international activity managed under one organizational structure—came into being. The expansion of multinational companies at this time was facilitated by improvements in communication technologies, from railways and steamships to the telegraph and the telephone. It was in the La Belle Époque and in the period immediately preceding it that many of the multinational corporations that remain household names today were born: Siemens, Bayer, Courtaulds, Kodak, Gillette, Dunlop, and the four enterprises featured in this study (Cameron and Neal 2003, 336–7; Wilkins 2005).

Companies take on a multinational identity when they establish a presence in another country to reduce transport costs; exploit its raw materials; take advantage of its labor force (because it may be cheap, or skilled) and its financial environment (for example, its tax or tariff regimes); and use it as a strategic base to enter its markets or those of nearby countries (Fieldhouse 1978, 10–11). Companies can also take on an international identity, without becoming multinational, simply by exporting to other countries.

In the second half of the nineteenth century, Britain's advanced industrial status and its established imperial networks provided a strong basis for both inward and outward investment. Foreign companies began manufacturing in Britain in the middle of the nineteenth century. Starting as a trickle, such investments became a flood in the years around 1900. The invasion from America was spearheaded by such names as Western Electric (1898), Kodak (1899), American Tobacco (1901), Columbia Gramophone (1905), Heinz (1905), and Ford (1911). Investments were also made by a large number of European companies, including Bosch (Germany, 1905), Osram (Germany, 1907), Ericsson (Sweden, 1903), Néstle (Switzerland,

1901), Hoffman La Roche (Switzerland, 1909), and Pirelli (Italy, 1914) (Jones 1988). Mirroring the growth of foreign investment *in* Britain, late-nineteenth century British corporations sought new global investment opportunities, both within and outside the Empire. By 1914 about 40 percent of British overseas investments were direct, rather than of the portfolio kind (Fitzgerald 1995, 505). While some complained that such foreign investments robbed English workers of employment (Wilson 1968, 98), the move towards substantial levels of overseas production by large British companies had by 1900 become irreversible.

The Information Management and Organizational Communication Revolutions

The increasing complexity and size of corporations prompted the implementation of rational management. As operations and transactions multiplied and became more intricate, control by owners over once easily surveyed domains inevitably and necessarily gave way to the rise of a professional cadre of managers more capable of making decisions regarding production, costs, labor, product development, marketing, and strategy.

In contrast to the enterprises of the proto-industrial era, those of the early twentieth century were characterized by an adherence to "system," wherein *information* systems of various kinds became a central feature of a company's operations and planning (Yates 1989). Constituting a pre-computer information management revolution, companies introduced new technologies and techniques of recording, compiling, copying, storing, organizing, retrieving, analyzing, and transmitting information (Beniger 1986; Orbell 1991; Yates 1989). A range of innovative *technologies* were deployed, including typewriters; addressing, punched-card, and duplicating machines; vertical filing cabinets; and telephones (the last of these critical to communication in the tall office buildings that were beginning to appear at the time). New *techniques* included formal office memoranda, minutes for management meetings, printed instructional manuals and protocols, circular letters, digestible graphic means of presenting data, central registries and filing rooms, in-house libraries and information bureaux—and, the subject of this discussion, in-house staff magazines.

In addition to radical changes in the way information was handled, the information revolution of the late nineteenth and early twentieth centuries was characterized by notable developments in organizational communication. Appearing long before this concept was formally articulated in management theory, early staff magazines were relatively sophisticated forms of such communication. The "classical" theory of the organization, closely associated with scientific management, was concerned essentially with organizational structures, not with people. This was superseded after World War II by the "human relations" theory of the organization which prioritized people-oriented over production-oriented management (Goldhaber 1974, 28–63). Central to the people-oriented approach was organizational communication that went beyond the mere "downward" communication of information. As Peter Drucker explained in 1973, the human relations school *enjoined listening*, that is, it was the task of the executive to find out "what subordinates want to know, are interested in, are in other words, receptive to" (Drucker 2008 [1973], 318). The human relations school sought a greater empowerment of workers and problem solving through cooperation. Listening did not mean surrendering authority. Control would be maintained but its nature would be *coactive* rather than *coercive* (Eisenberg, Goodall, and Tretheway 2007, 82).

Long before the rise of the human relations school, however, the futility of relying simply on the transmission of messages down the hierarchy—through such means as formal instructions, procedural manuals, and task protocols—was acknowledged by those who established and ran staff magazines. Managers supported staff magazines because they were a medium for conveying messages to workers *and* a means of receiving signals from them. Publications that emanated outside the executive, such as those produced by staff associations (formed when employees came together to pursue shared interests, though not of the kind prioritized by trade unions), also served in various ways to transmit information up the hierarchy; to make visible the thinking and behaviors of the workforce. Further, both kinds of magazine facilitated horizontal communication between peers or those occupying levels in the organization that were close to each other.[2]

Ford Motor Company

The quintessential example of an early multinational corporation adopting new methods of scientific and information and communication management is the Ford Motor Company. Ford was founded in Detroit in 1903. By 1906, even before the landmark innovations that Henry Ford instituted in car production—the assembly line, vertical integration, standardized parts, a sharp division of labor—Ford had become the largest car manufacturer in the United States, a position it consolidated with the introduction of the legendary Model T in 1908. By 1913 Ford had nearly 7,000 dealers in the United States. The magnitude of the company could be measured in terms not only of the size of its factories—the assembly-line Highland plant was opened in 1913, while work was started on the mammoth River Rouge plant in 1917—but also its international posture and ambitions. In 1904 Ford established a Canadian plant, in Walkerville, Ontario. This served as an initial gateway into the markets of Britain and the British Empire (Nevins and Hill 1954).

In 1907, having vastly reduced the number of European cars sold in the United States, Ford set its sights on the European market itself. A European representative,

---

2   Theories of downward, upward, horizontal, and even diagonal communication have been discussed by Koehler, Anatol, and Applbaum (1976), 41–9; and Lewis (1975), 36–45.

headquartered in Paris, was appointed. Contracts were signed with agents in Europe's principal cities, including London where Percival Perry, an automobile enthusiast and co-founder of the American Motor Car Company (1904), was given the responsibility of establishing sub-dealerships throughout Britain.[3] Perry won the rights from Ford's Walkerville plant in Canada to sell Ford cars in Britain, and by September 1910 he had established links with 61 other dealerships (Nevins and Hill 1954, 358–62).[4]

Before World War I, most Ford cars sold abroad were exported directly from the United States (Tolliday 2003, 153). This said, by 1914 the trend towards the true multinational status that Ford achieved after the war, with the company registering a physical manufacturing presence in multiple other countries, was already underway. The fact that foreign agents and dealers, under constant pressure to maintain lines of credit, were required to pay in full for cars before Ford released them for export affected sales negatively. The solution was to manufacture in the regions of the world to which cars would be exported. This would also avoid import tariffs charged by foreign states.

In Britain, in 1911, Ford established a factory in Manchester's Trafford Park, Britain's first business park (McIntosh 1991; Nicholls 1996). Strategically, this meant that Britain could replace Canada as the main portal for selling cars to its colonies (Tolliday 2003, 155). The Trafford Park plant was soon able to exploit the assembly-line methods Ford had developed in the United States, to the extent that between 1912 and 1913 production doubled to 6,000 cars per year (Ford Trafford Park Factory 2012). By 1914 Ford was the biggest car manufacturer in Britain (Jones 1988, 435).

Ford Times (Britain)

One aspect of managing the giant Ford operation was the founding of an in-house magazine, the *Ford Times*, first issued on April 15, 1908. The magazine described itself as being "in the interests of the Selling and Business organization of Ford" and the company's "branches, agencies, managers, bookkeepers, clerks, cashiers, and stenographers," and it aimed to "chronicle happenings at the factory and in the field." One of the major functions of the magazine was to extract information from, and disseminate it to, the vast Ford empire. Content included selling tips, details of unique advertising schemes, examples of good circular letters, and information about systems "for keeping stock, for supplying repairs, for keeping records, [or for] running the repair shop or office."[5] Readers of the magazine were urged to be active participants in its production: "Just because it happens in your town or store you do not regard it as news, whereas if it happened somewhere else you would

---

3   "Foreign Business," *Ford Times*, September 1, 1908.
4   "Home of the Model T at Leeds (England)," *Ford Times*, September 15, 1910.
5   "We Make Our Bow," *Ford Times*, April 15, 1908.

Figure 16.1   An advertisement for Ford in Australia, reflecting the corporation's global presence in 1910

Source: *Ford Times* (1 September 1910, p. 13). Reproduced with the kind permission of the Benson-Ford Research Center, The Henry Ford, Dearborn, Detroit.

read it with interest."[6] A regular column, appropriately entitled "Here and There," served as a conduit for information from dispersed sources.[7] The meteoric rise of Ford to global dominance was nicely illustrated in the *Ford Times* when in 1910 it reproduced an advertisement that had been circulated in Australia depicting a model T as heavenly body racing through space and keenly watched by observers on earth (figure 16.1).

A crucially important section of the magazine's audience, therefore, was Ford's extended family (it was not until 1917, with the appearance of another in-house magazine, *Ford Man*, that the home-plant, shop-floor readership was specifically targeted). To reach this dispersed family more effectively, the logical next step was for the company to publish magazines in those parts of the world where Ford had established operations. *Ford Times (Britain)* was launched in February 1913 (a Canadian edition of the *Ford Times* followed in the August of that year). *Ford Times (Britain)* contained the same diet of motivational, instructional, and knowledge-sharing messages that characterized its parent publication. Readers

---

6   "News Items Wanted," *Ford Times*, June 1, 1908.
7   "Here and There," *Ford Times*, June 15, 1908.

were treated to lists of aphorisms, such as "Seek out the automatic act of habit—do it better."[8] News was broadcast of the activities of dealers around the country and the network of workshops able to repair Ford cars.[9] There were tips on engine maintenance and road worthiness as well as letters from owners praising their cars and detailing the long-distance journeys they had undertaken.

However, one of the most striking features of *Ford Times (Britain)*, which was ostensibly a parochial publication, was its international flavor. Often embellished by photographs, stories abounded of the performance of Ford cars in exotic places around the globe. Titles included "With the Ford in Egypt," (February 1913); "The Alley Roads of Palma, Majorca," (February 1913); "The City of the Nut," (February 1913); "The Heathen Chinee," (December 1913); and "With the Ford in India" (December 1913).

The magazine's editors were keen to keep workers informed of the progress of Ford in the United States: "Today the Ford factory in Detroit is one of America's seven wonders—a model plant in every respect, equipped with the most modern machinery and devices," trumpeted the first issue.[10] Later issues carried large centerfold photographs of the Detroit factory such as the 1913 photographs of the 12,000 Ford shop-floor workers assembled outside the factory's main building and the company's five hundred executive, engineering, and office staff.[11] A panoramic photograph in that same year captured an entire day's output of cars from the plant, a thousand in total[12] while the Ford plant in Walkerville, Canada, was depicted in a line drawing.[13] Such images were powerful reminders of the globalized and globalizing nature of the company.

Lever Brothers

William Lever began his working life as a commercial traveller for his father's wholesale grocery business in Lancashire. With a flair for marketing, and eager to take advantage of the removal of taxes on soap that had occurred in the 1850s, Lever began to concentrate on selling only soap products (Bradley 1987, 174). In 1884 he registered the trademark "Sunlight Soap," the springboard for what was to become a giant international enterprise selling cleansing and other products worldwide. Lever began manufacturing soap in Warrington in 1885. His great innovation was to offer packaged and branded soap. Previously, customers had no option but to buy lengths of soap cut from anonymous blocks. Lever's soap bars were not only presented in bright cartons (an idea borrowed from the United States), they were also wrapped

---

8   "Thoroughness," *Ford Times (Britain)*, February 1913.
9   For example, "The Dealers' Department," *Ford Times (Britain)*, November 1913.
10 "Our Mammoth Factory," *Ford Times (Britain)*, February 1913.
11 *Ford Times (Britain)*, November 1913; and *Ford Times (Britain)*, December 1913.
12 *Ford Times (Britain)*, December 1913.
13 *Ford Times (Britain)*, December 1913.

in parchment paper to guard against the oxydization process that quickly turned soap rancid (Bradley 1987, 178). In 1890 Lever Brothers was registered as a private company. Conversion to a public company came four years later. By 1900 Lever was the largest soap manufacturer in Britain. It eventually became one of the first large companies to be the subject of an extensive history, the product of the emergent field of business history (Wilson 1968).

As early as 1888, the company opened an office in Sydney. A factory followed in 1900. This pattern—the shift from exporting via overseas agencies to manufacturing in foreign countries—was replicated by Lever in countries around the globe, so that by the twentieth century Lever also had a production presence in the United States, Canada, Germany, and Switzerland. Factories in other countries, the company explained, were mostly necessary to avoid import duties when exporting to them.[14] Lever preferred the word "affiliate" to "subsidiary" in describing its overseas companies; the former was said to be associated with fatherhood and was thus judged to be more in keeping with the company's paternalistic management style.[15]

After 1900, Lever factories sprang up in other European countries and in Japan. The company of Lever Brothers (South Africa) was registered in 1911 and a factory was opened in Durban the following year. In large countries like Canada and South Africa, multiple manufacturing units were required. To obtain secure flows of raw materials, a number of overseas properties and concessions were purchased in the Pacific and Africa, including plantations in the Solomon Islands (coconuts) and the Belgian Congo (palm oil) (Fieldhouse 1978, 27, 64–6, 102; Roach 2005, 20; Wilson 1968, 98, 159–210).

The center of Lever's operations was the industrial utopia of Port Sunlight on the Wirral that comprised a model factory and village in which had been established a range of welfare measures for workers. By 1902, there were 3,000 workers at Port Sunlight and 650 company houses.[16] Lever managers were attuned to the problems that accompanied corporate enlargement. Men and women operating in the large factory, William Lever observed, had been "forgotten and neglected." The "modern problem in considering industrial development," he stressed, "is merely one of size," and so good corporate welfare was required to correct this (Leverhulme 1917, 1).

Progress

The company's paternalism allied to its increasing size placed a premium on effective communication with the workforce. An in-house magazine was an important element in Lever's communication arrangements and at the same

---

14   As explained in "The Australian Tariff and Its Effects on Soap," *Progress*, January 1909.

15   "Subsidiary or Affiliate," *Progress*, October 1900.

16   "The Australian Tariff and Its Effects on Soap," *Progress*, January 1909.

reflected its paternalistic welfare provision. Lever established its first staff magazine in 1895, the *Port Sunlight Monthly Journal*. Although by the following year its cover was proclaiming that the magazine was for the company's workers in London, New York, and Sydney, the content was essentially parochial, focussing on life in the factory and village at Port Sunlight. A new magazine was launched in 1899, *Progress*. Unlike its predecessor, however, *Progress* developed geographical breadth in its reporting, in parallel with the company's increasingly multinational operations. Its international remit was evident in its declaration that it was:

> Printed and Published specially as a means of inter-communication between the HEAD OFFICE and WORKS at PORT SUNLIGHT, the BRANCH OFFICES in the UNITED KINGDOM, and the OFFICES, AGENCIES, OIL MILLS and AFFILIATED COMPANIES Abroad.[17]

By the early 1920s, *Progress* had secured a massive circulation of a quarter of a million worldwide, and was said "to appeal to the great Lever Brotherhood in every land."[18]

Progress, said Lever brothers, was a way of giving a "hearty hand-shake to all our staff at Port Sunlight and those scattered throughout the world." The company saw the magazine as important in an era when owners and executives were "not now able, as was the case in earlier days, to come into daily contact with all who are working with us." It had a double duty to fulfil: "to bring you [the staff] into contact with ourselves [the owners and executives] and with each other."[19] Two years into the run of *Progress*, the editor assessed that his magazine had made the world of Lever Brothers:

> larger and smaller at one and the same time. Larger in that our readers know more about the limitless market on which the manufactures of Lever Brothers Limited constantly figure that peculiarly large success born of sheer merit; smaller by reason of the fact that the great big world over which are now scattered broadcast the representatives of Lever Brothers Limited and the Associated companies, with their factories, agencies, and offices, has been made more familiar to our readers, and, therefore, less vast, less "foreign," and less un-neighbourly than it would have been had our popular little "means of communication" not existed.[20]

---

17   Stated on the title page of *Progress*, Vol. 1, No. 13 (October 1900) (upper case in the original).

18   As reported in its successor, *Port Sunlight News*, November 1922.

19   "Our Second Birthday: The Progress that Knows No Rest," *Progress*, October 1901.

20   "Our Second Birthday: The Progress that Knows No Rest," *Progress*, October 1901.

Progress was directed at both shareholders and employees. Regarding the latter, an important target audience was the army of district agents (known as DAs). District agents formed what was known as the "outdoor" sales force that brought Lever products to all corners of Great Britain. The magazine was a means of exhorting and advising district agents, as well as the workforce at Port Sunlight and in London (Wilson 1968, 42). As the months and years passed, however, although news about the British workforce and scene remained, foreign news began to compete with it for space. *Progress* eventually recognized itself as not only "metropolitan," conveying local news, but also "cosmopolitan," a form of "capital in the new social world that has begun to spring up on the ruins of the old."[21] From August 1901, home news and foreign news were organized into separately titled sections.[22]

The role of the magazine was to draw more closely together the many thousand units which comprised what *Progress* termed the global "Sunlight family."[23] A poem published in an early issue proclaimed Lever's cosmopolitanism:

> Oh, we are there; we are there
> SUNLIGHT SOAP is working wonders everywhere
> With Kruger and with Steyn
> Or with Bobs at Bloemfontein
> Where there's anything to clean
> We are there.[24]

Progress aimed to bridge distance:

> If any member of our staff has anything to say which he considers would come as light in darkness to some other member working perhaps two hundred, or even two thousand, miles away, we want him to feel that, in *Progress*, he has the right vehicle for his thoughts. Just peep into those mental storehouses of yours, and see what is stored away in the pigeon-holes of your brains.[25]

The magazine carried stories and photographs of Lever workers and installations in other countries, such as in Canada.[26] Writing from the vantage point of the company's British headquarters, the magazine's editor was confident that due

---

21   "Editorial Notes," *Progress*, January 1909. This perhaps a forerunner of the recent concept of social capital: that is, the social benefits that are derived from investments in structures and institutions that facilitate cooperation and trust between individuals and groups.

22   "The Progress of '*Progress*'," *Progress*, August 1901.

23   "Our Second Birthday: The Progress that Knows No Rest," *Progress*, October 1901.

24   "Sunlight's There," *Progress*, June 1900.

25   "One Year Old," *Progress*, October 1900.

26   "Our Toronto Salesmen," *Progress*, May 1903.

to his organ of communication, news of agents "whose ground extends far into the trackless forests of Canada is also as familiar with our doings here at Port Sunlight as if he worked at Head Office."[27] An agent in Nova Scotia recorded his delight at receiving *Progress*: "This publication is a grand medium, giving a chance to the large army of 'Sunlight' workers of exchanging ideas which could not be accomplished in any other way, and it will do a large amount of good."[28] News was circulated of the factory in Toronto, the "stronghold of the Sunlight Soap Company" in Canada.[29] A short article accompanied by two photographs celebrated the work of the factory in Philadelphia.[30] Photographs of shop windows decorated entirely with Lever products situated the company as a global enterprise that had meaning to people's everyday lives around the world.[31] The cover of the February 1904 issue of *Progress* depicted an Indian peddler with a trunk full of the company's wares. In 1906, mimicking, the magazine published an impressive map of the globe in which countries with a Lever Brothers presence were similarly darkly shaded. By 1906 Lever Brothers had established a Foreign & Colonial Department to oversee its international operation, the large extent of which were depicted in the impressive world map of the globe shown in Figure 16.2. This map in which countries with a Lever Brothers presence are darkly shaded mimicks world maps showing the British empire in red. The image is faded, but shading can just be glimpsed of North America, South Africa, Australia, parts of Europe and, of course, Great Britain.

Rowntree and Company

Quaker Henry Rowntree founded the confectionery business Rowntree and Company in 1862. In 1869 his brother, Joseph Rowntree, joined the company. When Henry died in 1883 Joseph took over as head of the company and it was he who oversaw the company's meteoric rise before World War I. Rowntree's base was the city of York which combined an ancient ecclesiastical and administrative heritage with industrial modernism. Despite fierce competition in the confectionery sector (Cadbury 2010), Rowntree's revenues and profits rose steadily from the mid-1890s onwards. Initially, most profit came from fruit gums and pastilles but in the early 1890s a new process for the manufacture of a purer cocoa powder—branded "Cocoa Elect"—was introduced ahead of competitors, resulting in vastly increased sales (Bradley 1987, 143–4). By the 1930s, assisted

---

27   "Our Second Birthday: The Progress that Knows No Rest," *Progress*, October 1901).

28   "From Nova Scotia," *Progress*, March 1900.

29   "Touching Toronto," *Progress*, January 1900.

30   "Philadelphia Works," *Progress*, September 1901.

31   "What Rotterdam Is Doing in the Way of Window Dressing," *Progress*, June 1900; "A Window Display in Quincy, Massachusetts," *Progress*, May 1905.

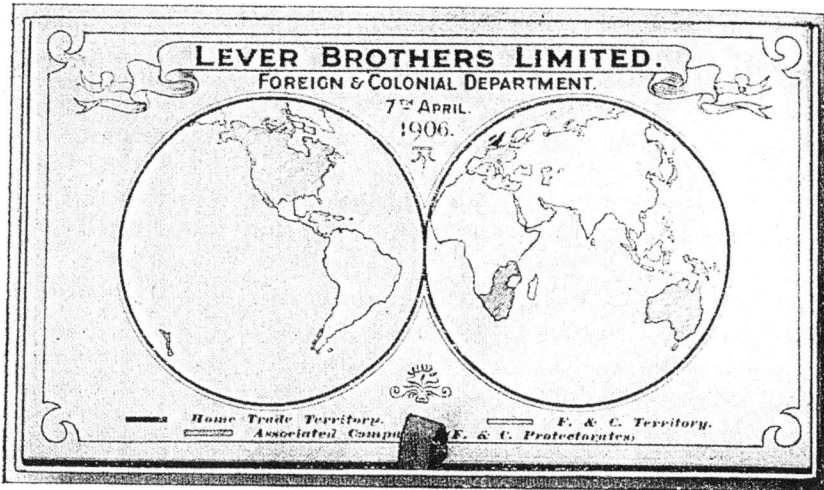

Figure 16.2   Lever Brothers foreign and colonial department

Source: *Progress* (June 1906, p. 172). Reproduced with the kind permission of the Unilever Archives.

by innovative marketing methods, Rowntree had become one of the world's largest manufacturers of confectionery, its products being supplied to a variety of countries, its supply chain very much dependent on cocoa, banana, and sugar production in the Caribbean (Fitzgerald 1995).

Run in accordance with Quaker ethics and a sensitivity towards new ideas of organization as well as an awareness of the power of mass consumption, Rowntree's success was based on a potent mix of paternalism and scientific management (Black 1994), a combination which has been theorized as corporate "welfare capitalism" (Sanford 1997). Dating from the late nineteenth century, corporate welfare capitalism entailed strategic action by large enterprises to protect their workers against risks associated with living and working in an industrial, urban environment where unemployment was a constant threat and where factory and home conditions could be dangerous and unhealthy (Sanford 1997, 4). Paternalistic welfare-minded employers like Rowntree provided sporting and other recreational facilities for social interaction between workers as well as good in-house medical and education services.[32] They were also at the forefront of the staff magazine movement.

---

32   Rowntree commenced free medical and dental services for its workers in 1904, a company supported pension fund in 1906, and education classes for girls and boys in 1905 and 1907, respectively (Bradley 1987, 147).

Cocoa Works Magazine

The *Cocoa Works Magazine* was inaugurated in 1902, produced by the company's Social Department and typeset and printed locally by Cooper and Swann, York.[33] The aim of the magazine was, in its own words, to tell of "the comings and goings of the people connected with the Works, doings of clubs, societies and classes, the recollections of those who took an early share in the building up of the business, the marriages and deaths and births in the families of our fellow-workers, and all the news of the life of the great and ever-growing population of the Cocoa Works."[34] According to the editor, the magazine sought a "cultivation of a spirit of unity and good fellowship."[35] In its first issue the company's founder, Joseph Rowntree, gave the magazine his blessing. From his perspective, the company had grown so large that there had been a loss of "personal intercourse" and "personal acquaintance" with the staff, which now numbered in excess of two thousand. The number of work sites and departments had expanded, meaning that those working in one room or section would often never see a fellow worker working in another. It was the job of the magazine, therefore, to repair this dislocation and revive a strong common purpose: "the entire body of workers must be animated with a common aim."[36]

Through the pages of the *Cocoa Works Magazine*, workers were made aware of the large international operation of which they were a part. Readers of the magazine were given educational lessons in the international origins and nature of the ingredients they used to prepare Rowntree products, such as the gum-producing acacia trees of the Sudan and Senegal and bananas from the West Indies.[37] A number of articles were devoted to the company's links with Jamaica where, in accordance with authentic Fordist vertical integration, the company had purchased banana and cocoa plantations and had established processing facilities.[38] In the case of Jamaica, the island's culture was explained with the help of anthropologically styled accounts of inhabitants' culture.[39] The company's workforce on the island, for example, was said to be made up of two ethnic groups:

---

33 "The Making of the 'Cocoa Works Magazine'," *Cocoa Works Magazine*, December 1911.

34 "The Making of the 'Cocoa Works Magazine'," *Cocoa Works Magazine*, December 1911.

35 "Editorial," *Cocoa Works Magazine*, March 1902.

36 "The Cocoa Works, York," *Cocoa Works Magazine*, March 1902.

37 "Gum Arabic," *Cocoa Works Magazine*, June 1904; "Bananas," *Cocoa Works Magazine*, October 1905.

38 Jamaica, *Cocoa Works Magazine*, February 1907. Interestingly, Lever also adopted vertical integration, in its Toronto factory where wooden and cardboard boxes were made and printed in-house: "Canada Is Awake," *Progress*, February 1902.

39 "Scenes from Jamaica I," *Cocoa Works Magazine*, January 1908; "Scenes from Jamaica II," *Cocoa Works Magazine*, February 1908; "Scenes from Jamaica III," *Cocoa Works Magazine*, March 1908.

"Negro" workers and East Indian "coolies."[40] News was published in 1907 of a serious earthquake in Kingston, the island's capital, and of the company's plan to distribute relief.[41] Coverage was also afforded of the island of Dominica where Rowntree had bought lime and cocoa plantations and ran a sugar mill.[42] As well as focussing on agriculture, the magazine reports, often embellished with exotic photographs, centered on the island's history, topography, wildlife, and native culture.[43] A photograph in one issue pictured a children's cricket match on one of the company's West Indian estates.[44] It was not uncommon for the magazine to adopt a patronizing tone towards the tenant farmers and laborers who cultivated its Caribbean estates.[45]

It sought to portray itself as a model company in pursuit of the latest methods of industrial organization and labor management. A number of articles featured the United States and its innovative production methods. A company executive reported on a visit to Pittsburgh, where he saw Carnegie's Homestead works and the factories of American Westinghouse.[46] The magazine gave considerable coverage of a visit to Canada and the United States by Arnold Rowntree, brother of Joseph Rowntree, to examine the quality of labor and of education and training systems.[47] In 1904 the magazine reported on the previous year's World's Fair in St. Louis, an exhibition twice the size of the World's Fair in Chicago a decade earlier; the giant Palace of Agriculture received special attention in the report.[48] The company's international self-identity was reinforced by an article explaining the attributes of the universal language Esperanto.[49]

British Westinghouse

The electrical engineering firm British Westinghouse was a subsidiary of the American company Westinghouse Electric and Manufacturing, established by George Westinghouse in Pittsburgh, Pennsylvania, in 1886 (Prout 2005[1921]). In the "electrical age" American Westinghouse grew rapidly and was chosen as

---

40 "Scenes from Jamaica II," *Cocoa Works Magazine*, February 1908.

41 "The Jamaica Earthquake," *Cocoa Works Magazine*, March 1907.

42 "A Trip to Dominica: I," *Cocoa Works Magazine*, May 1906; "A Trip to Dominica: II," *Cocoa Works Magazine*, June 1906; "A Trip to Dominica: III," *Cocoa Works Magazine*, September 1906.

43 "Blenheim," *Cocoa Works Magazine*, May 1904.

44 "Cricket Where the Cocoa Grows," *Cocoa Works Magazine*, September 1910.

45 "A Trip to Dominica: II," *Cocoa Works Magazine*, June 1906.

46 "Pittsburg and Its Industries," *Cocoa Works Magazine*, January 1905.

47 "Mr Arnold Rowntree in Canada and the States," *Cocoa Works Magazine*, November 1912.

48 "Our 'Special Commissioner' at the World's Fair," *Cocoa Works Magazine*, April 1904.

49 "What is Esperanto?," *Cocoa Works Magazine*, April 1904.

the leading supplier of electrical equipment at the Chicago World's Fair in 1893. Before World War I it established subsidiaries in France, Germany, Russia, and Britain. British Westinghouse was formed in 1899 and two years later moved into a new, purpose-built factory on a 100-acre plot it had purchased in Trafford Park (Manchester), where Ford was later to establish its British production facility. The firm was heavily involved in the scheme to electrify Britain's railways (British Westinghouse 1903) but despite the business this generated, in 1907 it went into receivership. As a result, American Westinghouse's control was weakened, although it maintained influence as a majority shareholder (British Westinghouse 2012).[50]

Westinghouse Gazette

In April 1912 British Westinghouse launched a staff magazine: the *British Westinghouse Gazette* (another magazine, *British Westinghouse Club News*, was started the following year, though its content was essentially social with no international dimension and thus of no immediate interest to this study). The *Gazette* was aimed primarily at staff but past and potential buyers of Westinghouse products were also included among the intended readership. Content was largely made up of what the magazine itself described as "semi-technical" articles,[51] "constructed that they may be read and digested with a minimum of effort."[52] The material was serious rather than social, reflecting the company's status as a research-based enterprise as well as its belief in the importance of applying science to industry.[53] Articles illustrated the use of Westinghouse electrical engineering in factories, printing works, shipbuilding, bakeries, water supply companies, mining, general electrical supply schemes, and railway construction and upgrading.

The majority of articles had a British focus, but the magazine was also keen to stress the international flavor of the company's work. Occasionally British Westinghouse's parent company in Pittsburgh was featured, as in the cases of the death of its founder George Westinghouse and a visit by a group of Trafford Park engineers.[54] The majority of stories with an international slant, however, covered industrial users of Westinghouse technology around the world. News was communicated to staff of activities in countries as far apart as Scandinavia, Australia, and Canada.[55] Progress and achievements abroad were featured in

---

50   "George Westinghouse," *British Westinghouse Gazette*, April 1914.

51   "Editorial," *British Westinghouse Gazette*, April 1912

52   "Editorial," *British Westinghouse Gazette*, May 1914.

53   "Mobilisation of Science in Regard to Industry," *British Westinghouse Gazette*, March 1916; "Some Features of British Westinghouse Research Work," *British Westinghouse Gazette*, January 1917.

54   "Professor Mike Walker," *British Westinghouse Gazette*, May 1912; George Westinghouse, *British Westinghouse Gazette*, April 1914.

55   "Norwegian Power and Industries," *British Westinghouse Gazette*, April 1912; "Hydro-Electric Power in Sweden," *British Westinghouse Gazette*, July 1912; "Water Power

annual reports termed "Retrospects" (digestible, end-of-year narrative accounts designed to keep workers informed of the company's technological innovations, sales, and projected business). The "Retrospect" for 1912, for example, reported on electrical turbines supplied to gold mines in Africa, to Melbourne Corporation in Australia, and to Napier Corporation in New Zealand.[56] The "Retrospect" for 1914 gave news of the supply of electrical material to the Ashanti Goldfields Corporation in the West African Gold Coast Colony. Like most other articles in the magazine this article was largely of a technical nature. However, it also addressed the labor issue in African gold mining, commenting negatively: "Although some intelligent natives were given responsible posts, most failed at work that wasn't routine and subject to constant and close supervision."[57]

Conclusion

The four British companies featured in this study were either subsidiaries with a foreign parent, as in the case of Ford (Britain) and British Westinghouse, or parent companies with foreign offspring, as in the case of Lever Brothers and Rowntree (although in Rowntree's case its Caribbean estates might be more accurately described as wards or adopted children). Through their staff magazines these companies expressed their internationalism in different ways.

In the *British Westinghouse Gazette* most international stories stressed not the company's relationship with its foreign parent but its success as a leading exporter in the sector. In the pages of the magazine the workforce was informed of the company's world-wide operations and status, conveying the message that they were globally connected and important, even if small, parts of an international organization that had originated in the United States. Whereas the international narratives that appeared in the pages of the *British Westinghouse Gazette* centered on exporting prowess and the beneficial effects of British Westinghouse products around the world, those that were printed in the *Cocoa Works Magazine* revolved around the sources of raw materials for importation that Rowntree had secured and developed. Rowntree did not enter the confectionery export trade to any significant degree until after World War I. Its international dimension was thus expressed through the description and discussion of its overseas supply operations. The *Cocoa Works Magazine* was laden with the imperial spirit, Rowntree's West Indian estates being situated in the British colonies of Dominica and Jamaica. It was a strain of corporate internationalism laced with Britain's imperial project.

---

in Tasmania," *British Westinghouse Gazette*, May 1912; "Random Notes on Electricity in Ontario," *British Westinghouse Gazette*, August 1913.

56   "Twelve Month's Work," *British Westinghouse Gazette*, December 1912.

57   "The Generation of Power for Gold Mining," *British Westinghouse Gazette*, December 1914.

Rowntree's international connections were characterized by vertical integration: specifically, the ownership of primary sources of raw materials. Buying directly into the supply chain was a major feature of early Fordism. So also was control of components in the distribution network. In the case of Ford (Britain) there was no direct ownership of the dealers or employment of the agents who sold and promoted its vehicles. Rather, influence was exercized through formal contracts and the informal medium of the *Ford Times (Britain)*. Through the magazine, corporate culture was transmitted, corporate loyalty strengthened, and corporate branding developed. The *Ford Times (Britain)* was itself an international phenomenon, being the organ of a subsidiary with an American master. But in addition, within its pages, it contained information that reminded readers not only of the grandeur and innovation of their American parent but also the extensive international reach of the Ford empire, of which they were seemingly an increasingly important part.

Of the two *British* companies featured in this chapter, only Lever Brothers established foreign subsidiaries before World War I. Its magazine, *Progress*, celebrated the company's strident internationalism. In celebrating the globalism of the company, the magazine also strove to surmount the barriers that arose from the extended lines of communication that were an inevitable product of corporate internationalization. Like the other magazines discussed in this study, a major aim of *Progress* was to serve as a "vehicle of inter-communication between all members of ... staff in all parts of the world."[58]

References

Beniger, R. 1986. *The Control Revolution: Technological and Economic Origins of the Information Society*. Cambridge, Mass.: Harvard University Press.

Black, A. 1994. "Information, Paternalism and Cocoa: 'Confectionery Fordism,' Northern Innovation and Library Provision at Rowntree and Co. of York before the Second World War." *Library History*, 10: 51–70.

Black, A. 2011. "'A Valuable Handbook of Information': The Staff Magazine in the First Half of the Twentieth Century as a Means of Information Management." In *Information History in the Modern World*, edited by T. Weller. Aldershot: Palgrave Macmillan, 130–54.

Black, A. 2012. "Organizational Learning and Home-Grown Writing: The Library Staff Magazine in Britain in the First Half of the Twentieth Century." *Information & Culture*, 47 (4): 487–513.

Bradley, I. 1987. *Enlightened Entrepreneurs*. London: Weidenfeld and Nicolson.

"British Westinghouse." 2012. *Wikipedia*. http://en.wikipedia.org/wiki/British_ Westinghouse (accessed January 28, 2012).

"British Westinghouse Electric and Manufacturing Company." 1903. *The Mersey Railway of Today: Being a Souvenir of the Introduction of Electric Power on*

---

58   To Our Staff, *Progress*, October 1899.

the Mersey Railway. London and Manchester: British Westinghouse Electric and Manufacturing Company.

Cadbury, D. 2010. *Chocolate Wars: The 150-Year Rivalry between the World's Greatest Chocolate Makers*. New York: Public Affairs.

Cameron, R. and Neal L. 2003. *A Concise Economic History of the World: From Paleolithic Times to the Present*. New York and Oxford: Oxford University Press.

Carlos, A.M. and Nicholas S. 1988. "Giants of an Earlier Capitalism: The Chartered Trading Companies as Modern Multinationals." *Business History Review*, 62 (Autumn): 398–471.

Chandler, A.D. 1990. *Scale and Scope: The Dynamics of Industrial Capitalism*. Cambridge, Mass.: Belknap Press.

Drucker, P.F. 2008 [1973]. *Management*. Revised edition. New York: Collins.

Eisenberg, E.M., Goodall, H.L. and Tretheway A. 2007. *Organizational Communication: Balancing Creativity and Constraint*. Boston and New York: Bedford/St. Martins.

Fieldhouse, D.K. 1978. *Unilever Overseas: The Anatomy of a Multinational, 1895–1965*. London: Croom Helm.

Fitzgerald, R. 1995. *Rowntree and the Marketing Revolution, 1862–1969*. Cambridge: Cambridge University Press.

"Ford Trafford Park Factory." 2012. *Wikipedia*. http://en.wikipedia.org/wiki/Ford_Trafford_Park_Factory (accessed March 2, 2012).

Goldhaber, G.M. 1974. *Organizational Communication*. 2nd edition. Dubuque, Iowa: William C. Brown.

Griffiths, J.R. 1999. "Exploring Corporate Culture: The Potential of Company Magazines for the Business Historian." *Business Archives: Sources and History*, 78 (November): 27–37.

Jacoby, S.M. 1997. *Modern Manors: Welfare Capitalism since the New Deal*. Princeton, N.J.: Princeton University Press.

Jones, G. 1988. "Foreign Multinationals and British Industry Before 1945." *Economic History Review*, 41 (3): 429–53.

Koehler, J., Anatol, K. and Applbaum, R. 1976. *Organizational Communication*. New York: Holt, Rinehart and Winston.

Leverhulme, W. 1917. *Standardizing Welfare*. Port Sunlight: Lever Brothers.

Lewis, P.V. 1975. *Organizational Communication: The Essence of Effective Management*. Columbus, Ohio: Grid Incorporated.

McIntosh, I. 1991. *Ford at Trafford Park: An Americanised Corner of Old Jog-trot England*. Manchester: Department of Sociology, University of Manchester.

Nevins, A. and Hill, F. 1954. *Ford: The Times, the Man, the Company*. New York: Charles Scribner's and Sons.

Nicholls, R. 1996. *Trafford Park: The First Hundred Years*. Chichester: Phillimore.

Orbell, J. 1991. "The Development of Office Technology." In *Managing Business Archives*, edited by A. Turton. Oxford: British Archives Council, 60–83.

Prout, H.G. 2005 [1921]. *The Life of George Westinghouse*. New York: The American Society of Mechanical Engineers.

Riley, S.G. 1992. *Corporate Magazines of the United States*. New York: Greenwood Press.

Roach, B. 2005. "A Primer on Multinational Corporations." In *Leviathans: Multinational Corporations and the New Global History*, edited by A.D. Chandler and B. Mazlish. Cambridge: Cambridge University Press, 19–44.

Tolliday, S. 2003. "The Origins of Ford of Europe: From Multidomestic to Transnational Corporation, 1903–1976." In *Ford: The European History 1903–2003*, Volume 1, edited by H. Bonin, Y. Lung, and S. Tolliday. Editions PLAGE: Paris, 153–76.

Tugenhat, C. 1971. *The Multinationals*. London: Eyre and Spottiswoode.

Wilkins, M. 2005. "Multinational Enterprises to 1930: Discontinuities and Continuities." In *Leviathans: Multinational Corporations and the New Global History*, edited by A.D. Chandler and B. Mazlish. Cambridge: Cambridge University Press, 45–79.

Wilson, C. 1968. *The History of Unilever: A Study in Economic Growth and Social Change, Volume 1*. New York and Washington: Frederick A. Praeger.

Yates, J. 1989. *Control Through Communication: The Rise of System in American Management*. Baltimore, Maryland and London: John Hopkins University Press.

Index

References to illustrations are in **bold**.

Abel, Frederick, Sir 268
Agence France Presse (AFP) 36
Agence Havas 18, 35, 36
    area of control 38
    independence 39
Angenot, Marc 137
Albert, Eduard 79–80
American Institute of Social Service 128
American Philosophical Society 116
Amis du Palais Mondial 222, 235
Anderson, H.C.L. 62
Anderson, Perry 110
Angell, Norman, *The Great Illusion* 183
Anglo-American Telegraph Company,
        North Atlantic map **26**
*Annales de l'Institut international de
        sociologie* 152
Année Sociologique 154
Annuaire de la Vie Internationale 14, 181,
        183, 184, 187, 188, 191, 192
    content 186
    internationalism 195
    publication problems 189
Anthropological Institute of Vienna 51
Antwerp, Universal Exhibition (1885) 259
arbitration
    and science 212, 216
    and war 216
Archives Slaves de Biologie 75, 76
Armstrong, Edmund la Touche 251, 255
    Brazier, antipathy 252–3
Associated Press 18, 36
    area of control 38
    independence 39
    and newsworthiness 41–2
    Western Union, association with 39
association and international association
    as evolutionary mechanism 161

information on 158
organicist definition 161–2
Augsburger Allgemeine Zeitung 37
Australasia
    Historical Society of 59, 65
    Library Association of 249, 251
Australia
    "9 by 5 Impressionism Exhibition" 246
    Aboriginal history, neglect of 64, 65
    Aboriginal languages, publications
        64–5
    archival material, dispersed 61
    in the *belle époque* 243
    Boards of International Exchanges 19,
        54–5
    Britain, telegraph cable 19, 50–51
    Commonwealth, formation (1901) 58,
        243, 244
    frontier thesis 246–7
    Heidelberg school of painters 245,
        247–8
    historical records
        gaps 60–61
        transcripts 62
        in UK, microfilming of 64
    historical societies, establishment 59
    history, importance of 56–7
    information seeking 50
    learned societies 51–2
    national identity, making of 19, 58,
        243, 244–5
    public libraries 51, 52–3
        see also Melbourne Public Library
    publication exchanges 54–5
    racist ideology 246
    utopianism 246
    see also Melbourne; New South Wales;
        Queensland; Tasmania
Australian Commonwealth Government,
        Historical Records of Australia 60